Christiane Stephan
Living with floods

T0139521

ERDKUNDLICHES WISSEN

Schriftenreihe für Forschung und Praxis

Begründet von Emil Meynen

Herausgegeben von Martin Coy, Anton Escher, Thomas Krings
und Eberhard Rothfuß

Band 165

Christiane Stephan

Living with floods

Social practices and transformations
of flood management in Chiapas, Mexico

 Franz Steiner Verlag

Umschlagfoto: Überflutete Straße am Fluß Usumacinta in Chiapas, Mexiko
© Christiane Stephan

Bibliografische Information der Deutschen Nationalbibliothek:
Die Deutsche Nationalbibliothek verzeichnet diese Publikation in der Deutschen
Nationalbibliografie; detaillierte bibliografische Daten sind im Internet über
<http://dnb.d-nb.de> abrufbar.

Dieses Werk einschließlich aller seiner Teile ist urheberrechtlich geschützt.
Jede Verwertung außerhalb der engen Grenzen des Urheberrechtsgesetzes
ist unzulässig und strafbar.
© Franz Steiner Verlag, Stuttgart 2019
Angenommen als Dissertation durch die Mathematisch-Naturwissenschaftliche Fakul-
tät der Rheinischen Friedrich-Wilhelms-Universität Bonn unter dem Titel „Living with
floods. Social practices and transformations of flood management in Chiapas, Mexico"
Druck: Hubert & Co., Göttingen
Gedruckt auf säurefreiem, alterungsbeständigem Papier.
Printed in Germany.
ISBN 978-3-515-12480-5 (Print)
ISBN 978-3-515-12481-2 (E-Book)

TABLE OF CONTENT

TABLE OF FIGURES

TABLE OF TABLES

LIST OF ABBREVIATIONS

Abbreviation	Original Term	Translation of Spanish terms to English
BID	Banco Interamericano de Desarrollo	Inter-American Development Bank
CADA	Campaña por la Desmilitarización de las Américas	Campaign for the Demilitarisation of the Americas
CDM	Clean Development Mechanism	
CENAPRED	Centro Nacional de Prevención de Desastres	National Centre for Disaster Prevention
CIASI	Comisión Intersecretarial para la Atención de Sequías e Inundaciones	Intersecretarial Commission for Adressing Droughts and Floods
CICC	Comisión Interministerial de Cambio Climatico	Inter-Ministerial Climate Change Commission
CIESAS	Centro de Investigaciones y Estudios Superiores en Antropología Social	Centre for Research and Superior Studies in Social Anthropology
CFE	Comisión Federal de Electricidad	Federal Commission of Electricity
CONAGUA	Comisión Nacional del Agua	National Water Commission
CONEVAL	National Council of Evaluation of the Politics of Social Development	Consejo Nacional de Evaluación de la Política de Desarrollo Social
DFID	Department for International Development	
DGV	Deutsche Gesellschaft für Völkerkunde	German Society for Ethnology
DIF	Sistema Nacional del Desarrollo Integral de la Familia	National System of Integrative Development of the Family
DRM	Disaster Risk Management	
DRR	Disaster Risk Reduction	
Eco-DRR	Ecosystem-based Disaster Risk Reduction	
ECOSUR	El Colegio de la Frontera Sur	
EZLN	Ejercito Zapatista de la Liberación Nacional	Zapatista National Liberation Army
FONDEN	Fondo Nacional de Desastres Naturales	Fund for National Disasters
FOPREDEN	Fondo de Prevención de Desastres	Fund for Disaster Prevention
GIS	Geographical Information System	

Abbreviation	Original Term	Translation of Spanish terms to English
GPDRR	Global Platform for Disaster Risk Reduction	
HFA	Hyogo Framework for Action	
IFRC	International Federation of Red Cross and Red Crescent Societies	
IIASA	Institute for Applied Systems Analysis	
INGO	International Non-Governmental Organisation	
INAI	Instituto Nacional de Transparencia, Acceso a la Información y Protección de Datos Personales	National Institute for Transparency, Information Access and Data Protection
IPCC	Intergovernmental Panel on Climate Change	
LGCC	Ley General del Cambio Climatico	General Climate Change Law
LGPC	Ley General de Protección Civil	General Law of Civil Protection
MCEER	Multidisciplinary Center for Earthquake Engineering Research	
NGO	Non-Governmental Organisation	
OMC	Otros Mundos Chiapas	Other Worlds Chiapas
PAOM	Plan de Adaptación, Ordenamiento y Manejo integral de las cuencas de los ríos Grijalva y Usumacinta	Plan for Integrated Adaptation, Planning and Management of the basis for the rivers Grijalva and Usumacinta
PAR	Pressure and Release Model	
PEMEX	Petróleos Mexicanos	Mexican Petrols
PNI	Programa Nacional de Infraestructura	National Infrastructure Programme
PRI	Partido Revolucionario Institucional	Revolutionary Institutional Party
RECOMA	Red Latinoamericana contra los Monocultivos de Arboles	Latin American Network against monoculture of trees
REDLAR	Red Latinoamericana contra las Presas y en Defensa de los Ríos, sus Comunidades y el Agua	Latin American Network against dams and in defence of the rivers, their communities and water
RETOS	La Red Trasnacional Otros Saberes	Transnational Network of the Other Knowledges
RICAM	Red Internacional de Carreteras en América Central	International Network of Highways in Central America
SDGs	Sustainable Development Goals	

Abbreviation	Original Term	Translation of Spanish terms to English
SFDRR	Sendai Framework for Disaster Risk Reduction	
SINAPROC	Sistema Nacional de Protección Civil	National Civil Protection System
SIEPAC	Sistema de Interconexión Eléctrica de los Países de América Central	System of Electric Interconnection of Central American Countries
UNAM	Universidad Nacional Autónoma de México	National Autonomous University of Mexico
UNISDR	United Nations International Strategy for Disaster Reduction	United Nations International Secretary for Disaster Reduction
UNU-EHS	United Nations University - Institute of Environmental and Human Security	
VCA	Vulnerability and Capacity Assessment	

SUMMARY

People around the globe deal with floods in various ways and evaluate flood risk and other risks based on general patterns they experience as part of their everyday lives. In the South of Mexico, floods along the river Usumacinta take place on a regular basis and interact with temporal and spatial dynamics in a complex socio-ecological system. Descriptions by people from villages along the Lower Usumacinta in Chiapas are manifold and represent highly ambivalent perceptions. Flooding is evaluated as a positive and at the same time negative dynamic in the cyclic pattern of social life. Perceptions and evaluations of floods along the Usumacinta influence options for flood management and long-term Disaster Risk Reduction as they form part of the social practices that make up flood management. These practices involve not only local people but connect different actors at different geographic scales who in different ways shape the conditions for the ways floods are managed.

In the state of Chiapas in the South of Mexico, flood management is a pressing issue given the fact that various processes of the physical environment and the socio-ecological systems interconnect. Processes of global climate change which are discussed to increase extreme rainfall events in the future and the effects of the *El Niño* phenomenon which may lead to both an increase in droughts as well as in floods shape hydrological conditions in the case study region (Landa et al. 2008: 13; IPCC 2001: 54). At the same time, a range of pressing social and political dynamics define and influence floods and flood management in the region. While climate change underlines the need for concerted action in disaster management on different geographic and administrative scales, flood events are currently approached in disaster risk science in a range of different and sometimes conflicting ways. Approaches informed mainly by natural sciences conceptualise risk as an objective entity and present quantitative methods for risk analysis based on probabilistic models (Bedford & Cooke 2001; Calvi et al. 2006). In contrast, a variety of social science disciplines put human activity at the centre of risk processes and the emergence of disasters (Renn 2008: 54f; Egner & Pott 2010a).

In a socio-cultural tradition of disaster risk science, this study understands flood risk as a result of societal construction and decision making. Based on foundations of social geography, the study tries to build an understanding of socio-spatial patterns of human activity through linking complex patterns of social life with relevant dynamics in the physical environment. This perspective entails a review of dominant concepts in disaster risk science and of locally used conceptualisations of floods, risks and disasters from Mexico. Investigating floods in this way requires involving local perspectives and conceptualisations of floods and most importantly, the multiple social practices that are created and performed around and as part of flood phenomena. Understanding floods as social phenomena with specific spatial relevance is accomplished using a specific version of social practice theory. The work of Theodore Schatzki (i.a. 2001b, 2002, 2003, 2011) allows the conceptualisation of social practices in a flat ontology of the social and describes, how practices are constituted mentally, bodily and materially. Linking social practice theory with Lefebvre's (1991) conceptualisations of space a specific conceptual and analytical approach is chosen. Building on the *riskscapes* concept presented in social geography by Müller-Mahn and Everts (2013), this study develops one version of a *riskscape* conceptualisation that accounts for the complex and interwoven character of flood related social practices.

The methodology designed for empirically studying social practices of flood management in Mexico, is informed by ethnomethodology and audio-visual methods in qualitative research. Based on grounded theory and influenced by two "alternative" epistemologies, field research gives an active role to research partners from case study villages and to a co-researcher from Chiapas, thereby integrating local conceptualisations and performances in a dialectic process of conceptual and empirical work. Empirical research carried out in a period of nine months in 2014 and 2015 involved in-depth qualitative research in selected case study villages in the municipalities Catazajá and Palenque in the North of Chiapas. In addition, qualitative interviews were carried out with other actors relevant for flood management and its larger context in the South of Mexico, among them representatives of civil protection, government agencies, NGOs and research institutions. The different actors involved in this research are selected in their relevance as carriers of specific practices that form part of flood management, which allows the identification of a complex pattern of interrelated performances. The use of audio-visual methods, especially participatory photography and video workshops carried out in rural settlements along the river Usumacinta, provides novel information on the characteristics and dynamics of social practices. The visual as a highly sensitive medium for patterns in the material world, presents how material objects "tell a story" about the social practices which they are or have in the past been part of and might be in the future. The visual medium allows the analysis of different temporal dimensions which coexist in the physical world in the present. Visual methods are presented in this study as an important component of qualitative social research, that equip social geography with an additional approach to enquire questions of materiality and embodiment of human activity in spatial settings.

As a result of this study patterns of social practices in flood management are presented, which are of relevance within and beyond the case study region. Linking empirical results with social practice theory in the *riskscape* approach enables to focus not on the actors primarily but on the practices performed. Practices of living with the flood and practices of anticipation that are identified in case study villages can be contrasted to practices performed by external actors like those that try to transfer knowledge on floods, introduce a development discourse or change local patterns of preferences and behaviour. The specific *riskscape* approach developed in this study provides the necessary analytical view to identify specific types of interactions between social practices. Diverse types of interrelations between social practices are explored, while the types of interaction range from co-existence and support to conflict and competition. As the empirical example of a projected dam on the Usumacinta shows, different types of interrelations between social practices show different spatial repercussions and beyond current consequences carry along additional repercussions in the near or far future. The analysis of interrelations of social practices are relevant both conceptually as well as practically. On the one hand, the empirical results provide information on which a more in-depth understanding of social practices can be developed. On the other hand, underlying conflicts, negotiations and capacities are identified that influence flood management and other spatially relevant phenomena in the case study region. Making changing patterns of relations between social practices and the transformation of discourses in flood management visible, this thesis provides detailed insight into ongoing processes of social transformation in the South of Mexico. These processes reconnect to and reflect dynamics that foster social inequalities on a global scale, among them processes of promoting discourses of development and technological modernisation. Presenting scientific analysis of the socio-spatial practices that guide social transformation, the study tries to provide one part of an analytical and methodological perspective and apparatus, which is regarded necessary to elaborate more sustainable practices of flood management in the future.

ACKNOWLEDGEMENTS

Carrying out a doctoral research and writing this thesis would not have been possible without the support of a large range of people and institutions to which I am very thankful. To begin with, I want to thank DAAD for funding my research stays in Mexico in 2014 and 2015.

My sincere thanks go to Caroline Blankenagel, Henrik Junius and Celia Norf, who gave their constructive feedback on the first draft of this thesis and to Eva Bogdan, who has been a great counterpart in reflecting on practice theory. I am grateful to my colleagues at TH Köln, who reflected with me on theoretical and methodological questions and brought me back to reality after hours of interview coding. Special thanks go to Alexander Fekete who supported my idea to carry out a doctoral research from the beginning on and provided more than the necessary environment, resources and motivation to express my ideas.

I am deeply grateful to my supervisor Detlef Müller-Mahn, who inspired my conceptual ideas, who encouraged me to walk unconventional ways in research and who shared the idea with me that a dissertation is a marathon that can only be run step by step. I owe many thanks to my colleagues in the Department of Geography at University Bonn as well as in the DELTA project at University Köln, at Freie Universität Berlin, University Bielefeld and University Erlangen, as they inspired me to develop further my ideas and to keep on going.

There are many colleagues and partners in Mexico, to whom I am thankful because they revealed new ways of thinking and doing research to me. Among them are Fernando Briones, Laura Herrero Garvín, Anaïs Vignal, Fernando Brauer and Ursula Oswald Spring. Special thanks go to Xochitl Leyva Solano, who invited me to be part of "creando saberes". My gratitude is with Noemi and David, who supported my research in 2014 with artistic skills and empathy. My highest gratitude goes to Xuno, who has been a trustful and inspiring research partner in the co-research experience and taught me what an "epistemology of the heart" can look like.

Without my family, I would not have been able to carry out this study. I am deeply thankful to them for encouraging and believing in me in the difficult times of this journey. I am grateful to my colleagues and friends who have been great intellectual and emotional counterparts and much more than that. Moreover, I thank my brothers and sisters at HaF, who helped me to believe.

I am deeply grateful to the inhabitants of the case study villages in Chiapas, who opened the doors of their houses and hearts to me, spent time with me in their *milpas*, on the river as well as in the backyards of their houses, and allowed me to participate in their day-to-day activities. I am especially grateful to the brave women who live along the Usumacinta and other waters. May God bless you.

PROLOGUE

Agua de las primeras aguas, tan remota,
que al recordarla tiemblan los helechos
cuando la mano de la orilla frota
la soledad de los antiguos trechos.

Y éste es el canto del Usumacinta
que viene de muy allá
y al que acompañan, desde hace siglos, dando la vida,
el Lakantún y el Lakanjá.

Excerpt from "El canto del Usumacinta"
by Carlos Pellicer, May 9th, 1947

Water of the first waters, so remote,
that remembering them, the ferns start to tremble
when the river bank rubs his hands
on the solitude of the old stretches.

And this is the song of the Usumacinta
which comes from far away
and whom accompany, since centuries, giving life,
the Lakantún and the Lakanjá.

[own translation]

1 INTRODUCTION

People around the globe deal with floods in a variety of ways. While some people who live in flood prone areas take decisions to leave their settlements, others decide to or have to stay in a flood-prone area and deal with floods as part of their everyday lives. "Waters that come, waters that go". While this description of a flood may appear simple, it introduces one way in which locals in the South of Mexico have described floods for centuries. Floods are studied in different scientific disciplines with a focus on the hydrological dynamics they entail (Brown 2016; Pedrozo-Acuña et al. 2014; Kundzewicz et al. 2013). However, floods are understood here as complex social-ecological dynamics, which involve different processes of the physical environment and the human and non-human beings who live and interact within it. In this thesis floods are approached from a social geographic perspective that places flood dynamics within a myriad of human performances. One relevant dynamic is flood management, a pattern of social practices performed by different groups of people, among them inhabitants of flood-prone areas, civil protection staff and government representatives. Looking into flood management from a social geographic perspective not only enables to learn about a natural phenomenon and human-nature interaction but it also gives insight into constructions of reality, into social interactions and into constructions and reproductions of material and immaterial spatialities. While as a researcher in social geography I am interested in understanding social patterns of human activity, it is relevant to connect patterns of social life with relevant dynamics in the physical environment. It is for this reason that floods are studied here as social patterns which link and interact with global socio-ecological dynamics.

The relevance to study flood management can be underlined by processes of global prevalence discussed in different scientific disciplines: Climate change is argued to be likely to increase flood risk in selected world regions through the increase in extreme rainfall events in the future (IPCC 2001: 29). In the South of Mexico, it is especially the effects of the *El Niño* phenomenon (ENSO= El Niño Southern Oscillation) which may lead to both an increase in droughts as well as in tropical storms and extreme precipitation events followed by floods (Landa et al. 2008: 13; IPCC 2001: 54). While processes of climate change show the need for concerted action of disaster prevention on different geographic scales, flood events are approached in disaster risk science in a range of different ways. Approaches informed mainly by the natural sciences conceptualise risk as an objective entity, to be measured and calculated with quantitative methods and expressed in probabilistic models (Bedford & Cooke 2001; Calvi et al. 2006). In some accounts of natural science, a flood is characterised as a hazard that has its origin in physical processes and it is argued that these physical dynamics impact the lives of humans (Smith & Ward 1998). In contrast, a variety of other scientific disciplines have identified human activity as playing the central role in the generation of risk and the emergence of disasters, among them psychology as well social science disciplines and geography (Renn 2008; Egner & Pott 2010a). Social science and human geography perspectives describe a social construction of risk and largely detach enquiries on risk from objectivist approaches (Weichselgartner 2001). Flood dynamics in this sense are understood as a part of social processes that involve decision making based on aspects like perception, values and needs (Ibid: 64, 110).

The overview on risk perspectives brings up the following question: If floods are embedded in such complex social processes, how can we as researchers account for them in an adequate manner? One possible approach is represented in social practice theory as this line of theory provides relevant conceptual linkages from flood events on one side towards practical activities and on the other side towards larger social dynamics including inequalities in access to flood management resources. Practice theory initially entered social geographic research mainly through the work of Pierre Bourdieu who made important contributions to the conceptualisation of the linkages between social structures and spatial relations and the ways in which human activity is part of larger societal structures (Etzold 2014: 38). In this thesis, conceptualisations made by Theodore Schatzki (i.a. 2001, 2002) on social practices and their constitution through and as part of *the social* are used in order to look into flood related dynamics. His version of a conceptualisation of social practices enables to identify major performances and local conceptualisations relevant for and in flood management. Based on an analytical approach derived from this theory I argue that it is relevant for research on floods to recognise and represent water, land, human beings and animals as more than variables in an equation of flood risk. These entities are closely tied to each other through the social practices of human beings, among which we have to identify mental and bodily performances of humans, abstract concepts and tangible materilities as well as social and ecological processes.

It is believed to be highly relevant for social geography to investigate floods and social practices in a rural area in Chiapas in the South of Mexico. Relevance is stated for a number of reasons: The empirical analysis of social practices of flood management in the study region helps to understand and refine general conceptual features of social practices and practice patterns. Moreover, it can be recognised that floods are no phenomena to be regarded independently from other phenomena and social dynamics. Therefore, floods and flood related social practices can point research towards general patterns of social life in which they are embedded. The Mexican case reveals dynamics which are specific to the case study region or the socio-cultural, political and even physical-geographic context of Southern Mexico. It thereby gives a detailed insight into a geographic region for which only few scientific accounts on the social and ecologic processes in a historical overview are found (Wilkerson 1991; Scherer & Golden 2012). Furthermore, the case study reveals patterns of social practices which link different geographic scales, different world regions and different temporalities. It is in the Mexican case that reflections of globally relevant social practices can be recognised, including global projects for economic and infrastructural development and dominant discourses, among them risk discourses. The reader of this thesis is invited to follow the analysis of how these practices are informed by and reproduced in different groups of people, thereby forming what are called *riskscapes*. The *riskscapes* of flood management, namely the socio-spatial dynamics of flood and risk related social practices, help to understand socially and spatially relevant inequalities and to deconstruct inherent conceptualisations of risk and flood.

Given the amount of works published on floods from social science and related scientific fields, it is relevant to argue what new insights can be expected from this thesis. It is acknowledged that relevant conceptual and empirical contributions have been presented by a range of scholars in the last 15 years (i.a. Crate 2011; Krause 2016; Oslender 2002; Rohland et al. 2014). These works underline that there is a lack of analysis which serves academia, practitioners and local people alike. What is needed are publications that result from long-term engagement with local people in flood-prone areas and at the same time involve processes of collective research and collective practical action. This study intends to contribute to an ongoing broadening of conceptual and methodological horizons through a specific way of thinking and doing research. It is a way in which people's everyday practices in dealing with floods are addressed through a qualitative approach that gives a strong role to visual and audio-visual methods. The design of methodology follows two major purposes which are interlinked. In the use of visual methods, it is possible (1) to incorporate people actively into the creation of empirical material for the research process, e.g. through participatory workshops and processes of reflection on and representation of visual material and (2) to produce a large set of visual empirical information to be included in the analysis of social practices which complements and extends other sets of empirical data. The interpretation of visual material demands procedures of analysis – in this case the documentary method – which have not been broadly used in human geography yet, but are worthwhile exploring. In the visual approach, the researcher included local people from the case study villages as well as a co-researcher from Chiapas. This provided a unique option for active interaction between all involved partners. Moreover, producing pictures and videos with a camera allows researchers and research participants to express thoughts, research questions and perceptions in a different language than the languages of spoken and written text. Reflections stimulated from these collective processes of video and photography production and their revision are invaluable for this study and the long-term relationship between all involved research partners from Mexico and Germany.

The partly unconventional conceptual and methodological approaches chosen are closely connected to the main motivations and objectives I follow in this study: The main motivation to carry out a doctoral thesis on flood phenomena in the South of Mexico has been to learn about floods from people who experience them on a regular basis. Believing that people develop conceptualisations as part of their daily lives stimulated the decision to carry out research among flood-affected people. This aspect is linked to the motivation of supporting processes of self-reflection and awareness raising among local people in Chiapas on various socio-spatial patterns and transformations that take place in the case study region. Research can serve to reflect on processes that people are aware of but which they have not exchanged information on in a communicative process. Reflection takes place in a bi-directional way: the researcher offers a different view into the lives of research partners upon which they can reflect and vice versa the research partners provide views that help the researcher to reflect on her research question, academic processes and life (personal communication Leyva Solano 07.08.2014, Field notebook 1/2014). In this research, "mirroring" as the process is called by Mexican social anthropologists (Ibid) provides the option to make explicit those inequalities perceived by research partners and to support social practices of transformation that can be of use to reduce inequalities. Understanding the field of human geography as a space not only for academic research but for active involvement in knowledge production, representation and social transformation, this thesis can hopefully make a small contribution to make a step from knowledge and understanding towards sustainable social transformation. It is believed that it is not only the inhabitants of case study villages who can benefit from the research results and interactions initiated throughout field research. Moreover, the different actors from the South of Mexico addressed in this research, among them civil protection on regional and state level in Chiapas, municipal governments, research institutions and NGOs as well as INGOs are invited to continue the trustful interaction with the researcher in order to improve flood management along the Usumacinta River. Based on the motivations and objectives, I suggest that quality and relevance of this thesis may be evaluated by scientific criteria as well as through requesting the social relevance of the research presented.

This thesis is divided into ten chapters. While *chapter 1* introduces the study and its relevance as accomplished in the paragraphs above, *chapter 2* describes the research problem of this thesis and develops the research questions. The principle features of floods and flood risk management in Mexico and in the state of Chiapas are presented, deriving specific points of interest by the author as well as critical aspects that make a social geographic investigation relevant. The points of entry into the specific research problem and interest are thereafter developed in the formulation of a major research question. This main question is subsequently broken down into sub-questions that help to underline the specific focus of this thesis and to operationalise conceptual ideas into the empirical approach.

Chapter 3 introduces the case study region by means of a historical overview of selected socio-spatial dynamics in Chiapas and Mexico. This overview charts major processes since early Mayan settlement in the Usumacinta region until today, while among a large amount of historical processes those events and processes are selected which are regarded to be of relevance for current flood management. A focus is set on population dynamics in the South of Mexico, processes of economic exploitation of natural resources and labour along the river Usumacinta and adjacent lands during colonial times and major political events like the Mexican Revolution and Agrarian Reforms as these processes have fundamentally influenced the socio-spatial patterns and challenges we find in the case study region today.

Chapter 4 lays the theoretical basis of this thesis. It presents the paradigmatic and epistemological basis of this thesis and introduces the three main strands of theory used to approach flood phenomena. Social practice theory is presented as the principal theoretical basis among which the work of Theodore Schatzki is chosen to guide the theoretical examination of flood related human action. This is complemented by risk theory among which social science perspectives are underlined and theories of space, most especially the triadic conceptualisations coined by Henri Lefebvre (1991). The chapter concludes by weaving together the three theoretical lines into a specific concept of use to analyse the social practices of flood management: the *riskscapes* concept.

Chapter 5 presents the methodology designed for this study. It starts with a consideration of positionality, ethics and questions of ownership which are presented as highly relevant aspects in empirical research. Building up on principles of ethnographic research, the methodology incorporates qualitative methods of social enquiry, especially those developed in social anthropology and human geography. Moreover, a range of visual and audio-visual methods are presented as part of the methodological design. Finally, this chapter gives a short overview on the different methods used for the analysis of data and information and the processes of interpretation.

Chapter 6 is the first of four results chapters which highlight different aspects among the social practices identified in the analysis of the empirical material. This chapter presents an overview and in-depth insight into the *social practices of living with the flood* in the case study villages. This involves practices directly and visibly contributing to flood management in the case study village as well as those social practices that indirectly relate to them, e.g. practices of retelling history, practices of fighting for land or practices of reproducing collective identity.

Chapter 7 presents an identification and analysis of additional social practices through the use of visual empirical material and analysis with the documentary method. Moreover, the chapter links the identified practices with some of the social practices identified in the foregoing chapter. Thereby a comprehensive analysis of the *social practices of anticipation* is presented, amended by additional insight into the materiality of flood management in the case study region.

Chapter 8 opens a different perspective into the social practices of flood management as here a large range of social practices of other actors is presented and analysed. Those practices performed at different geographic scales are selected that show repercussions in the local practice patterns of the case study region. Here, insight is won into the different discourses, interests, and risk perspectives that transpire through the social practices identified.

Chapter 9 consolidates the empirical analysis presented in chapters 6, 7 and 8 through the (re)formation of relevant practice patterns. This is accomplished through the conceptual approach of *riskscapes* which redirects the focus from the actors towards practices and the interrelations between practices. Different types of interrelations are identified and visualised with the *riskscapes* approach. This leads to a presentation of critical points of practice interrelations, where dominant practices are identified to stand in contrast and partly rule out other practices, (re)producing inequalities and creating new risks. As a result of this overview, selected points are described where relevant dynamics of social transformation are recognised.

Chapter 10 draws together the lines of this thesis in a conclusion. After the formulation of final statements concerning the results of this thesis, the chapter presents an outlook on future research in the fields of social practice theory and riskscapes also involving questions of methodology, empirical field research and epistemological perspectives. This is complemented by selected recommendations for practical work and strategic orientation in disaster risk management that addresses amongst others the specific use of social practice perspectives.

2 TOPIC AND RESEARCH QUESTION

Yearly flooding and flood risk in a region of vested interest in territorial property and user rights; this can be described as the first and outer layer of a multifaceted situation along the Lower Usumacinta River and its adjacent strips of land. The processes described involve the inhabitants of many small and widely dispersed villages as well as a range of other actors, e.g. from government, the economic, humanitarian, and NGO sectors that interact with the people, the river and the land. Consequently, a situation is addressed – in a topical as well as topographical sense – which is of interest due to its relevance for science and society alike.

Within the interest of this investigation in social geography, a range of topics have been identified in the study area as highly relevant. These range from the topics of disasters and risks linked to processes of flooding, to concepts of nature and culture discussed in social anthropology, as well as processes that involve large infrastructural projects in different world regions. The specific interest in these topics is evoked because analysis of the relevant scientific literature and other documents shows that they have not been analysed profoundly in this constellation and for this world region with its specific social, ecological and political context. On the one hand the topics presented in more detail below are relevant in the sense of the necessity to gain more scientific insight through advancements of theoretical and methodological approaches. On the other hand, they are regarded by the author of this study as relevant in a social and socio-political sense. Several of the topics addressed here directly or indirectly touch aspects of basic human rights. As described in the introduction, the aim of this study is to analyse socio-spatial dynamics of flood and risk management in Chiapas. This is exemplified by empirical cases located along the river Usumacinta in the municipalities of Catazajá and Palenque. Figure 1 presents the location of the case study region in the larger topographic context of Mexico, highlighting the state of Chiapas, as well as the two other states (Campeche and Tabasco) whose water bodies form part of the hydrologic network of the Usumacinta River. In figure 2 (see colour plate p. 136) the two municipalities of Chiapas where empirical research was carried out are put into focus, displaying the location of the municipal capitals, Palenque and Playas de Catazajá, and the location of the case study villages. Alongside the development of new links between theories and the identification of relevant patterns of social dynamics it is crucial to point towards those dynamics and patterns that are involved in the creation of social vulnerability relevant in the everyday lives of people. This includes general and structural processes of economic and political marginalisation but also specific derogatory action as for example the deprivation of land property and of basic human rights.

In the following section, the main research problem is defined. Following this, the
major topics identified in the research problem are described, including an overview
on the state of the art in research and current challenges in the disciplines. This
serves as an entry point into a specific situation in the field study region, which will
be further developed and conceptualised throughout this study. The definition of
the research problem and problem context as well as available knowledge and challenges leads towards the development of a research question, which is presented
below (chapter 2.4). The broad topics touched here are linked with specific theoretical concepts which are specified further in the theoretical outline of this study
(chapter 4).

Figure 1: River Usumacinta in the Mexican states of Chiapas, Campeche and Tabasco
Source: author´s elaboration using ArcGIS (version 10.1/2012) (edited by Johannsen)

2.1 PHENOMENA AND PROCESSES OF INTEREST

In the focus of this research are social actors in Southern Mexico, that face two
different types of challenges: People in villages along the river Usumacinta and the
adjacent water bodies (lagoons and small streams) live in a region where yearly
floods are common in the months of June to October, which is the main rainy season
in this climatic region (Reyes Barrón 2012; Fernández Eguiarte et al. 2012; CONA-
GUA 2014). In recent times however, extreme rainfall events have manifested a
significant increase in frequency and severity which are linked to the effects of climate change (Arreguín Cortés et al. 2011: 22). In Mexico, severe droughts as well
as storms and extreme precipitation events can be linked to the phenomenon of
ENSO (El Niño Southern Oscillation) as well as more generally to global processes

of climate change (Landa et al. 2008: 13; IPCC 2001: 54). It is estimated that ex-
treme rainfall events will increase in frequency as well as severity in the future
(Elsner et al. 2008: 92 cf Schroth et al. 2009: 606). For the case of the state of
Chiapas, recent events of drought in 2015 are linked by scientists to the ongoing
ENSO event (CONAGUA 2015a: 14). The last severe flood events happened in the
region of the Lower Usumacinta in the years 2007 and in 2010 (Guha-Sapir & Ph.
Hoyos - EM-DAT 2016) and two smaller flood events were reported by village
inhabitants and civil protection staff for 2014 in the months of May and June, which
are months in which floods are unusual in this region. The 2007 flood event is linked
to the major flood in the neighbouring state of Tabasco, in which around 62% of
the area of the state were covered with water from the rivers Grijalva and Usumacinta (Perevochtchikova & Lezama de la Torre 2010: 73). In contrast to the
floods, in the year 2015 the northern parts of the state of Chiapas faced a severe
drought mainly from August until October (CONAGUA 2015b, 2015c, 2015d).
According to the general climatology of Southern Mexico and Gulf of Mexico region, these months usually are among those with the highest amount of yearly precipitation (Fernández Eguiarte, et al. 2012: 107).

Besides a link of flooding to processes of anthropogenic climate change, a
growing vulnerability of people towards flood events can be described for the Mexican state of Chiapas (Boyd 2013: 8). While the term vulnerability is described in
more depth in chapter 4, the term is used here to summarise characteristics of social
groups that are a precondition for a flood event to become an event of risk and of
possible damage. For the case of Chiapas, major factors that are described to contribute to high levels of social vulnerability against floods and other events are population growth, poverty, urbanisation, social injustice, conflicts and inadequate economic strategies (Gobierno del Estado de Chiapas 2013: 46). The dynamic interplay
of climate change and conditions of social vulnerability can be estimated to result
in increasing economic losses and damage in Southern Mexico in the coming years
and, if no adequate measures are developed to reduce vulnerability and effects of
climate change, also for the long-term future.

The second type of challenge identified for the South of Mexico is the fact that
the territories where rural population lives are included in a set of large-scale development projects. These include economic and social projects on local and regional level as well as large national and international infrastructural projects ongoing or prospected for the coming years. Mexico forms part of a large infrastructure project called *Proyecto Mesoamérica* (PM). Since the inception of the anteced-
ent programme, the "Plan Puebla Panamá", in 2008, a range of transformations have
taken place related to economic and social projects (Fromm Cea 2015). One large
infrastructural project prospected in Chiapas and Tabasco is a project involving the
construction of a number of dams on the river Usumacinta for the generation of
electric energy. The dam project is planned to be constructed at Boca del Cerro, a
geologic formation located near the city of Tenosique, Tabasco, upstream of the
case study region (see figure 2, colour plate p. 137). The construction could create
a range of different problems for the population of the study region, including a

permanent fall of water level in the Usumacinta River and the connected water bodies like lagoons and small river branches. This would severely affect local small-scale agriculture, which is the livelihood base of most local people. Moreover, in a dam system which is used for the purpose of energy-generation, generally a high water level is kept in the reservoir (Costa et al. 2014: 1). This is a challenge for flood management especially in tropical areas where extreme weather events occur with high variability (Ward et al. 2013: 3).

It is assumed here that both processes – the increased flooding and the putting forward of development projects – create new risks and new uncertainties for the population. New uncertainties arising for the people living in the villages have provoked a process of multiple changes and transformations. These transformations are not only to be found in visible and materially evident processes, e.g the frequent census of population by government agencies, soil assessments and mapping of the region by engineering companies. At the same time transformation processes take place in the everyday lives of the population (e.g. change of economic activities and employment patterns). While the region of this study and its´ inhabitants are most frequently described to suffer political marginalisation as well as economic and socio-cultural inequalities, there is one important issue that must be highlighted when talking about floods on the Usumacinta: People have lived in the region for various decades and have experienced floods for generations. We can suppose that people have developed locally adapted strategies of flood risk management.

In the study region, a range of different actors are involved in activities related to flood risk. As an exploratory study performed by the researcher in Mexico in 2013 indicated, the types of involvement of actors in flood risk dynamics are of a diverse nature. Whereas direct involvement can be stated for the local population and for those authorities in charge of the implementation of flood protection measures, namely the civil protection units at municipal and regional level, a range of other actors and persons is involved indirectly. Apart from structural measures and support during floods, the actors involved directly or indirectly articulate different interests in the resources and people located along the river Usumacinta in the relevant municipalities of Palenque and Catazajá. One pressing issue linked to these interests are plans to relocate local people from those villages that lie in direct vicinity of the river banks. The different actors involved hold arguments in favour or against relocation plans. They promote these ideas by carrying out certain activities and by linking their arguments with specific discourses and practices.

It is believed that the ways of involvement and the levels of action and interaction with the people in the villages reflect to some degree the distribution of power in the region. However, it is believed that power as a dynamic entity is under constant negotiation and reorganisation. Flood risk management in Chiapas appears to be a process where greater societal challenges crystallise and where new dynamics evolve. It is therefore of high relevance for the understanding of dynamic and conflict-loaded processes to take a closer look at the various parts involved in flood risk management with a focus on the local and regional level.

2.2 STATE OF THE ART – FLOOD RELATED RESEARCH

The following section introduces the main domains of research that deal with topics such as hazards, disasters, risks in general and with flood dynamics more specifically. Before elaborating the theoretical basis of this thesis (chapter 4), it is necessary to identify challenges and lacks in current research on the topics and processes presented above. As a starting point, dominant topics, questions and current or long-term challenges in the scientific disciplines are presented. Research results and challenges which have relevance for Mexico and the specific case study region are emphasised. Focus is laid on those aspects that are relevant from a perspective of human geography.

2.2.1 Research on hazards, disasters and risk

A major phenomenon addressed in this study is flooding on the river Usumacinta. Floods are often described as hazards because they can have major impacts on people settling in or in vicinity of flood prone areas (Jonkmann 2005: 168; Smith & Warde 1998). Flooding, just as other socio-physical dynamics, can provoke a disaster to occur, when dealing with them becomes a task that local population cannot perform without substantial support from government or other external bodies (Wisner et al. 2004: 61). Geography addresses hazards and disasters from a range of different ways. Integrating perspectives with natural science and social science focus, disaster research in geography comprises different scientific paradigms, research traditions as well as methodological approaches (Egner & Pott 2010a). Three selected key terms that reflect not only different paradigms but also the historical development of the research field are the terms *hazard*, *disaster* and *risk*.

The origins of *hazard* research in geography lie in research carried out in the USA in the 1940s, which build on approaches of social and human ecology from the 1920s (Felgentreff & Dombrowsky 2008: 15). The hazard perspective takes into view a part of the earth's surface and attributes certain hazards to it. These hazards result from a specific interaction of humans and the environment. As Pohl (2008: 51) points out, the early concepts of human ecology argued in line with a strong contrast between human and nature, which are revised in more recent approaches of human ecology. Therefore, early research on natural hazards did not consider the aspect of dynamic interaction adequately and overemphasised technical solutions in a natural and engineering sciences approach (Felgentreff & Dombrowsky 2008: 17). Moreover, hazard research in geography is based on an utilitaristic concept of the human through the model of the *homo oeconomicus* (Pohl 2008: 52). Today, the terms *hazard* and *natural hazard* are still predominantly used in the natural and engineering sciences (Felgentreff & Dombrowsky 2008: 14). The most important institution for natural hazard research up to today is the Natural Hazards Research and Applications Information Center (NHRAIC), which was founded by the geographer G. F. White at the University of Colorado in Boulder in 1976. In recent years, it has moved its focus more to societal dynamics and man-made hazards, which reflects a general orientation towards social science perspectives (Pohl 2008: 50).

For long periods *disaster* research has faced a challenge in finding a consistent answer to the question "What is a disaster?" (Quarantelli 1998; Oliver-Smith 1999; Perry & Quarantelli 2005). As Quarantelli (1985) showed in a comprehensive analysis of literature from three decades, research defines disasters in a great variety of ways. The variety which spanned from e.g. disaster as a physical agent to disaster as the social disruption following a detrimental physical event and even to the definition of a disaster as the social construction of reality which does not necessarily involve a physical event, made it appear impossible to come to one consistent definition (Ibid cf Oliver-Smith 1999: 21). The definitional debate, which was mainly carried out between the scientific field of sociology and geography (Oliver-Smith Ibid), reveals no purely scientific challenges. The relevance of this definitional debate results from the fact that acts of defining a disaster has crucial socio-political and economic repercussions. National and state governments around the globe specifically address the responsibility to support society in cases of disaster and catastrophe. Moreover, structures on the global level have been built with the specific aim to "coordinate effective and principled humanitarian action [...] in disasters and emergencies" (UNDAC 2013: 7). While a scientific definition that emphasises the physical processes attributes responsibility of damages that affect society to the natural environment, other definitions make explicit that a disaster can only happen when mechanisms installed for the protection of society and in society fail (Carr 1938: 211 cf Felgentreff & Dombrowsky 2008: 24). Research on disasters and catastrophic events has in large parts turned away from the specification of disasters as *natural disasters* and emphasised that it should be referred to as "social disasters" (Felgentreff & Glade 2008). Despite its´ controversial use in science, the term *natural disaster* is used frequently in media coverage (e.g. New Zealand Herald, El País, The New York Times) in science-journalistic accounts like e.g. the journal National Geographic (n. Y.) as well as in scientific publications up to the current date (Tseng & Chen 2012; Nix-Stevenson 2013; Neumayer et al. 2014).

Risk is a term discussed with similar controversy in scientific discussion and policy making as the term disaster. This results in the challenge to find an adequate terminology to communicate within and between scientific disciplines up to the present day. An attempt to take into account the different research traditions is made in the definition presented by the United Nations International Strategy for Disaster Reduction (UNISDR) (Felgentreff & Dombrowsky 2008: 19). UNISDR terminology presents a definition of risk as "the combination of the probability of an event and its negative consequences" (UNISDR 2009: 25). In science, the term risk is not only relevant in the context of hazards or disasters, but has a much broader field of application spanning various disciplines. As an example, portfolio theory developed in finance modelling can be named (Markowitz 1952). Different scientific disciplines have elaborated risk definitions, linked to specific conceptual vocabulary that cannot be easily transferred from one disciplinary field to another. The strong contrast between quantitative approaches and qualitative approaches to define or measure risk represents general contrasts between major scientific paradigms. The term *risk* is a key term of this study and is highly relevant in social as well as interdisciplinary geographic research. Theoretical accounts on risk are elaborated and discussed in more detail in chapter 4. As the risk concept is however crucial for the development of the research question, some basic accounts on the way the risk concept is addressed by the researcher is given here. Risk can be seen as a factor external to society that comes into being when there is a hazard and an exposure to this hazard with an attributable probability (Deleris et al. 2004). Other perspectives present risk as "an inevitable by-product of the pursuit of benefits that are important to societies and communities" (Tierney 2014: 31). In this study however, risk is regarded as the result of a primarily and fundamentally social process. Dynamics in which risk is produced involve organisational processes just as cultural and social-psychological ones and must be regarded as long-term and dynamic ongoing processes (Ibid: 40, 46). Resulting from this processual understanding it is necessary to highlight the social and political relevance of the term risk and specific power dynamics that reflect in processes of risk definition. This is revealed in a series of research results from case studies around the globe, including Mexican cases (i.a. Bankoff 2001; Collins 2009; Frerks & Hilhorst 2004; García 2005; Hörnqvist 2010; Pelling & Manuel-Navarette 2011). Building on these findings, relevant power dynamics involved in risk making in Chiapas are of interest in this study.

It has been argued among scientists from diverse disciplines that disasters and disaster risks – whether emphasis is given to natural or social dynamics – need to be investigated in an interdisciplinary manner (Gelman & Macías 1984; Schenk 2007: 18). Interdisciplinary research comes with challenges as it brings together not only different research questions, methods and foci, but especially because of the different paradigms involved. Moreover, the paradigmatic basis and assumptions on scientific truth and epistemologies are not always reflected upon in a transparent manner (Lincoln et al. 2011). Comprehensive literature review has shown that research has in the past decades largely remained discipline-specific (Gall et al. 2015: 263). Furthermore, Tierney (2007: 517) argues that interdisciplinary research has resulted in high transaction-costs for social scientists which can even undermine or weaken scholarly research. Notwithstanding, funding schemes and national research programmes largely require research projects on hazards, disasters and risk to be set up in interdisciplinary consortia.

The three terms presented in this section frame a growing research field that involves researchers and research cases from around the globe. Prominent journals such as Natural Hazards, Disasters, or the International Journal of Disaster Risk Science, reveal that the terms are well-accepted and known in English-speaking scientific circles. Gall et al. (2015: 255) point towards a predominance of European and North American researchers and institutions in this research field. Moreover, the authors reveal that the dominance of journals and papers in the English language pose a challenge to non-native speaking researchers and practitioners to identify and utilise relevant research results from and for their study areas (Gall et al. 2015: 260). For the case of research carried out in Latin America, the language barrier however also exists in the other direction. Research results on disaster related topics published in Mexico are hardly taken up in international knowledge platforms unless they are written in English.

The three key terms and their specific definitions are also reflected in strategies and action taken in the political arena, from the local to the global level. All three terms are common while translations, definitions and operationalisations can differ significantly from country to country. In 2009, UNISDR published the suggestion for consistent terminology in various languages (UNISDR 2009). Relevant for the case of Mexico is the translation into Spanish, with the terms "desastre" (=disaster) and "riesgo" (=risk) that can be directly translated from English. The term *hazard* can be translated as "peligro" (Cambridge University Press n. Y.), while the Spanish word can also mean danger or risk and therefore is less specific. Documents published by public actors with influence on civil protection and climate change adaptation refer to these terms and related concepts, also including terms such as *vulnerability* and *resilience* that are used increasingly today (CENAPRED 2014; DOF 2015). It is questionable however, if a transfer of terms and concepts developed in one geographic region to other parts of the world through formal translation presents the best option. As will be revealed throughout the chapters of this study, local and colloquial terms used to describe dynamics and processes in the social and physical environment comprise relevant information about local and regional conditions. It is highly relevant to take up these terms in order to gain access to relevant data and to have the necessary context information to interpret these data. An open approach which is capable of identifying relevant terms beyond the predefined, such as *hazard*, *disaster* or *risk*, can establish an adequate basis for communication in qualitative research, and ensure that scientific contributions go beyond what is already known and discussed.

2.2.2 Research on floods and climate change with relevance for Mexico

In current discourse on climate change and global risks in general, flooding is described as one of the fundamental perpetrators of loss and damage in present times as well as possibly for the future (World Economic Forum 2014: 13). From a global perspective, floods are discussed among the disasters that cause the highest losses of human lives (Jonkmann 2005: 172). However, it is also the loss of land and livelihood which are severe consequences of many floods around the globe. Research on floods is carried out in a range of different scientific disciplines in most diverse paradigmatic and epistemological frames. A wide range of countries and world regions present empirical case studies on floods. Scientific literature distinguishes different types of floods concerning the origin and effects of a flood (Ashley & Ashley 2008; Berz et al. 2001; Jonkmann 2005). Three types of floods mentioned frequently in scientific literature, government documents and reports by organisations dealing with disasters and humanitarian aid are river floods, coastal floods and flash floods (Berz et al. 2001: 458; The Sphere Project 2004: 371). Moreover, there are many different typologies for floods, whether they take place in an urban or rural environment, linked with permanent or temporal water bodies or whether they happen at a lake, a river or the sea. It is with a more frequent occurrence of floods in countries of the Global North that the necessity to realise a transfer of knowledge between academic disciplines and practitioners has been pronounced repeatedly (Weichselgartner 2008: 334). At the same time, countries in the Global South, especially the densely populated countries of South Asia, suffer strongly from floods in terms of human and economic losses (Doocy et al. 2013: n. p.).

In Latin America, an abundance of investigations has been carried out on floods and consequences of flood events, including various cases from Mexico (i.a. Briones 2012; Lammel et al. 2008; Romero Rodríguez 2011). In Mexico, the National Centre for Disaster Prevention (CENAPRED), defines a flood as

> "any event which – due to precipitation, waves, storm surge or failure of any hydraulic infrastructure – provokes an increase in the open water surface of rivers or the sea and generates an invasion or penetration of water in areas where usually there is no water and which causes damage to population, agriculture, livestock or infrastructure" (CENAPRED 2004: 5, translated by the author).

As indicated in figure 3, in a national comparison over a period of 50 years, Chiapas is shown to be one of the states that have in the past experienced a significant number of floods (Ibid). Considering global and regional climatic dynamics, the variability of frequency and severity of flood events in the study region is likely to increase in the future. The increase in extreme meteorological events, especially those related to the ENSO phenomenon (El Niño Southern Oscillation) in recent years, has given rise to a range of practical research cases. Empirical findings relate ENSO phenomena to the loss of livelihood (Olesen 2010: 113; Nielsen 2010: 134) as well as to flood events that have caused severe structural damage (Rossing & Rubin 2010: 65f; Duncan Golicher et al. 2006). While research on the dynamics of the ENSO phenomenon still is under development and predictions of the severity of an ENSO period can hardly be made, researchers describe that for Central and Southern Mexico the *La Niña* years result in higher than average rainfall (Endfield 2008: 11; Olesen 2010: 113). Moreover, a link can be described between ENSO and the frequency of hurricanes (Endfield Ibid), which can cause extreme precipitation. Besides the ENSO phenomenon, Mexico experiences a range of recent meteorologic dynamics that are attributed to global climate change. These can mainly be observed in changes in temperature and rainfall patterns. Although changes can be observed throughout Mexico, there is a spatial variance concerning different climate change patterns (Karmalkar et al. 2011). Before describing recent changes in climate, the general classification of climates in Mexico and the study region is presented.

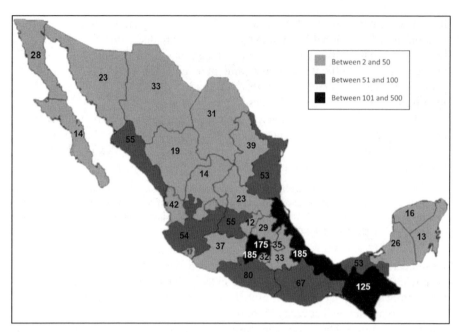

Legend:
- Between 2 and 50
- Between 51 and 100
- Between 101 and 500

Figure 3: Floods registered in Mexico between 1950 and 2000
Source: apadted from CENAPRED 2004: 5 (legend translated by the author)

Mexico is characterised by great climatologic diversity and includes strong contrasts, e.g. between predominantly arid zones in the Northwest and humid zones in the South of the country. The climate is mainly under the influence of the characteristics given by latitude, the distribution of the land mass, altitudes and surface configuration of land as well as by surrounding oceans and influencing currents (Vivó Escoto 1964: 188f). An important influence of atmospheric circulation on Mexican climate are the Northeast Trade Winds, which – as they do not only involve horizontal but also vertical movement of air masses – cause instability of air masses and can result in heavy rainfalls as well as thunderstorms (Ibid: 192). According to García (1988), Mexico can be divided into different climate zones, adapted from the Köppen classification. The *Atlas Nacional de Mexico 1990–1992* (UNAM 1992) presents a map by García which describes eleven different climatic regions in the country. The study region is located in the humid tropics, corresponding to the Af and Am climates in the Köppen classification (García 1988: 20f). In these climates, the mean temperature lies above 18 degree Celsius and there is an abundance of rainfall throughout the year (Ibid: 19). More precisely, the region is located at the border between two main climatic regions described as characteristic in the South of Mexico (figure 4, colour plate p. 138).

The Gulf of Mexico region is characterised by i.a. tropical cyclones in summer and autumn and north-winds ("nortes") in winter; the Southwest region, that is in the zone of intertropical convergence (ITC) during parts of the year, is characterised by summer monsoon winds and tropical cyclones (UNAM 1992). The mean temperature in the study region lies between 26–28 degree Celsius and the amount of precipitation ranges between 1500–3000 mm per year (INEGI 2010a). Whereas the temperature shows little variability throughout the year, there are significant differences in the distribution of rainfall (Ibid). The month with most precipitation is generally the month of September and the lowest amounts of rainfall occur in March (CONAGUA 2014: 29).

Long-term data on the mean yearly temperature in the state of Chiapas from 1901–2000 show a general increase in temperature. Whereas the highest increase has been of 1.8 degree Celsius in the Fronteriza, Soconusco and Sierra zones, others have experienced an increase between 1–1.4 degree Celsius (Ramos Hernández et al. 2010: 33). Data on precipitation in the period 1951–2000 reveal an atypical increase in the amount of rainfall in the Soconusco and Sierra zones, in parts of the Selva zone as well as in parts of the North of the state (Ibid: 32). While in some zones of the state rainfall has decreased in the years between 1901–2000, parts of the North and Selva region have experienced an increase of 100–300 mm (Ibid: 34). Under the influence of *El Niño*, rainfall patterns in the main rainy season are described to be below normal, including droughts, while in years of *La Niña* influence, rainfall amounts are in the normal range or above normal (Magaña et al. 2004: 29ff). As the study region lies in the Selva region of Chiapas, the information presented in the report prepared for the Government of Chiapas (Ramos Hernández et al. 2010) indicates that recent changes in the study region may be attributed to climate change. This includes the changes in mean temperature and precipitation but also in the extreme events linked to flood and drought. The Catazajá and Palenque municipalities are among the few regions in the state of Chiapas where heat waves have been recorded (Ibid: 41f). The fourth assessment report (AR4) of the Intergovernmental Panel on Climate Change discusses regional climate change projections for Latin America (IPCC 2007: 50). There is growing consensus among scientists that climate change will increasingly interact with other environmental and natural resource dynamics around the globe, including disaster risks (Ibid: 70). For Latin America, an increase of the exposure to climate-related disasters is discussed (Rossing & Rubin 2010). Accordingly, in Mexico an increase in several disastrous events is expected which includes floods and droughts just as a rising intensity of tropical storms (Verner 2010: 11).

In the study region, a diversity of problems exists related to the availability of water (Oswald Spring 2011); however, the phenomenon of flooding is highlighted in this investigation due to a lack of knowledge about the temporal dynamics of flood processes and related socio-ecological dynamics. The study region being located in direct vicinity of the river Usumacinta and the larger Usumacinta water system experiences seasonal floods of varying intensity year over year. Generally, floods are likely to occur in the months between September and November following high rates of precipitation (CONAGUA 2014: 30). These are the months with highest amounts of precipitation in the region which saturate soils and overflow water bodies. Moreover, the water level in the river is mainly influenced by rainfall in the upstream regions and the adjacent rivers in Tabasco, the Lacandon forest and the highlands of Chiapas as well as in the highlands of Guatemala, where the river originates. It is only by a combination of high rainfall in the study region and a high water level in the river coming from upstream regions, that severe floods occur in the Lower Usumacinta region. It is still a challenge for the natural as well as for the social sciences to model and understand the complex processes that take place in a widely dispersed socio-ecological network including the physical river network and human as well as non-human entities.

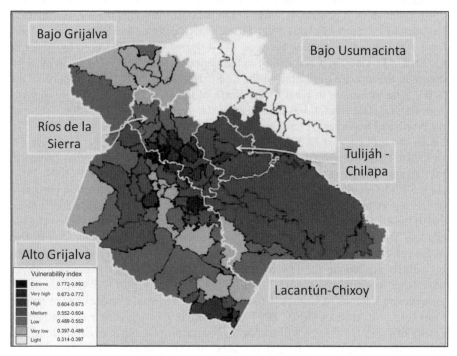

Figure 5: Vulnerability index Grijalva-Usumacinta
Source: adapted from BID 2014 (legend translated by the author)

The Usumacinta also is part of a larger water system called the Grijalva-Usumacinta rivers system. This hydrologic system and the people living within the administrative regions are considered to be highly vulnerable towards climate risks (BID 2014: n. p.). Vulnerability is attributed to the region due to the fact that its inhabitants suffer high levels of marginalisation as well as social and economic inequality (BID 2014: n. p.). Figure 5 shows the spatial distribution of vulnerability in the different municipalities of the states of Chiapas and Tabasco in a map elaborated by the Inter-American Development Bank (BID). The Palenque municipality, in which one of the case study villages is located, shows high levels of vulnerability, while the Catazajá municipality is attributed with a low vulnerability level (BID 2014: n. p.). Large parts of the population in Tabasco, as well as the population in the study region in Chiapas, live in settlements which are located at or below 50 metres above sea level.

The Plan for Integrated Adaptation, Planning and Management of the basis for the rivers Grijalva and Usumacinta (PAOM), elaborated by the Inter-American Development Bank (BID) identifies five priority regions for future intervention with regard to climate change dynamics. The region under investigation in this study is included in one of these (zone 5 in figure 6, colour plate p. 138) (BID 2014: n. p.). The priority regions were identified due to the following characteristics and assets (BID 2014: n. p.):

1. Consideration of the population which is most vulnerable towards climate change
2. Strategic infrastructure with a high exposition towards climate risks
3. Key ecosystems and environmental services for the building of resilience against climate change
4. Existence of economic activities which are highly vulnerable towards climate risks
5. Political and social feasibility which allows for the implementation of measures and their sustainability over time.

For several decades, research on floods in Mexico has been dominantly carried out by representatives of the natural sciences and engineers, especially in the disciplines of hydrology, hydro-engineering, meteorology and physical geography (i.a. Domínguez & Sánchez 1990; Soriano Roque 1988; Vega 1980). The case of Mexico reflects a general tendency in science and public to regard floods as related dominantly to natural phenomena (Felgentreff & Dombrowsky 2008: 14). Following the dichotomy opened up between nature and human beings, the so called "natural disasters" have for a longer period in the twentieth century primarily been studied in disciplines of the natural sciences (Elverfeldt et al. 2008: 32f). With a turn in scientific discussion towards concepts like vulnerability in the 1980s, the social factors interacting with physical processes of the environment were highlighted and a more comprehensive model of risk was developed (Ibid: 33). This tendency has shown repercussions in research on floods in Mexico which has also experienced a growing amount of contributions from the social sciences and human geography in more recent years. These contributions were especially realised in interdisciplinary research projects on topics of climate change, disaster management and resources management (i.a. García Acosta et al. 2012; González-Espinosa & Brunel Manse 2015). Contributions from social anthropology have emphasised that there are key social processes involved in the construction of risk and have presented case studies from Mexico as well as from other parts of the world (García Acosta 1996, 1997; Lammel et al. 2008; Briones 2012). Furthermore, a range of postgraduate thesis and journal papers from Mexico have addressed the topic of floods in an interdisciplinary manner (i.a. Angulo 2006; Gómora Alarcón 2014; Reyes Barrón 2012; Sandoval-Ayala & Soares-Moraes 2015; Valadez Araiza 2011). Besides the characteristics of the physical environment, aspects such as social vulnerability, livelihood and political participation have been included more and more in scientific research on floods (Valadez Araiza 2011). However, exchange between different disciplines and actors which study disastrous events like floods, droughts and other phenomena and their dynamic causes seems to be in an early stage and linked to a small group of researchers.

The government of Chiapas highlights the risk that is linked with both the hydro-meteorological phenomena and the high levels of vulnerability of the population of the state (Ramos Hernández et al. 2010: 58). Ramos Hernández et al. (2010: 28) highlight a range of major causes that can provoke floods in Chiapas:

– persistent rainfall in a region (storm) over an extended period of time
– very strong rainfall out of the rainy season
– abrupt rise of the high tide in coastal zones due to storm surge or thunder storm,
– obstruction of river bed due to landslides or earthquakes
– sudden burst of a large dam wall which can be provoked by overloading of amount of water or by an earthquake.

This list shows that a general awareness exists concerning the variety of factors that cause or aggravate flood events. It is especially with the Law on National Waters (DOF 2016) and the legal mandate held by CONAGUA for the management of national rivers and water bodies that the government tries to address river related floods. Moreover, the challenges posed by climate change have set impulses for the creation of a new legal environment in Mexico. The General Climate Change Law (LGCC) published in 2012 and amended in 2014 and 2015 addresses the task to reduce vulnerabilities towards the effects of extreme climatic events (UNDP 2014: 45). An Inter-Ministerial Climate Change Commission (CICC) was founded in 2005 which involves key units of the civil protection system (Ibid). While the connection between scientific results and political action in the field of floods is still a challenge, the creation of an Inter-Ministerial Commission to Address Droughts and Floods (CIASI) can be seen as a starting point to manage different risks together in a holistic manner (Ibid).

Following the brief overview on state of the art research and open questions, the next section takes a look at current trends in global DRR as well as in the civil protection system and disaster risk management in Mexico. This is believed indispensable in order to better understand how hazards, disasters and risks are defined and dealt with on an international level as well as by the authorities in the country. Additionally, specific challenges in flood management are presented that relate to processes of DRR as well as to current developments in other public sectors in Mexico and to broader global strategies.

2.3 CHALLENGES IN DISASTER RISK MANAGEMENT IN MEXICO

Disaster risk reduction (DRR) is a term that was coined on the international level and has gained prominence through various reports and activities of the United Nations International Strategy for Disaster Reduction (UNISDR) since the year 2000 (UNISDR n. y.). Since the 1960s various disasters had raised attention and called for action on a global level. This call was especially pushed forward with the announcement of the International Decade for Natural Disaster Reduction from 1990-1999. With the installation of UNISDR significant levels of awareness for pressing issues of disaster risk have been gained. Disasters are not only linked to natural hazards but a perspective has been increasingly developed towards the different levels of vulnerability of population groups around the globe. Following the 2005 World Disaster Reduction Conference, a first framework was developed with the goals to describe in more detail the contribution and action that is required by different stakeholders and to substantially reduce disaster losses: The Hyogo Framework for Action 2005–2015: Building the Resilience of Nations and Communities to Disasters (HFA) (UNISDR 2005). This first framework was superseded by the Sendai Framework for Disaster Risk Reduction 2015–2030, that presents a more comprehensive set of tasks and has been developed in line with a general post-2015 development agenda (UNISDR 2015a).

Mexico is one of many countries that has actively pushed forward action for global Disaster Risk Reduction on the international level, especially since the inauguration of the Hyogo Framework for Action (OECD 2013a: 207). Having experienced the devastating effects of the 1985 earthquake in the capital Mexico City, a densely populated megacity located in a physical environment prone to various hazards (earthquake, flood, volcanic eruption and others), the government strongly revised and modernised the national system of civil protection, which resulted in the creation of two mechanisms. The Mexican system integrates risk management into the different levels of government and civil society. With the installation of the National Civil Protection System (SINAPROC), the coordination of emergencies from the local up to the state level and finally to the federal level is outlined (SINAPROC n. Y.). Moreover, SINAPROC is the national platform for Mexico collaborating with UNISDR. The other entity created after the 1985 earthquake is the National Centre for Disaster Prevention (CENAPRED). Supported by the Government of Japan and with scientific advice by experts of the National Autonomous University of Mexico (UNAM), it was created in 1988 (CENAPRED n. Y.). CENAPRED is the technical and scientific branch of the national system supporting SINAPROC. The government of Mexico manages two national funds to finance prevention and recovery operations for disasters. The Fund for Natural Disasters (FONDEN), mainly used for emergency responses such as food and water supplies as well as reconstruction was installed in 1996. The other mechanism is the Fund for Prevention of Natural Disasters (FOPREDEN) which was created in 2004 and mainly finances prevention activities such as risk identification, early warning systems and disaster risk awareness. One example for action is that since 2004 FOPREDEN has been implementing a strategy to give incentives to all Mexican states to elaborate a risk atlas (OECD 2013b: 9). Besides the financial mechanisms other initiatives include an integrated hydrology programme in the state of Tabasco. This programme was developed as a consequence of severe damage caused in this flood-prone state in a 2007 flood which had directly affected around 70% of the state's territory (CONAGUA 2012: 7). Mexico disposes of a General Law of Civil Protection (LGPC), which was published in 2012 and amended in 2014. In the National Programme for Civil Protection a range of national goals is presented, among which especially preventive action against disasters is highlighted and concrete actions are promoted such as the elaboration of risk atlases at all relevant levels including all municipalities in the Mexican states (SCT 2016: 11f).

Today, Mexico is a country that emphasises the necessity to further develop strategies for disaster risk reduction on the international level. In a commemorative event 30 years after the 1985 earthquake, Ms. Wahlström, head of UNISDR, addressed the role of Mexico as follows: "We know that we can count on Mexico as an ally and friend in contributing decisively to the implementation of the post-2015 development agenda, and in making the world more resilient to disasters" (UNISDR 2015b). Despite intense international cooperation and action intensified under the Hyogo Framework for Action and a modern national system of civil protection, disaster prevention is still in its inception on the local and regional levels in the country. The national system has revealed major shortcomings in large disasters such as the 2007 flood in Tabasco as well as in 2013, when landslides and severe floods affected more than 75% of the country's territory following the tropical storm Manuel in the Pacific and Hurricane Ingrid in the Gulf of Mexico (Pedrozo-Acuña et al. 2014: 295). On the international level, it is argued that in Mexico a shift is needed "from a reactive approach to disaster to a proactive disaster risk management approach" (The Worldbank 2013). In a programme operated by the Worldbank in 2012 and 2013, the Government of Mexico was supported in the development of a comprehensive disaster risk management framework. Focus was laid on a comprehensive strategy of risk finance, on measures to bring experience in risk financing from the national to the state level but also and maybe most importantly, from a perspective on local and regional challenges on activities to mainstream DRR in land use and urban planning (The Worldbank 2012: 41ff).

The civil protection system in Mexico is organised and works on three different governmental levels which are the federal, the state and the local level (UNDP 2014: 47). As a report by UNDP indicates however, there are still major shortcomings concerning the overview on risks and specific vulnerabilities on the local level. The authors of the report call for the collaboration of academia, state governments and the civil society in order to accomplish the compilation of state risk atlases and the integration into a national atlas (Ibid). An even greater challenge exists in the collection and integration of data from marginalised municipalities in the Centre and in the South of Mexico (Ibid: 46). With this data available an elaborate system of development planning and the integration into ecological and land use regulations would be possible.

The state of Chiapas relies on a strong system of civil protection. Chiapas as the Southernmost state of Mexico and due to its geographic location and characteristics of the physical environment, is regularly exposed to various hazards such as tropical cyclones, extreme rainfall events, droughts and wildfires as well as earthquakes and volcanic eruptions. The state government has developed a specific civil protection strategy in order to manage risks and disastrous events in a comprehensive manner. With the Civil Protection Law for Chiapas presented in 2014, the organisation of civil protection from state level to municipal and local level is outlined (Art. 8 cf Secretaría General del Gobierno 2014: 17). It also describes the formation of the Institute for the Integral Management of Disaster Risk for the state of Chiapas (Art. 33 & 34 cf Secretaría General del Gobierno 2014: 27). Closely linked to this institute, in 2013 Chiapas founded the first school for civil protection in the country in order to carry out applied research for disaster risk management as well as to assure a high-level vocational training for paramedics, fire fighters and civil protection staff in various fields of expertise (Secretaría de Protección Civil 2015a). Article 162 highlights that special responsibility is taken by the federal and state level for disastrous events taking place in rural areas of Chiapas that are caused by climatological events (Secretaría General del Gobierno 2014: 55). The state atlas of dangers and risks is in place and can be accessed online at any time (Secretaría de Protección Civil 2015b). However, when data is to be obtained from specific municipalities, the lack of local data becomes obvious. Besides the availability of data, a range of challenges exist at the regional as well as at local level, which include challenges of access to resources and staff, lack of capacity building as well as other constraints. These challenges lay part of the basis for this study and are within the main research interest.

Another challenge for flood management in Mexico is related to the energy and infrastructure sector. Current dynamics in these sectors which aim towards an increase in energy production and distribution involve the planning and implementation of hydro-energy projects in Chiapas. With this they are part of diverse water related infrastructure projects currently planned and implemented around the globe (Boelens et al. 2016: 6). In general, hydro-energy projects are designed to make use of river waters for the production of electric energy, which in many cases involves the construction of dams that alter the course of river networks in substantial ways. Environmental and social impacts of hydro-energy projects that involve dams are broadly documented and discussed (Postel & Richter 2003; World Commission on Dams 2000; Richter et al. 2010). Nevertheless, the technology to use the natural potential of water is often described as environment friendly and the energy produced is labelled in the group of renewables (Darmawi et al. 2013: 215). In general, dams have been constructed around the globe since historical times for purposes like flood control, supply of drinking water, for irrigation management as well as for energy production (Tahmiscioğlu et al. 2007: 760). In this case study the discussion of dams, dikes and hydro-energy projects in general is relevant due to a range of strategic decisions that are taken at the national and international level, which might have strong effects on the case study region. For Central America, an expansion of hydropower capacity is predicted for the decades to come, as national needs for electricity supply are on the rise (Esselmann & Oppermann 2010). Mexico started with the installation of hydro-energy projects in the year 1889, only few years after the very first small projects had been implemented in the UK, France and the USA (Ramos-Gutiérrez & Montenegro-Fragoso 2012: 103). With the formation of the Federal Commission of Electricity (CFE) in 1937, more and more projects were implemented around the country and substantially transformed Mexico´s water and energy infrastructure (Ramos-Gutiérrez & Montenegro-Fragoso 2012: 103). Today, Mexico administers a total of 4000 dams of varying size (figure 7). Chiapas holds a significant part of the total amount of water volumes in the Mexican territory with the Grijalva-Usumacinta system representing around 30% of national water volumes (BID 2014: n.p.). The Usumacinta River, that originates in the highlands of Guatemala, is the river with the highest water volume in the country (BID 2014: n.p.). In Chiapas four large hydro-energy projects exist on the river Grijalva (figure 8), which have an impact on the extended Grijalva-Usumacinta river system.

The first hydro-energy project built on the river between 1958 and 1966 is the Malpaso dam (also known as Netzahualcóyotl) that was installed in a general development process implemented in the Southeast of Mexico in the 1960s (Ramos-Gutiérrez & Montenegro-Fragoso 2012: 116). The second project, La Angostura (also known as Belisario Domínguez), constructed between 1969 and 1976, is the largest dam project on the Grijalva, with a basin that covers around 18,000 square kilometres (Domínguez-Mora 2009 cf González Villarreal 2009: 12). The other two dam projects realised on the Grijalva are Chicoasén which started commercial operation in 1980 and Peñitas, which as the smallest and most recent project started operation in 1987. The national importance of this comprehensive hydro-energy infrastructure on the Grijalva is underlined by the fact that in 2008, the dam-system operated 42.3% of the total hydroelectric capacity in the country (Ramos-Gutiérrez & Montenegro-Fragoso 2012: 111).

The latest hydro-energy project on the river Grijalva is the project Chicoasén II. Construction work has been started in 2014 and is planned to be finished in 2018. However, large protests by the public as well as by the workers reveal highly critical issues of labour and human rights and show an impact on the progress of construction (Noticias Voz e Imagen 22.02.2016). On the river Usumacinta no hydro-energy project or other large infrastructural project has been realised so far. However, plans to construct dams on this river have existed since the 1970s and have been discussed for their possible political, socio-ecological and archaeological impacts (Wilkerson 1991: 119f). Recent dynamics on the national as well as international level indicate that new plans exist to realise hydro-energy projects on this transnational river. A hydro-energy project called Boca del Cerro or Tenosique, that is planned near the town Tenosique in the state of Tabasco, appears in current plans by CFE (2014) as well as in development plans of the national government, most prominently the National Infrastructure Programme 2014–2018 (DOF 2014: 162). While no official documents are available to the general public, journalistic work by the newspaper

Tabasco Hoy in 2015 has revealed that a transnational agreement between the governments of Guatemala and Mexico has been signed that would allow for the construction of hydro-energy projects on the Usumacinta (INAI 2015).

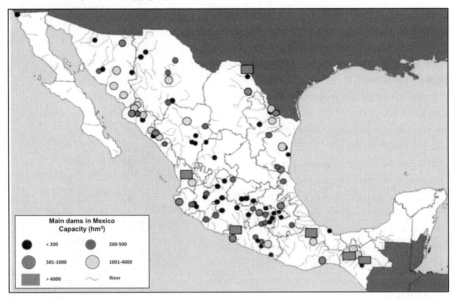

Figure 7: Major dams in Mexico
Source: adapted from CONAGUA (2010) cf OECD (2013)

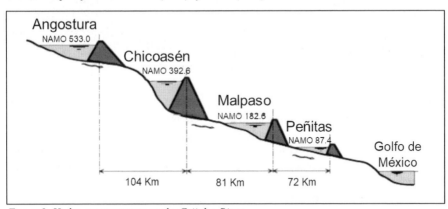

Figure 8: Hydro-energy system on the Grijalva River
Source: adapted from CFE cf González Villarreal (2009)

Table 1: Major challenges of dam and hydro-energy projects

Challenges of dam projects	Ecologic	Economic	Social
Limitation of sediment feeding into downstream river	x		
Change in water-soil-nutrient relations downstream (change in areas of former seasonal floods)	x		
Increase in erosion (human activity; permanent increase in the water turbidity)	x		
Increase in earthquakes (possible as an effect of filling of big dam reservoirs)	x		
Increase in evaporation loses as a result of the increase in the water surface area	x		
Microclimatic changes related to changes in air moisture percentage, temperature and movements	x		
Change of water quality (e.g. vertical change of temperature, salt and oxygen distribution; returning water from irrigation schemes; discharge of toxic matters)	x		
Threat to fish species (no access to egg lying zones; death when passing through floodgates, turbines, pumps)	x	x	
Increases in water sourced illnesses like typhus, typhoid fever, malaria and cholera		x	x
Effects of resettlement processes on the social, cultural and economic structure of the region and on resettled population groups		x	x
Changing agricultural patterns due to change in soil fertility downstream (need for additional fertilizer)		x	x

Source: author's elaboration based on Tahmiscioğlu et al. (2007)

The construction of a dam and hydro-energy-project can have a range of positive and negative effects, concerning people, the economy and the environment. As Tahmiscioğlu et al. (2007: 763) highlight a range of possible effects exist, which must be distinguished concerning their short- or long-term effects. The major benefits are flood control, provision of water, land improvement through irrigation and a major economic benefit made through the generation of electric energy (Ibid 764). Some of the major challenges are listed in table 1. Among them are an increase in erosion and possibly even an increase in earthquakes which is a highly relevant aspect given the tectonic conditions in Mexico. Furthermore, health aspects such as a possible increase in water-sourced illnesses (e.g. typhoid fever and malaria) as well as detrimental effects on the agricultural characteristics due to a change in soil fertility in downstream regions are discussed (Ibid: 762f).

The abovementioned aspects relevant in the consultation about and planning of hydro-energy projects on the river Usumacinta are of high interest in this study because of the various possible effects this infrastructure and subsequent activities in the region could have on the people and environment in the case study region. The topic of hydro-energy projects however does not stand in isolation; rather, it interacts directly and in various ways with each of the topics of this research mentioned throughout this chapter.

One of the benefits of a dam system listed above is the possibilities it creates for water management. This includes options to store water in times of high precipitation and to distribute it in times of drought. Moreover, floods and landslides can be prevented through diverse techniques, such as through the deviation of water. However, dams planned for the generation of electric energy do not always offer safe conditions for water retention in times of flooding. In a dramatic case this proved to be true on the river Grijalva in 2007, when gates of the dam Peñitas had to be opened in order to prevent the threat of destruction of the main dam wall caused by extraordinarily high amounts of water in the river system (Gracida Galán 2011: 55). The meteorological conditions as part of the cause and the disastrous flood events that stroke Tabasco have been discussed above. However, it must be pointed out here that the management of the Grijalva dam system was not prepared for this amount of water in the river following an extreme precipitation. Although the dam system had had the capacities to control water levels in other events, a comprehensive strategy to reduce the risk of flooding under extreme conditions was not in place. The 2007 case of the Grijalva system shows the possible disastrous effects that can arise when extreme precipitation events affect the water level in a river system that is interrupted by dams that operate mainly for the purpose of production of electric energy. With climate change, scenarios are discussed for the South of Mexico that could reduce the amount of available water in the river systems (González Villarreal 2009: 161). At the same time, scientists discuss a growing likeliness of extreme weather events, including extreme precipitation in times of the year that had not been characterised by large amounts of rainfall in the past. Data on rainfall and river discharges however are still not easily accessible from CONAGUA, which is the institution in charge for river management (OECD 2013a: 87). For the management of hydro-energy systems, the lack of reliable data to anticipate extreme rainfall and water level rise in large river systems, will continue to be a major challenge in the coming years if no adequate measures, including the installation of an elaborate early warning system, are taken. The example from 2007 also shows that DRR strategies and prevention plans by the civil protection system in Mexico have not yet adequately taken into account hydro-energy projects.

As presented in this section, geographic research and other disciplines that deal with the physical and social aspects of hazards, disasters and risks have identified a range of open questions for the investigation of flood dynamics. Processes like ENSO and climate change as well as other social, ecologic and economic dynamics with effects on the global as well as on local and regional levels underline that addressing these questions is highly relevant. Different scientific disciplines need to contribute to a better understanding of the interrelated processes in which flood events are embedded. In the following section, specific challenges identified in the different lines of research and DRR and other sectors in Mexico and on a global scale are selected. Selection takes place based on what it is believed social geography can and should make a contribution to. This selection leads to the elaboration of the major research question and sub-questions of this study.

2.4 RESEARCH QUESTION

The complexity of the research field and problem presented above is high as the topics presented span various global and local phenomena and processes as well as different research disciplines and traditions. It is therefore necessary to point out the major tasks identified as relevant for geographic research in this field and to present a consistent research design to address these tasks. For the case of this doctoral thesis, a major research question is presented which connects the various topics addressed in a new way. This question is broken down into various sub-questions which indicate the scientific and practical directions towards which this research proceeds.

Risks and disasters are prevalent around the globe and have been discussed from global, regional and local level perspectives. As mentioned, this research addresses the topic of risks and disasters with the specific case of flood events in a Mexican river system. Floods are among those dynamics that have influenced and interacted with humans along the history of mankind. However, as it has been shown above, new dynamics that involve changes in atmospheric just as in social conditions, result in increasing losses of property and damage to human life. It is of political relevance and high scientific interest to find out, why loss and damage increase and how societies and their physical environment could be designed to prevent an increase in detrimental effects. From a social geographic perspective, it is especially the study of processes of decision making and of everyday life in social groups that are believed to provide substantial insight into the basic patterns of society with relevance for long-term flood management. Floods happen around the globe, while the type and severity differs strongly according to factors including topography, climate or infrastructural conditions. Moreover, the societal, cultural and economic specifics in locations and the larger geographical regions that create and encounter flood events are diverse. Research on floods needs to take into account these specifics as they have been found to be important for the analysis of differences in strategies, policies and measures. By choosing a case study in the South of Mexico, it is believed that conditions can be identified which are characteristic for the specific institutional, political and socio-cultural patterns in this region. The interplay between relevant actors who collaborate or compete on the different political levels is regarded as highly relevant for the understanding of flood risk management and spatial dynamics. Bringing together the abovementioned aspects, the major research question is the following:

How do different actors construct and negotiate flood risks in the region of the Usumacinta River in Chiapas, Mexico and what consequences do conflicting social practices of flood (risk) management have on the local spatial level in villages of Palenque and Catazajá municipalities?

The fact that the research question includes the notion of "construction and negotiation of risks" shows that the author choses a constructivist perspective towards the phenomena of this study and argues that risks are socially constructed. In the research question the term "flood (risk) management" puts the word "risk" into brackets as the concept of risk is considered ambivalent and requires further analysis in this study (chapter 4). The main research question implies a scientific approach as well as a specific research interest in selected patterns in the social and environmental complex encountered in the South of Mexico. In order to make this scientific approach explicit, four sub-questions are formulated that guide this study and help operationalise the main research question:

1. What social practices do local people and other relevant actors carry out in flood (risk) management?
2. How can the analysis of social practices and the materiality of practices help to better understand flood (risk) management?
3. How do different power relations reflect in the social and spatial construction of risk, such as in flood (risk) management (e.g. protection and prevention measures) and in larger social phenomena (e.g. discourses)?
4. Which role does the planning of hydro-energy projects on the Usumacinta River play in spatial dynamics of flood, flood risk and other risks for the future?

While the main research question makes use of the concept of the social construction of risk, in the sub-questions, further key terms of conceptual relevance are presented. The terms social practices, materiality and spatial dynamics point towards a specific conceptual orientation. Taken together, this configures the general theoretical outline of this study presented in chapter 4, linking (1) social practice theory, (2) theories of risk and (3) theories of space.

The main research question aims towards the explanation of flood related dynamics relevant in the South of Mexico. There are different relevant levels or scales implied in this research question. This comprises the local level, represented by the people who live in villages along the river Usumacinta and the regional level, which comprises government authorities and civil protection staff who are responsible for flood management and other measures that influence flood vulnerability (spatial planning, land rights, provision of basic services, employment and education, etc.). However, the research question is designed in a way that leaves analysis open to other actors on other levels that might interact with the practices on the local and regional level.

The relevant dynamics and spatial processes brought up in the research question require on the one hand a thorough and in-depth analysis of practices within social groups. On the other hand, it is also necessary to analyse complex patterns of practices across social groups. This demands to develop an understanding of those processes that feed into shaping and reproducing social practices related to flooding. It is believed that in order to accomplish this task it is necessary to connect empirically with the *life-worlds* of those people who shape their lives along the Usumacinta River. In order to introduce the larger context of the *life-worlds* to be addressed conceptually and empirically, the following chapter presents an overview on selected geographic and historical characteristics of the case study region.

3 INTRODUCING THE CASE STUDY REGION
– SELECTED SOCIO-SPATIAL DYNAMICS
IN A HISTORICAL OVERVIEW

In this study, an in-depth research in a delineated geographical region is carried out. A case study area has been identified in the state of Chiapas, in the South of Mexico. In order to describe the general characteristics of the larger region and the specific characteristics of those villages, where empirical research has been carried out, this chapter provides a presentation of the region starting from a short presentation of the case study villages before presenting the historical context of the larger geographical region. This chapter develops from a historical, descriptive perspective towards a more detailed and analytical approach. Following this approach, it is possible to take into consideration different relevant processes in the past and in the present and to identify and elaborate on the specific challenges encountered in the case study sites today. In the description of the research problem in chapter 2, general features relating to climatology, hydrology and especially flood dynamics in the research region have been discussed. This chapter in contrast emphasises topics of historical events and development, migration dynamics as well as cultural, economic, political and social processes and characteristics. The reconstruction of historical processes and events lays the basis for a conceptual analysis of the territory and processes of territoriality. Approaching the case study region from a perspective of territoriality allows the identification of first key actors and characteristic dynamics in the case study region. The introduction into a conceptual abstraction of historical and recent processes at the end of this chapter also builds the connection to the major theoretical concepts that are presented thereafter.

The larger region of relevance in this study is the Lower Usumacinta region. While the state of Chiapas has been chosen as the larger political and administrative entity to focus on in this study, the Lower Usumacinta region is an area that has its main extension in the states of Tabasco and Campeche (see also figure 1). The economic, organisational and political characteristics differ from state to state, however cultural, historical and social characteristics are shared beyond administrative borders. Therefore, patterns and dynamics that are presented as relevant for Chiapas may also be of relevance and present a challenge to future developments in other Mexican states. The Mexican state of Chiapas is located in the South of Mexico, sharing borders with Guatemala to the South, the Mexican states of Veracruz and Oaxaca to the Northwest and West and Tabasco to the Northeast and East. In the West, Chiapas has a coastline that connects the state with the Pacific Ocean. In 2010 Chiapas had a population of 4,796,580, which is around 4.3 percent of the total population of Mexico (INEGI 2010b). With 46.3 percent of the population at poverty level and 11.4 percent in extreme poverty in 2010, Chiapas is the state with highest poverty rates in Mexico (SEDESOL 2012). A graphic elaborated by the National Council of Evaluation of the Politics of Social Development (CONEVAL) (figure 9, colour plate p. 139) presents that in almost every of the 118 municipalities of the state, large parts of population suffer from detrimental living conditions (CONEVAL 2010). These conditions include economic poverty, lack of access to basic education, health care, social security and living conditions in houses built with poor construction material (SEDESOL 2012). The capital of the state, Tuxtla Gutiérrez (yellow in figure 9, colour plate p. 139), builds an exception from the general conditions of poverty. In Chiapas, 51 percent of the population live in urban areas, while 49 percent reside in rural settlements, which are those settlements with a population below 2,500 inhabitants (INEGI 2010b). The most important cities in the state are Tuxtla Gutiérrez (604,891 inhabitants), San Cristóbal de Las Casas (203,387), Tapachula (351,165), Palenque (120,882), Comitán de Domínguez (156,143) and Chiapa de Corzo (97,660) (Gobierno del Estado de Chiapas 2014: 1ff). The 118 municipalities are grouped into thirteen larger administrative entities, called *regiones*.

The case study villages are located in the municipalities of Catazajá and Palenque, which both belong to the socioeconomic region *Región XIII "Maya"* (CEIEG 2015). This region is located in the North-eastern and extreme Easter part of Chiapas and shares a border to the North with the state of Tabasco, while the other part of the region shares a border with Guatemala (figure 10, colour plate p. 140). The borderline in this part of the state is built mainly by the Usumacinta river. In 2010, Catazajá municipality which has the size of 631,764 square kilometres had a population of 17,140 and the considerably larger Palenque municipality which covers an area of 2,897,443 square kilometres counted a population of 110,918 (SEDESOL 2013a, 2013b). The two municipalities are characterised by a large share of rural population. While in Palenque municipality, around 43 percent is urban population and 57 percent is rural, the contrast is higher in Catazajá municipality with 21 percent of urban and 79 percent of rural population (Gobierno del Estado de Chiapas 2014: 12f). The case study villages are located in remote parts of the municipalities, characterised by difficult road access and lack of some basic services (e.g. health service and education). The remoteness of the villages is a crucial characteristic for the case study sites as specific challenges for assistance and prevention mechanisms for flood events can be assumed to result from the remote location and its side effects.

To begin with an overview of the history of the case study region, it is relevant to mention that the geographic region of concern covers little space in textbooks on Mexican history. As the study region lies in the border zone between today´s states of Chiapas, Tabasco and Campeche, the history books of the states describe the lands along the river mostly as remote areas and do not include much information about the ecological and social characteristics (Martínez Assad 1996; Mendozah 1955; Thompson & Poo 1985; Zebadúa 2010). In spite of the lack of written documents, it can be retraced that the case study region has been part of relevant historical processes that have influenced how people in the larger region and in the three states settle, work, live together and go about relevant everyday practices. A social geographic approach is chosen here in order to introduce the case study region, which links historical perspectives with socio-spatial perspectives. These perspectives allow for the identification of a range of processes that have shaped the case study region. These processes are presented in a chronological order, from pre-Columbian times until today. As no comprehensive historical reconstruction is intended in this study, key processes and events that have been identified as representative of the major development lines of the region are selected for this account. The literature accounts are few but represent influential works by famous authors as John Kenneth Turner (1910), Frans Blom (1954) and Jan de Vos (1987).

3.1 STRATEGIC IMPORTANCE OF THE USUMACINTA RIVER IN MAYAN TIMES

The river Usumacinta is one of the largest water systems in the country. It is not only the length but also the richness in water volume that presents the extraordinary value of this river. As mentioned in chapter 2, there is high economic interest in the water and its potential for hydro-energy production. At the same time, biologists and other environmental experts highlight the importance of the Usumacinta because of its high biodiversity and its large number of flora and fauna (Amezcua et al. 2007: 20). The qualities and strategic characteristics of this river have been valued since the rise of Mayan cultures and even before. The first population group from which traces have been found in the South-eastern part of the country, the Gulf of Mexico and presumably along the lower part of the Usumacinta, belonged to the Olmec culture which dates back 4,000 years (Hudson et al. 2005: 1031). Early settlements in the Maya Lowlands and along the Lower Usumacinta have been dated back to the periods of the Maya Classic (250–900 AD) and the Post Classic (900 AD – arrival of the Spaniards) in paleo-environmental studies (Englehardt 2010: 58; Solís-Castillo et al. 2013: 270). It is assumed that human settlements in these periods were established on the flat alluvial river banks which were richer in humus and nutrients than soils in the uplands and thus provided a solid basis for agricultural production (Solís-Castillo et al. 2013: 285). As Ruz (2010a: 10) argues, early population may have belonged to the Mayan group of the Chontales, who at a later time merged with other ethnic groups. Ancient Mesoamerican narratives documented in scriptures such as the *Popol Vuh* or *Chilam Balam* attribute an important role to water for the world view or *cosmovisión*, which includes the cyclical destruction and renovation of the world induced by events like floods, fires or hurricanes (Martínez Ruíz et al. 2007: paragraph 7). Besides the mythological importance of water in the Mayan culture, one of the strategic values of settlements along the Lower Usumacinta lies in the fact that this region geographically connects the Yucatán peninsula, the Highlands of Chiapas and Guatemala and further important regions like the Highlands of Central Mexico (Ruz 2010a: 10). The region was located along important trade routes in Mayan times. In the Early Classic (150–550 AD) the Usumacinta River and the San Pedro River attracted people looking for communication routes and with the installation of these routes along the Lower

Usumacinta the population in the area increased significantly (Solis-Castillo et al. 2013: 272). The name Usumacinta reveals the different cultural influences in the region. Whereas the region was mainly influenced by Mayan culture and language, the name Usumacinta originated in Nahuatl, an Aztec language (Ruz 2010a: 7). "Ozomatli" or "osomahtli" being the word for monkey, the Usumacinta is referred to as the "River of the Sacred Monkey" (Canter 2007: 1). Frans Blom, renowned archaeologist and expert in Mexican history and especially Chiapas, argues that the river Usumacinta served as the "principal arteria of Mayan culture" (Blom 1956: 9). Remains of sacred sites such as those of Piedras Negras, Yaxchilán, Pomoná and others are testimonies of societies with high cultural activity and a complex political organisation. However, Compañ Pulido (1956: 10) argues that the river Usumacinta had been forgotten for a period of almost one thousand years until rediscovery in the twentieth century. Archeologic sites in places like Emiliano Zapata and Chablé document the Mayan influence in the direct vicinity of the case study villages (Canter 2007: 23). Moreover, the name Catazajá, which is the name of the lagoon system and the municipality, is of Mayan origin and can be translated as "Waters that come, waters that go" ("Ha"=water and "Cat/Catha"= coming and going) (Latournerie Lastra 1985: n.p.).

3.2 ECONOMIC EXPLOITATION AND ECOLOGICAL EFFECTS IN COLONIAL TIMES AND AFTER

At the beginning of colonial rule, the region did not immediately attract the *conquistadores*, as the climatic conditions of the wet tropics and dense vegetation made life hard for Europeans. Since the 16th century, the region has served as a refuge for population groups from Campeche and Yucatán, who escaped inhuman labour and living conditions at plantations (Ruz 2010a: 13). Economic interest however caused the Spanish crown and other Europeans to open up pathways into the Lower Usumacinta region. The *Palo de tinte* or Campeche wood (bot. *Haematoxylum campechianum*) was a resource of major interest to Europeans. Whereas it had already been used in prehispanic Mexico, the *conquistadores* found interest in the wood and its qualities as dying agent as early as 1561 (Villegas & Torras 2014: 79). Trade with the different colours produced from *Palo de tinte* flourished due to increasing demands from textile producing nations such as France, England and the Netherlands (Ibid). The river was the main infrastructure for transport and commerce and it linked various places along the widespread river and lagoon system (Torras Conangla 2012: 10f). In the 19th and 20th century, the economic potential of the river and the adjacent lands was explored and exploited mainly by traders of precious woods. The species most appreciated included mahogany (bot. *Swietenia macrophylla*) and cedar (bot. *Ceiba pentandra*), which are among the characteristic tree species of large height (25 – 50 m) in the region (Amezcua et al. 2007: 20). Mexican traders mainly originating from the state of Tabasco organised the felling of the trees and their transport on the river to the coast (Vos 1987: 73). This timber was mainly shipped to Europe, where it was used in the production of furniture or in the construction of ships (Benjamin 1981: 509). Important national as well as international timber companies built factories in the towns along the river, e.g. Tenosique in Tabasco, and expanded their concessions from the Lower Usumacinta into the Lacandón Forest in Chiapas as well as Guatemala in the last decades of the 19th century (Vos 1987: 73f). Another natural product was exploited along the river Usumacinta: the *chicozapote* tree (bot. *Maninkara sapota*) (Arrivillaga Cortés 1996: 362). *Chicle*, the natural gum extracted from the tree, again a process that had already been applied by Mayan people, served as a basis for chewing gum which was at that time mainly produced in the USA. The way into the exploitation of *chicle* was opened through the wood industry, because it was the forest experts working in the region that best knew about the qualities of this tree (Ibid: 366).

The massive use and exploitation of forest resources along the Usumacinta had a vast temporal as well as geographic extension. A long period characterised by the unsustainable extraction of *palo de tinte* under colonial dominance as well as the shorter period of the extraction of precious woods and of *chicle*, the latter also involving the installation of plantations in the 19th and early 20th century, had substantial effects on the forest vegetation and settlement structure along the river and the wetlands. Over the decades this has resulted in a range of ecological and social challenges. Economic processes in the 20th century intensified environmental degradation in the region. González Pacheco (1995: 37) describes three major processes that have caused transformation and degradation of virgin or secondary forests in Mexico (table 2).

Table 2: Major processes contributing to transformation and degradation of virgin and secondary forests in Mexico

1.	The growth and modernisation of industrial activity derived from silviculture
2.	Development of extensive cattle breeding in forest zones, especially in areas of the humid tropics
3.	The construction of infrastructural mega projects in the sectors of hydro-energy, fisheries, petrochemical industry, tourism etc.

Source: adapted from González Pacheco (1995)

All three processes prove to be relevant in the case of the study region. While industrial activity derived from silviculture represents the earlier phases of degradation described above, the study region exemplifies the process of transformation of large parts of agricultural land, forest area and marsh lands into grazing land for cattle, which has been intensified in the lowlands of Chiapas and Tabasco since the 1970s (Voisin 2012: 42; Eche 2013: 81). The third process of forest degradation was initiated by infrastructural mega projects which were implemented mainly from the 1960s (Ramos-Gutiérrez & Montenegro-Fragoso 2012: 198f). Besides the general trends of deforestation in Mexico, some regional interventions have contributed to substantial deforestation along the Usumacinta River. One of these was the logging of around 440,000 hectares of forest by timber companies which was concessioned after a study carried out by PEMEX on the local oil reserves in the Lacandón forest (Howard 1998: 362).

In the last 70 years, extensive growth of the cattle production has taken place in Mexico with high amounts of international loans given for its development (González Pacheco 1995: 40). This system of extensive use of lands for cattle herding is responsible for the transformation of forest in the humid tropical zones, especially in the states of Chiapas, Tabasco and Veracruz and has taken place even in the coastal zones, strongly affecting coastal lagoon systems and economic activities such as fishing (Ibid). Deforestation is a process that has changed Chiapas and the South of Mexico in general throughout history. As Eche (2013: 81) reveals however, the largest part of deforestation in Chiapas occurred between 1990 and 2012.

Many of the abovementioned changes in the environment, the settlement patterns and economic activity have caused a range of ecological challenges. These challenges are in direct or indirect interaction with flood prevalence in the physical environment as well as with patterns of social interaction and use of resources. Deforestation of river banks and basins has strong influences on the hydrological cycle of regions and the options for water flow regulation (Chakravarty et al. 2012: 16). In the vast Usumacinta river network, it can therefore be assumed that major changes in the water cycle have occurred following the massive destruction of the vegetation cover. In the Lower Usumacinta River, recent research in the hydrological system focuses on indicators such as the absorbing capacity of the soil or the amounts of runoff and of flow accumulation in the river, which mainly influence the flooding of the wetlands zones (Tapia-Silva et al 2015: 1508). With deforestation, the absorbing capacity of the soil decreases leading to a more rapid saturation of a soil in times of high water flow or precipitation (Chakravarty et al. 2012: 16). Moreover, current agricultural practices in the study region interact with flood dynamics. Diemon et al. (2006: 206) argue that agricultural practices, including unsustainable farming as well as herding leads to soil erosion and compression. The soils thus have less capacity to retain the abundance of water which is necessary to prevent flooding.

Whereas the river Usumacinta was an important infrastructure for trade and the transport of people in Mayan times just as in the times of colonial rule, today the river is not the main transport infrastructure in the region. In the villages along the river, transport on motor boats is important for daily life and economic activity, especially during the times of floods. However, transport on roads has become more important and frequent in everyday activities. The strategic character of the study region today materialises e.g. in an interstate highway network, that passes in close vicinity to the villages. The highway connects Chiapas, Tabasco and Campeche but it also is a major connection hub for transport from the centre of the country to the Yucatán Peninsula, including legal transports but also illegal trafficking of drugs, weapons and even humans.

3.3 POPULATION DYNAMICS, FIGHT FOR LAND AND MAJOR POLITICAL DYNAMICS IN THE 20TH CENTURY

The different cultural and ethnic backgrounds of the groups that live in the area reveal the large population dynamics that have materialised in the study region throughout centuries. Ethnic backgrounds are hardly documented in statistics for this region today. Taking up the point of Torras Conangla (2012: 11) it can be argued that the many different origins have in some way been made invisible or reinterpreted. What is known in general is, that the area of the Lower Usumacinta is inhabited by people of different ethnic backgrounds. Among them people of Mayan, European as well as African ancestry can be named (Ruz 2010a: 9, 21). A multitude of processes has influenced the population dynamics and the subsequent social and economic processes that characterise the region up to today. They are relevant in this study as they may build an important basis for the identification and analysis of key social practices encountered in the region today, especially those linked with flood management. Some of these processes are historically documented, other processes however lack important information concerning the local and regional consequences in the study region and therefore have to be complemented by information accessed through field research. While a comprehensive historical reconstruction of population dynamics is beyond the scope of this study, an overview of the different processes in the 20th and early 21st century is presented here (table 3).

The practices of large-scale forest exploitation along the Middle and Lower Usumacinta have shown great repercussions in the settlement patterns in the study region. The demand for workforce in the forest related activities led to an immigration and a subsequent growth of settlements in the places of forest extraction as well as along the river (García 2001: paragraph 8). A range of settlements grew in strategic importance and size during the time of the timber and *chicle* extraction in the 19th century (Ruz 2010b: 80). There is little reliable information available concerning *monterías*, the regional hubs for timber exploitation and trade (Benjamin 1981: 507). Among the most famous accounts on conditions of work and property in Yucatán is the work by US-American author John Kenneth Turner in his book *México Bárbaro* from 1910. He describes working conditions on plantations all over Mexico underlining the most precarious conditions in Yucatán (Turner 1973[1910]: 11). The henequen industry which had its major boom between 1876 and 1910 is one example for exploitation of human labour (Alston, Mattiace & Nonnenmacher 2008: 2). The legal framework in Yucatán at that time made it possible to use debts to tie labour to *hacendados,* the plantation owners (Alston, Mattiace & Nonnenmacher 2008: 7). In Turner´s account it is described that among the people working in the henequen industry there were eight thousand people from the ethnic group of the *Yaquís*, who had been brought to Yucatán from the state of Sonora in the Northwest of Mexico, three thousand people denominated as "chinos" or "coreanos" and around 100 to 125 thousand Mayan people (Turner Ibid: 13). While no numbers of the labourers who fled the *haciendas* are documented, it is described that those who fled often found refuge in the forests to the South of Yucatán "controlled by rebel

Maya" (Alston, Mattiace & Nonnenmacher 2008: 19). Oral history in the study region includes the narratives by grandfathers and other ancestors that fled from henequen plantations and found refuge along the remote strips of land along the Lower
Usumacinta River (personal communication La Sandía Digital 13.05.2015, Field
notebook 1/2015).

Table 3: Relevant population dynamics in the 20th century of influence in the study region

	Description of process, events or dynamics	Geographic dimension	Time dimension
1.	Growth of settlements under exploitation schemes	along the river Usumacinta and its afluents (from Frontera until Flores, Guatemala)	Starting under colonial rule
2.	Workers fleeing inhuman working conditions in timber and chicle industry	Yucatán Peninsula, Tabasco, Chiapas	1876 – 1910 (Porfiriato)
3.	Regional effects of the Mexican revolution	in different regions of Mexico, regional processes in the Usumacinta region	1910 – 1915
4.	Agrarian reform and fight for rights of ejido (communal land titles)	in different regions of Mexico, regional processes in the Usumacinta region	1916/1917 – 1950s
5.	Immigration strategy of the Mexican government	national	1950 – 1960s

Source: author´s elaboration based on literature cited in this chapter

The Mexican revolution in 1910 was a dramatic event in Mexican history that
brought about drastic changes all along the country. The armed conflict that took
place in Mexico from 1910 until 1917 showed local repercussions, as did the Constitution, passed officially in 1917, which included a list of social and political reforms (Collier & Lowerry Quaratiello 1994: 28). In Chiapas, the Mexican revolution started in 1914 upon the arrival of Venustiano Carranza who abolished labour
conditions of debt (Collier & Lowerry Quaratiello 1994: 28). While this resulted in
the freeing of workers in the central parts of Chiapas, plantations and *haciendas* in
the East of the state did not experience this wave of liberation (Collier & Lowerry
Quaratiello 1994: 28). In the remote parts of the zone *Los Ríos*, which comprised
the borderland of Tabasco and Chiapas, a revolutionary group called "Brigada Usumacinta" occupied strategic settlements including Balancán and Montecristo (today
Emiliano Zapata) (Martínez Assad 1996: 151). While the direct effects of the revolution in the case study region are hardly documented in written text, photographs
such as the one presented below (figure 11), taken of a steam boat on the river

Usumacinta depicts a group of revolutionaries that returned to Villahermosa in 1916.

Figure 11: Photography of the Brigada Usumacinta from 1916
Source: Dominguez Vidal (1955)

The Mexican revolution brought about important developments in favour of small holder agriculture and autonomy. However, it can only be described as a temporal success, because the political reforms taking place in the years after the revolution initiated a process of industrialisation of agriculture (Pichardo González 2006: 46). Alongside with the rise of cattle breeding, called "ganaderización" and the rise of monoculture production, changes were promoted in those laws that had been protecting land rights (Ibid: 47). There was no equal redistribution of lands to those persons who traditionally had been disowned of their lands. While the main goal was to prevent the development of extensive *latifundios*, large estates in the hands of powerful landowners were not redistributed or shared (Ibid).

It was shortly before the times of the Mexican revolution that conflicts between the states of Tabasco and Chiapas over the territories in the borderlands along the Usumacinta River came up (Carvajal 1951: 9). This conflict allows the assumption that strategic importance was ascribed to the Usumacinta lands by both state governments. Part of the case study region is located in an area that changed from belonging to Tabasco to belonging to Chiapas and vice versa several times throughout the twentieth century. It is documented that in 1897 the finca [...] (name anonymised), was officially registered as part of Palenque municipality in the state of Chiapas (Ibid: 139). While written documents on exact location of the borderline are rare, some indications on the borderlines can be retrieved from historical maps. Figures 12 and 13 represent a selection of maps that are part of a historical atlas of Tabasco. This historical atlas includes maps from the time period of 1570-1981. The first map on which the case study village is charted is a map from the year 1921. In this map, the village is located east of the borderline between Tabasco and Chiapas, which documents that in this year it was located in Tabasco (highlighted in figure 12, colour plate p. 141). The second map from 1952 shows that in the case study region the river Usumacinta is the borderline between the two states. Although no settlement on the side of Chiapas is charted in this map, the territory left of the river Usumacinta, downstream from the town Emiliano Zapata, is represented as part of the state of Chiapas (figure 13, colour plate p. 142).

Land and border dispute is not a singular case for the example of Palenque municipality. Rather, it stands in line with general struggles for territory that occurred in the area in a period shortly after the Mexican revolution and in the years of the Agrarian reform that started in the year 1916 (Eckstein Raber 1966: 30). This reform and the passing of the Mexican Constitution in 1917 opened up a window of opportunity for landless workers to rise against owners of large estates and gain access to property. Paragraph 27 of the constitution is the most discussed paragraph, as it lay an important basis for private property of land. The paragraph also clearly promoted the development of small-scale land ownership and splitting up of large estates (SEGOB 1917). Following the agrarian reform, e.g. Yucatán experienced the formation of an agrarian movement under General Salvador Alvarado, whose actions included the liberation of peasants of Mayan ancestry from inhuman working conditions and debts and released many indigenous people from jails (Huizer 1969: 118). It can be assumed that it was people from Yucatán as well as labourers from other areas of the South of Mexico, who subsequently took the chance to fight for land property in the region. It was within the so-called *ejido* land title system that groups of men were given the opportunity to gain access to land. The *ejido* is a collective land title that has its antecedents in Spain as well as in former Aztec land systems (Silva Herzog 1964: 28). While the land title is collective, the cultivation can be performed individually or collectively, according to the agreement of the *ejido* members (Eckstein Raber 1966: 1).

The processes of continuous struggle for land in the borderlands of Chiapas and Tabasco when Tomás Garrido Canabal was governor of Tabasco (1919–1934) is of specific interest for the case study region. Even though it is not officially documented when the first ancestors of today´s village inhabitants settled in this place, the investigation into the narratives of local people by the researcher assume that struggles for land started in the time of rule of governor Garrido Canabal. Official documents reveal that the first request for the endowment of ejido was made by villagers of today´s village in the year 1939 and was granted in 1954 (SEGOB 1954: 5). In 1954, the ejido was given to 35 families who each were attributed eight hectares of land for agriculture as well as place within the urban settlement zone and a zone of communal user right (SEGOB 1954: 6). Subsequent extensions and amendments to the territory and population have followed up to today.

Although the agrarian reform brought about important steps for a more just and equitable distribution of lands, in the long run it did not meet the goals of a social and just development under subsequent regimes of modernisation (Entrena Duran 1990: 153). In light of this argument it is relevant to review another historical process for the region that took place in the 1950s and 1960s. In these years, the Mexican government implemented a large-scale migration scheme in the country. In order to address high population densities in the central parts of the country and to intensify agricultural production, population groups were resettled (Stevens 1968: 81). The larger strategy to develop the Grijalva and Usumacinta River basins in Chiapas and Tabasco was initiated with the creation of the *Comisión del Rio Grijalva* in 1951 (Barkin & King 1986 cf Pérez Sánchez 2007: 1). Problems like floods, the lack of population settlements in zones of fertile soils were mentioned as the justification of large-scale technical interventions into a system of specific social and ecological characteristics (Pérez Sánchez Ibid). A well-known infrastructure scheme carried out under the auspices of the Grijalva Commission was the *Plan Chontalpa*. The major objective of this plan started in 1963 was to develop the large wetland area of the Chontalpa region in Tabasco in a process of modernised agricultural production (Ibid: 43). The major activities of this large scale infrastructural programme comprised drainage of wetlands, clearing of tropical forest, construction of roads and the foundation of new settlements (Ibid: 20). Another infrastructural project was the *Plan Balancán-Tenosique* initiated in 1972, which mainly developed into an agricultural scheme for the intensification of cattle breeding in the municipalities of Balancán and Tenosique. Just as the *Plan Chontalpa*, activities pushed forward under this development scheme were the clearing of forests, the drainage of wetlands and the creation of new *ejidos* as settlements of smaller size (Ibid: 44). It can be assumed that the *Plan Chontalpa* and the *Plan Balancán-Tenosique* showed local repercussions in the case study region.

3.4 RECENT PROCESSES OF INTERNATIONAL ECONOMIC INTEGRATION AND DYNAMICS OF RESISTANCE

A series of examples is at hand to underline the ongoing processes of technical modernisation of agriculture and an export orientation in Palenque and Catazajá municipalities. While oil extraction in the Gulf of Mexico is the most prominent and economically most relevant of these processes in the larger region, increasing palm oil production is a local example for more recent forms of land transformation and degradation (Castellanos-Navarrete & Jansen 2013: 12). Beyond agriculture-based production, recent plans for the economic development in the region highlight the growing demand for infrastructural development. Recent development plans for the larger region are officially known as the Plan for the Integration and Development of Mesoamerica – in short the *Proyecto Mesoamérica* (PM). The projects envisaged in this cross-national scheme comprise interventions in the areas of social development, environmental sustainability, infrastructure, technical assistance and the commercial exchange of goods and services (SRE 2012: 6). While a comprehensive discussion of the project is beyond the scope of this study, two processes which are part of the scheme can be highlighted in regard to the topic of this study. Besides a range of other initiatives, the PM includes an initiative to increase the production and interconnection of electric energy. The other relevant process is the initiative for the prevention and mitigation of natural disasters (Capdepont Ballina 2011: 142). Within this initiative a system of territorial information in Mesoamerica (SMIT) has been built up with the objective to improve the identification of disaster risks (Socialwatch 2011: 8). Economic development, energy production and disaster risk reduction are merged in a scheme that follows a discourse of development. Critical voices from civil society argue that the PM is of neoliberal orientation and promotes measures which contradict the improvement of living conditions (Capdepont Ballina 2011: 146). The PM is closely linked to the *National Infrastructure Programme* (PNI), which was installed in 2014 and comprises e.g. the construction of new dams in different parts of the country (DOF 2014).

Another important historical and ongoing dynamic with repercussions in the case study region is related to the Zapatista movement. As a reaction to increased efforts by state government to channel the distribution of solidarity resources, described as "neoliberal restructuring" by Harvey (1998: 233), local people started to oppose the government. The official birth of the movement took place in 1994, when the Zapatista National Liberation Army (EZLN) rose up in an armed conflict against the Mexican government and proclaimed autonomy. Until today its programmatic outline focuses on the objection of neoliberal projects thereby changing the course of a long history of exploitation and capitalism (Subcomandante Marcos 1994: chapter 4f). A leading figure of the movement, Subcomandante Marcos, presents the perspective towards ongoing spatially relevant dynamics by referring to the widely-known publication *The open veins of Latin America* by Eduardo Galeano (2004 [1971]):

> "Chiapas is bled through thousands of veins: through oil ducts and gas ducts, over electric wires, by railroad cars, through bank accounts, by trucks and vans, ships and planes [...] to various destinations: the United States, Canada, Holland, Germany, Italy, Japan – but all with the same destiny: to feed the empire" (Subcomandante Marcos 1994 cf Slater 1997: 270).

The conflict between EZLN and the national government is called a war of low intensity (Zaga 2015: 2). It is characterised by a range of direct and indirect measures by military and paramilitary groups, among them road cuts, crop damages or even physical violence as well as benevolent interventions in deprived communities in Chiapas to prevent new coalitions with EZLN (Ibid: 10). The Zapatista movement continues its work of resistance and autonomy from National government, supported by individuals and groups within the country as well as through a vast international network of solidarity. While the Zapatista movement is not physically present in the region of the Lower Usumacinta, there has been a discourse about the aims of the movement and the measures taken by the conflicting groups. Especially the fight for land and autonomy proclaimed by EZLN link to the historical processes and current dynamics of land ownership which are pressing issues in the case study region even today.

This chapter has presented selected historical dynamics in the case study region and larger region of South Mexico from Mayan times until recently. Among the processes presented in this selective overview there are relevant population dynamics like migration and refuge, economic patterns like plantation schemes and modernisation, changes in the ecological systems through deforestation and intensified land-use, as well as a range of political dynamics ranging from the Mexican revolution over agrarian reforms to current struggle for autonomy by EZLN. Moreover, the cultural backgrounds of the people who inhabit the case study region shape everyday lives and dynamics today. The processes are selected as they form the larger socio-spatial context in which perceptions of floods and performances of flood management are embedded. It is possible to assume that a process like the fight for land, which is a recurring pattern throughout history is a pressing issue today. Current socio-spatial dynamics in the case study region however cannot be adequately accounted for by a historical review alone. The historical review helps to underline the relevance of studying flood management in this specific case study region. However, in order to address the research question it is necessary to identify conceptual entry points from social theory and to analyse current dynamics in the light of these.

4 SOCIAL PRACTICES, RISK AND SPACE
– THEORETICAL BASIS

It is the aim of this chapter to introduce the theoretical basis of the study. The main theoretical interest of the research concerns practice theory and the question how it is related to theories of risk and of space. By introducing the main characteristics of practice theories, insight is given into the specific viewpoint these theories allow. Through a more in-depth dedication towards the perspective of practice theory pursued by the social theorist and philosopher Theodore Schatzki this study tries to clarify an understanding of what practices are, how they are thought of as key constituting elements of *the social* and how they connect to risk, space and time.

This chapter starts with a short introduction into the constructivist field of social theories that is dealt with in this study. The tradition of thought shaped by Berger and Luckmann (1967) gives orientation, how (scientific) reality is conceptualised in this study and which kind of contribution the theoretical basis of this work is believed to make. This general clarification is followed by an introduction into different theories of risk and theories of space, which are of relevance to this research and more in general to the field of social geography. Subsequently the entrance into practice theories and the specific orientation of Schatzki´s work (2001, 2002) is presented. The concept of *riskscapes* introduced by Müller-Mahn (2013) is then described and contrasted towards the assumptions made by Schatzki. Derived from this conceptual contrasting, a specific way to address practices in relation to everyday life, risk, space as well as to notions of materiality and the body is presented.

4.1 PARADIGMATIC AND EPISTEMOLOGICAL BASIS

The ground-breaking account on *The social construction of reality* by Berger and Luckmann presents essential observations and concepts for a sociology of knowledge which has not lost importance since its first publication in the year 1966. It is by describing the connection between the concepts of 'knowledge' and 'reality' in a process of social construction (Berger & Luckmann 1967: 15) that their work has influenced scientific reflection in sociology as well as far beyond the discipline. They describe the specific interest of sociology in these concepts based on the following argument: "Sociological interest in questions of 'reality' and 'knowledge' is thus initially justified by the fact of their social relativity" (Ibid). Relativity is one major consequence of the social construction of reality and it summarises the ways in which different people have different perceptions of reality, or more precisely, that and how different people create different realities. This line of thought presents a major entry point for the epistemological and empirical approach of this study.

Social constructivism as a specific direction among constructivist perspectives present in the social sciences today objects objectivist notions and highlights intersubjective aspects as part of the social and cultural embeddedness of *knowledge* (Reich 2001: 366). Moreover, it also regards scientific as well as non-scientific accounts on *nature* or the environment as constructed from a specific cultural context (Ibid).

The epistemological orientation of this study is based on these social constructivist assumptions. Social geography adopts a focus on social theory in dealing with the relationships between society and space (Smith et al. 2010: 5f). While many different paradigms are present in social theory in general, it is in the theoretical realm of social constructivism that this study is based and that it draws its basic epistemological assumptions from. More precisely, it is believed that everyday knowledge is drawn from the encounters of people with the physical world and other people and that social scientific knowledge is created in a process of reinterpretation of this everyday knowledge using specific language or concepts (Blaikie 2010: 95). The other consequences when buying into social constructivism is that the researcher's position is believed to influence research in a fundamental way and that all observation is seen as happening in an environment influenced by theory and existing concepts (Ibid). A social constructivist approach in the discipline of geography however must also have clear implications for the theoretical perspective towards space. It is in line with the main authors of social geography that the notion of space is addressed in this chapter.

Basic assumptions made on concepts like knowledge, reality or space are linked closely to academic traditions and schools of thought. Moreover, it is important to take into account basic epistemological assumptions present in the cultural background (e.g. derived from specific cosmologies or beliefs) of the people living or working in the larger research area. The collaboration with a co-researcher from the Mexican state of Chiapas in the empirical field work of this study makes it indispensable to include some of the basic assumptions he introduced into the research. Two versions of what are called "alternative epistemologies" are presented here and used in the conceptual and empirical outline of this study: *epistemologies of the South* as presented by Sousa Santos (2015: 12ff) and *epistemology of the heart* developed by the co-researcher Xuno Lopez Intzín.

Epistemology of the South is a research paradigm that tries to establish new ways of creating knowledge and of research in general, originating in the work of Latin American researchers and activists. It is a paradigm based on two basic assumptions, formulated by Boaventura de Sousa Santos:

"The understanding of the world is much wider than the occidental understanding of the world [...] [and] the diversity of the world is infinite, a diversity that includes very different modes of being, thinking and feeling, of conceiving time, the relation between human beings and between humans and non-humans [...]" (Sousa Santos 2015: 13).

De Sousa Santos and other authors of the compilation *The other practices of knowledge(s) – Amidst crisis, amidst wars* (translation from Spanish by author) (Leyva Solano et al. 2015) emphasise that there are two major ways to implement an *epistemology of the South*: "*ecology of knowledges* and *intercultural translation* (emphasis in original; translation from Spanish by the author)" (Sousa Santos 2015: 13). Ecology of knowledges points out that various knowledges exist (Sousa Santos 2015: 22). Leyva Solano argues that there are certain practices of knowledge currently used in research in countries of the Global South that "have been, bit by bit, allowing us to denaturalise the given and like this to start an attempt to decolonise our systems of thinking and of life [...]" (Leyva Solano 2015: 27). Exchange and collaboration with Xochitl Leyva Solano allowed for an introduction into the basic principles and practices of an *epistemology of the South*. This exchange made it possible to gain access to novel research perspectives and to be part of a process of research and practical action, involving knowledge generation and a constant critical reflection of knowledges.

The other "alternative" epistemological assumption in this research comes from a concept of the social group of the Maya-Tseltal people. The co-researcher, who is a representative of this group and a researcher in sociology, introduced this concept to me and we used it together in the collaborative research phase. It is a term called *yo'taninel snopel* (López Intzín 2015: 185) that is translated as *sentipensar* (Ibid) into Spanish language and as *feelingthinking* (translation by the author) into English. In this line of thought, to make sense of a phenomenon involves acts of thinking as well as acts of feeling. The co-researcher expresses this assumption as follows

> "the interaction of the heart and the mind – love, passion and reason – are more than a dichotomy in dispute but they are parts that complement each other and that form the Maya-Tseltal rationality" (López Intzín 2015: 185, translated by the author).

In line with the co-researcher, it is argued here that an adequate epistemological approach for this study can incorporate both strands of perception that give us access to relevant empirical material. It is the access to data through acts of thinking as well as through acts of feeling that create a comprehensive approach advisable to capture social practices and their meaning to the individuals and groups who carry them out. This concept called *epistemology of the heart* by López Intzín (personal communication 15.06.2015, Field notebook 1/2015) is a concept found in various accounts on research in the social sciences, mainly presented by Latin American authors. More recently, Guerrero Arias (2011: 21f) has presented an approach to integrate the aspect of *corazonar* into anthropological science. The Spanish word *corazonar* is a neologism formed by the two words "corazon" (=heart) and "razonar" (=to think/ to reflect). In her anthropological research into Maya-Tseltal cosmology Pérez Moreno (2012: 11) investigates the concept of *O'tan* in Maya-Tseltal culture, highlighting the importance of including the heart into research. It is necessary to reflect on the word "heart": Even though the heart is a physical organ in the human body, the concept of heart involves emotional perceptions that are often ascribed to the soul of a person. Using emotional perceptions as one entry-point into empirical research in the social sciences may be a contested undertaking. Maya-Tseltal conceptualisations contribute to a specific understanding of an *epistemology of the heart* in this study. Three local concepts can be highlighted as they are broadly discussed among researchers and activists in Chiapas who subscribe to "alternative epistemologies": *Ich'el-ta-muk'*, which can be translated as the reconsideration of the greatness of every human being and *ch'ulel*, a term that conceptualises soul, spirit, conscience and heart as one entity that gives access to perception and experience (López Intzín 2015: 194). The third expression expresses the core aspect of *epistemology of the heart*: *yo'taninel snopel*, which can be translated as "feeling-thinking" (from the Spanish expression "sentipensar" translated by the author) means that reflections, thoughts and knowledges do not only come into existence in the mind but that they are also developed in the heart (Ibid: 184). These concepts make explicit that bodily perceptions and those perceptions that originate in other sources than the mind are not perceptions of second order in qualitative empirical research. Rather, they open up the view to a new field of relevant information. The term *epistemology of the heart* was used by psychologist Rosemarie Anderson (2004: 308) to describe a key process in practicing transpersonal psychology and enquiry in scientific research. It is assumed here that what different qualitative researchers have called "intuition" (Sperber 1985: 93; Anderson 1998; Janesick 2001) during field research is closely connected to those levels of perception, which are not mainly part of a structured cognitive process of data collection. Due to its sensitive nature and the ethical implications linked to research involving personal information, all processes that involve emotional and bodily perceptions have to be handled with special care. If done so, *epistemology of the heart* is an invaluable source of empirical research and analytical insight.

Methodological aspects of how to perform and document emotional perception in qualitative research can be found in accounts on ethnographic research; however, a clear-cut separation between emotional perception and perception through other senses (e.g. vision, hearing, etc.) and the ways in which information won through these perceptions is processed is hardly possible. Therefore, it is argued here that any enquiry in the social sciences that performs an empirical approach involves the thinking and the feeling of a researcher. Making use of these powerful tools in combination and making transparent the contributions emotional perception makes to the research is considered a reasonable approach by the author. The way in which this epistemological perspective is put into empirical practice in this research is further described in chapter 5.

Having introduced the main epistemological basis for this study, the following paragraphs present the principal theoretical basis of this work. In this specific study of flood management in the South of Mexico, conceptual focus lies on the relationship between risk and space, especially in the mutual dependencies that exist in the construction of space and in the construction of risk through the main pattern of social life which are social practices.

4.2 SOCIAL PRACTICE THEORY

The main body of theory that guides this study is social practice theory. Social practices provide a valuable conceptual entry point into the understanding of everyday life and social phenomena. In order to understand flood management, it is argued here that it is necessary to take a closer look at the patterns of social practices that are part of flood management and which form the larger socio-cultural context in which flood management occurs. This tradition of social theory is attributed to a range of different authors from the twentieth century. The most prominent practice theorist, Pierre Bourdieu, first presented a theoretical account in his *Outline of a theory of practice* in 1972. This work is based on considerable empirical analysis, mainly from his work on Kabyle society in Algeria (Bourdieu 1972). The key terms of his work, *habitus*, *capital* and *field* have had substantial impact on theory development as well as on empirical research, putting his practice theory among the *Grand theories* (Reckwitz 2003: 282). With a theory of practice, he provides an important basis for the understanding of the relational character of social space (Etzold 2014: 38). Bourdieu´s conceptual strengths as well as strong empirical foundation underline the importance for practice theories developed in decades following his main work on practice theory. His concepts build major points of reference in sociological research and in geographic research up to the present day. The sociologist Anthony Giddens (1979, 1984) is another main author contributing to our current understanding of practice theory. Highlighting the importance of people´s everyday actions, he argues for a duality of structure in which "structure is both medium and outcome of the reproduction of practices" (Giddens 1979: 4). With their different versions of practice theories, Giddens and Bourdieu shared the attempt to build a conceptual synthesis between structuralist and interpretative approaches within the social sciences (Spaargaren 2011: 815). Both authors are named as belonging to the *first generation of practice theorists* (Postill 2010: 6). Foucault is also mentioned among the first generation of practice theorists (Ibid). Reckwitz recognises an orientation towards practice theory mainly in his late work (1984 cf Reckwitz 2002: 243). Postill (2010: 9) lists a range of other authors among the *second generation of practice theorists*. Among them are Karin Knorr-Cetina (2001), Sherry Ortner (1984), Andreas Reckwitz (2002), Theodore Schatzki (1996, 2001) and Alan Warde (2005). Before turning to the specific social practice theory made use of in this study, it is necessary to identify some of the basic characteristics of practice theories.

Reckwitz argues that the description of "'ideal types' of theories" (2002: 244) although simplifying the variability of perspectives, allows for the delineation of the main aspects that distinguish social practice theories from other sets of social theory. Practice theory as one version of cultural theories, is clearly distinct from those theories that use the concepts of *homo oeconomicus* and *homo sociologicus* for the description of how humans act and how social order emerges (Reckwitz 2002: 245f). More generally practice theory dismisses other prominent perspectives towards human action; among them are individualism, structuralism, systems theory, poststructuralism and others (Schatzki 2001a: 10f). Furthermore, practice theory opposes the dominant role given to the mind over the body in other cultural theories like mentalism, intersubjectivism as well as textualism by highlighting the fact that it is bodily and mental processes in conjunction that form practices and thus influence the social (Reckwitz 2002: 252; Schatzki 2001a: 20). A critical discussion on the idea of a practice turn and recent versions of practice theory is provided by Bongaerts (2007). He questions that approaches of practice theory can clearly be distinguished from other sociological theories and argues that for the case of the theoretical account on social action presented by Alfred Schütz a clear delineation towards concepts used in practice theory is not adequate (Bongaerts 2007: 251). Moreover, he highlights that in comparison to Bourdieu's conceptual apparatus on habitus a clear theoretical structure is missing among recent works in practice theory (Ibid: 255). While the point that Schütz as well as Schatzki make explicit reference to Wittgenstein's concept of *Tätigkeit im Vollzug* (Ibid: 251) is elucidating, the general discrepancies between phenomenological accounts and practice theory are believed to outweigh similarities. The decision for one theoretical pathway, in this case for practice theory, which belongs to the realm of cultural theory, is at the same time the decision against other theoretical pathways. Ongoing processes in the empirical case region could as well be framed as subliminal conflict situations that demand for the analytical tools offered by political geography. Geographical research on conflicts as applied and conceptually determined by Reuber (2012) lies an epistemological focus on intentional behaviour of actors (Reuber 2012: 119f). This intentionality can be described as one of the dividing lines between conceptualisations used in geographical conflict research in the tradition of political geography and the theories of social practice used in this thesis. It is a fine line, which has to be further focused on in order to make visible the strengths of the specific approach used here. In this research, the main points of interest are not consciously intended actions by actors with clear-cut roles. Of interest are the diverse everyday practices of actors in long-term situations of contingencies and inequalities. Actors make use of established social practices and adapt or change them according to dynamic patterns that involve e.g. interests, knowledges, beliefs and needs. Instead of cutting boundaries between different actor groups and their intensions in this research, the unclarity between intentional actions directed towards a pre-defined goal on one end of the spectrum and sub-conscious routine actions which are part of a larger context of values, practices and knowledges on the other end is highlighted.

A social practice is a routine way in which actions are carried out, e.g. the way an object is moved or an argument is put forward by a human being (Reckwitz 2002: 250). As Reckwitz (Ibid) argues, to delineate a practice as a social practice is a tautology insofar as all practices can be regarded as social. At the point of argumentation where the essence of *the social* is discussed, it is crucial to delineate the conceptual path within the scope of practice theories that is chosen to take in this study. The theoretical outline on social practices given by Schatzki and the specific social ontology he makes are believed to offer a valuable and adequate conceptual pathway for the social enquiry undertaken here (Schatzki et al. 2001; Schatzki 2002,

2003). In *A new societist social ontology* Schatzki (2003: 184) emphasises that (individual as well as collective) action presupposes the social. Any given practice is
essentially social because it involves various people and because a practice is always part of a larger set of socially ordered actions (Schatzki 2002: 87f). Therefore,
it is practices and social orders in conjunction that constitute social life (Schatzki
2002: 59, 116f). One of Schatzki´s arguments and a typical example of the vocabulary he uses is displayed in the following sentence: "Social orders are the ensem-
bles of entities, through and amid which social life transpires—the arrangements of
people, artefacts, organisms, and things that characterize human coexistence"
(Schatzki 2002: 38).

Building on Wittgensteinian thought, he defines social order as arrangements
rather than as regularities (Schatzki 2001b: 51). *Arrangements* link people, organisms, artefacts, and things and relate each of them to another in a specific way (Ibid).
This conceptualisation of social order meets the dynamic character of social life and
describes the relation of humans to other entities. Taking a look at practices themselves, the reader of Schatzki´s work is pointed towards the definition of practices
as "doings and sayings" (Ibid: 58). Sayings are doings that involve language and
say something, just as the action of shaking a hand or sending a greeting card
(Schatzki 2002: 72). All practices are carried out by the body (Ibid) and ordered in
the minds of people (Schatzki 2001b: 58). The mind however is not an external
entity that determines behaviour but it is constituted socially (Ibid: 50, 57). It is
rather the mental states as part of the social that are expressed in behaviour, by
"determining what makes sense to people to do" (Ibid: 57).

To make this point clear the action of hugging can be taken as an example. A
hug is performed bodily and it is a *doing* that *says* something (e.g. it shows interest,
affection, or it is done in compliance with certain social traditions). For a person to
hug another person, it involves the mental state of hugging, meaning that it makes
sense to the person to hug the other. This mental state is socially mediated, as hugging is only an accepted action in certain social settings or situations. The example
of hugging therefore shows that an action is performed bodily and at the same time
it is connected with certain mental processes. Larger sets of mental states and processes are relevant parts of practices.

Following Schatzki (2002: 77) a practice is organised in the mind by linking
the relevant *doings and sayings* through four entities: (1) practical understandings,
(2) rules, (3) a teleoaffective structure, and (4) general understandings. *Practical
understandings* are abilities of a person to carry out certain actions that are part of
a practice (Ibid). It is however more than merely the knowledge and ability to perform these actions but it includes also knowing how to identify these actions and
moreover how to provoke and to respond to these (Ibid). Although similarities exist
with the concept of habitus (Bourdieu) and the concept of practical consciousness
(Giddens) in describing a capacity needed for an activity, Schatzki (Ibid: 79) argues
that the two other concepts cannot determine particular actions as they ignore what
makes sense to people to do. *Rules* have a specific definition in his work as "formulations interjected into social life for the purpose of orienting and determining
the course of activity" (Ibid: 80). They present one of the normative aspects that

define what makes sense to human beings to do in a certain context or situation. Another aspect including normativity is called *teleoaffective structure* (Ibid: 80). It is a structure built up of certain ends (telos) and emotions or moods (affect), all of which carry a normative connotation (Ibid). This structure, just as the two above-mentioned aspects contribute to the distinction of what makes sense to somebody to do. *General understandings* are those structures that can be given specific importance as they guide the larger set of actions carried out not only in one specific practice but in various practices of everyday life of a social group (Ibid: 86). For the purpose of this study, figure 14 presents a visualisation of the organisation of practices as *doings and sayings* on the micro-level, involving the bodily and mental processes involved in a human being performing a practice.

The work of Schatzki purveys a valuable conceptual perspective as well as sophisticated analytical equipment for this study. Investigating the dynamic nature of social practices and orders, his work approximates the basic patterns of social life. Furthermore, in the empirical analysis of the Shaker medicinal herb production Schatzki (2002) offers a meticulous identification of the four abovementioned assets that order and determine social practices. Arguing at the same time that practices are "materially mediated nexuses of activity" (Schatzki 2001a: 20) he highlights the relevance of materiality and emphasises that social relations among humans involve nonhumans (Schatzki 2002: 40f). However, he objects posthumanist accounts that conceptualise non-human entities at the same level as humans in guiding action (Schatzki 2001a: 20).

Besides the main pillars of practice theory in Schatzki´s work, that are referred to above, this study makes use of the specific social ontology he presents. In line with Taylor (1985 cf Schatzki 2001b: 53), he reveals that practices are more than activity, but include "a site, or context, where activity occurs". In his 2003 essay *A new societist social ontology* Schatzki revisits individualist ontologies and earlier societist ontologies identifying their blind spots in order to describe sociality in an adequate manner. It is argued that any analysis of social life needs to take into account ontologies and to deal with them in a reflective manner (Schatzki 2003: 189). In his critical review of the most prominent ontologies in science of the 20th century, he points out that individualist ontologies are blind to the core of social phenomena while most *societist ontologies* (e.g. wholism, Durkheimian sociology, and structuralism) are inappropriate because they reify the social (Ibid: 175). The social, according to his account, is a "mesh of human practices and material arrangements" (Ibid: 195). What seems to be his main point is expressed as follows: "Something´s *site* is […] that realm or set of phenomena (if any) of which it is intrinsically a part" (Ibid: 177, emphasis in original). He demarcates his conceptualisation of site from an exclusively spatial understanding but describes that spatial aspects are included in a site among other aspects (e.g. time and teleology) (Ibid: 176). He presents three different ways in which a site can be understood: As the location where something takes place, as the broader set of phenomena in which it occurs and as the set of phenomena of which it constitutes a part (Ibid: 176f). The notion of location highlights that any human activity is located "in time and objective space" (Ibid: 176). Therefore, time and space are an important context for activity, not only being the background to an activity but having important interdependency with it. Beyond objective space, Schatzki also describes another version of space which he names "activity-place space" (Ibid). It is assumed here that Schatzki describes a notion made explicit in the concept of spatiality, described in postmodern geography (Soja 1996). Moreover, there is as well a teleological location that embeds an activity within a range of ends and tasks (Schatzki 2003: 176). The notion of the broader set of phenomena shows that one activity has to be regarded as one among many others (Ibid). It is within a specific set of activities that an activity becomes part of a practice. The third notion describes that a site is a set of phenomena of which it constitutes a part (Ibid: 177). In order to understand this point, it must be clarified that a site is not a context in the sense that it only builds an external frame for activities and practices. The site reveals a double nature, as it is a context for and is at the same time constituted by the practice. In this study it is argued that in order to understand social life and processes of transformation, the ontological assumptions on *the site of the social* and their relevance for the analysis of social practices are of specific value.

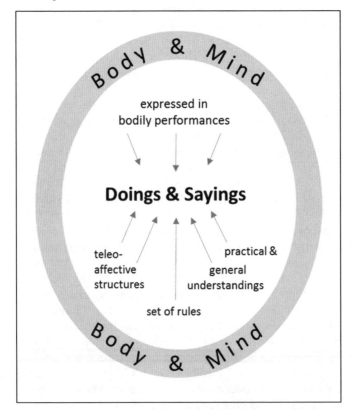

Figure 14: Organisation of practices on the micro-level
Source: author's elaboration based on Schatzki (2002)

4.3 THEORETICAL ACCOUNTS OF RISK

Theories on risk are known for the multiplicity of different conceptualisations and definitions of risk. It is the aim of this section to briefly retrace the main different scientific traditions in risk research and corresponding bodies of theory. Thereafter, the conceptualisation of risk and related terms which are regarded appropriate in this thesis are discussed. A historical perspective towards the development of risk as a relevant concept in social science reveals a dominance of the economic perspective which does however not grasp the complexity found in current societies (Kron 2013: 56f). While research on risk, including concepts of vulnerability, resilience and adaptation, has grown in recent decades, various authors highlight the importance of taking into account notions of uncertainty (Kasperson 2008; Pidgeon 2009; Jeschke & Jakobs 2013; Kron 2013). Pidgeon (2009: 349) points towards the fact that a research focus on uncertainty can include the analysis of relevant political aspects.

German sociologist Luhmann (1991: 28) argues that in late modern society, the term *risk* has developed to contrast to the term *Sicherheit*. Zachman (2014: 5) refers to Luhmann pointing out that the concept of risk helps to solve problems of uncertainty. In the discipline of economy, a clear distinction is made between decisions taken in conditions of risk and decisions taken in conditions of uncertainty (Schwarz 1996: 125). Risk implies that the probability of the occurrence of a specific event can be determined, while no information on probability can be given in conditions of uncertainty (Ibid). Zachmann (2014: 5) emphasises the need to reconsider uncertainty as an important feature of all societies and he argues that a specific historical relationship between uncertainty and risk exists.

> "Uncertainty, however, is a fundamental anthropological experience. People in all societies have had to deal with uncertainty in one way or another. Thus, if we want to understand the significance of risk in our present society, we need to explore the following questions: when did the attitude toward future uncertainties change so that the understanding of uncertainties became narrowed down to risk?" (Ibid).

In this study it is relevant, that risk can be identified as a concept closely linked to development of modern society in the Western world (Bankoff 2001: 29). Different conceptualisations of risk developed at different points in time and are still present and exert influence in society to a varying degree today. Therefore, a specific risk concept used in one world region or social group cannot be easily equated with concepts from other world regions or social groups and other ways to deal with related phenomena (Zachmann 2014; MacGregor & Godfrey 2011; Bonß 1995; Joffe 1999). Processes of weighing different risks or threats within different societies reveal important patterns that guide decision making (Rohland et al. 2014: 194f).

While all concepts of risk include the notion of "contingency of human action" (Renn 2008: 50), the specific role ascribed to humans and other entities in theory varies. Two main traditions of research on risk can be distinguished according to their epistemological and theoretical basis: In the natural sciences risk is often conceptualised as an objective entity, attributed with probability, measureable by technical instruments and calculable in a quantitative manner (Egner & Pott 2010b: 16; Renn 2008: 51; Mythen & Walklate 2006: 1). In line with the development of large technical facilities, procedures of risk analysis were developed that promote an understanding of risk as a product of a hazard (or its probability) and an estimated damage (Thoft-Christensen & Baker 1982; Hauptmanns et al. 1987: 4). Natural hazards research has developed risk indices that are calculated on the basis of standardised numerical values, e.g. the natural hazard risk index for megacities (MunichRe 2004: 42). Cognitive psychology as a natural science discipline designs models of human behaviour at an individual level (Lupton 2013: 28f). The psychometric model has been a dominant approach in psychological research, focusing on quantitative methods to analyse risk attitudes and risk perceptions (Slovic 1987: 281). This paradigm is based on the understanding that "perceived risk is quantifiable and predictable" (Slovic 1987: 282).

In the social sciences, risk is generally conceptualised as a phenomenon that includes a social context, a social process of risk definition and negotiation (Lupton 1991: 5; Egner & Pott 2010b: 18). Lupton (2013: 43) distinguishes different perspectives on risk which have in common that risk is conceptualised as "never fully objective or knowable outside of belief systems and moral positions" (Lupton 2013: 43). Social science research on risk has gained specific prominence following the publication of *Risikogesellschaft* by Ulrich Beck in 1986. His work strongly influenced other disciplines and provoked comprehensive discussion about larger social processes like globalisation, social inequalities and environmental management (Jarvis 2007: 2; Matten 2004). Another important contribution to the body of risk theory is presented by Giddens (1990, 1991) who shares Beck´s focus on larger societal dynamics of modernity that are relevant for the construction of risk in society (Lupton 1999: 1). Another relevant group of works is inspired by Foucault´s conceptual accounts on *gouvernmentalité* (i.a. Foucault 1994, 2007). Although theoretical contributions on risk have been made from a range of other authors, this chapter involves those works that have mainly influenced socio-cultural perspectives in risk research. Lupton (1991: 1) highlights the relevance of accounts by Beck, Giddens and Foucault as they emphasise the larger social, cultural, and (as in the case of Foucault) historical contexts which concepts of risk rely on. All three argue that risk has become a concept used in a cultural as well as political sense in order to organise society (Lupton 2013: 37). Beck gives a detailed account on the development of society in late modernity. The *new modernity* is characterised by processes of risk distribution and individualisation as well as by an increased reflexivity on the process of modernisation (Beck 1986: 251f). He describes a risk society, in which the activity to define risk is a way to exert political power and to regulate economic development (Ibid: 362f). Social inequality is intensified in the world risk society in which "Western governments or powerful economic actors define risks for others" (Beck 2006: 333). The risk society Beck describes "is an inescapable *structural* condition of advanced industrialization" (Adam et al. 2000: 7, emphasis in original). In *The consequences of modernity* Giddens argues that risk can be described as a "threat towards desired outcomes" (Giddens 1990: 35). While risk is related to people's actions, it is not in all cases something people are aware of (Ibid). However, risk is not mainly a result of individual action but a collective phenomenon taking place in certain "environments of risk" (Ibid). Giddens contrasts these environments of risk for *pre-modern* cultures on one side and *modern* cultures on the other (Ibid: 102). Threats and dangers in pre-modern society are described to have their origin in phenomena such as diseases, climatic conditions or floods, which are all categorised under the term "nature" (Ibid). In contrast, in a modern condition threats and dangers are a result of the "*reflexivity* of modernity" (Ibid, emphasis in original). In highlighting that *reflexivity*, i.e. the process of constant reflection, also involves the adaptation of social practices, Giddens (Ibid: 38) lays out an important basis for the consideration of practices in the analysis of larger social phenomena, like in this case, phenomena of risk. The theoretical contributions to risk research as well as to larger theory development made by Beck and Giddens are of unquestionable value for risk scholars up to the present day. However, authors revising their work critically note that it lacks a thorough empirical research base (Lupton 2006: 21; Wilkinson 2006: 27). This critique is relevant insofar as major challenges in the communication of results in risk research arise from the different possible entry points into understanding real-life social phenomena linked to risk.

Foucault's main works centre around questions of the exertion of power within society and not specifically on risk; notwithstanding, his work on governmentality has been taken up by various authors who discuss risk as a major political instrument in modern society (Ewald 1991; Dean 1998; Bunton 1998 cf Mythen & Walklate 2006: 83). Foucault is referred to in the connection between governmentality, territorial strategies and security (Elden 2007), governmentality and spatiality (Huxley 2007) or in the analysis of risk discourses and governmentality under neoliberal conditions (Kauppinen 2012).

An elaborate cultural theoretical perspective within risk research is given by Mary Douglas (1992) as well as in the account by Douglas and Wildavsky (1982). Douglas objects approaches of cognitive psychology and highlights a cultural rather than individualistic notion of risk (Lupton 2013: 53; Mythen & Walklate 2006: 13). Social principles are described to play a major role in decisions which risks to take and who should take them (Douglas & Wildavsky 1982: 6). Risk is described as a cultural strategy including beliefs and practices that are applied in order to stabilise social groups (Lupton 2006: 13). It is regarded as important to include evaluations of risk made by lay people into scientific analysis as they give access to specific cultural conceptualisations of risk (Lupton 2013: 54). Douglas (1992: 31) emphasises that actors compare risks and do not regard them in isolation from other risk items. Instead of questioning the absolute existence of risk, as strong constructionist perspectives would do, Douglas and Wildavsky (1982: 30) put an emphasis on the fact that risks are strongly political insofar as they are selected deliberately. The account of Douglas and Wildavsky has been criticised for reducing culture to the "organisational structure of groups" (Elliot 1983: 892) and for inconsistencies in the main typology of group formation (Boholm 1996 cf Tansey & O'Riordan 1999:

77). However, the contributions made by the two authors to understanding risk as a socio-cultural matter and the strong influence on anthropological and inter-discipli-nary risk research also inform the research presented here.

In risk research, especially in those areas investigating into disasters and haz-ardous events related to processes in the physical-natural environment, the concept of *vulnerability* has gained importance in the last two decades (Zehetmair 2012: 274). While in disaster risk management the aspect of vulnerability was neglected for a long time (Fekete 2010: 22), mainly development research as well as research on risk, hazard and disasters have increasingly taken up the concept (Zehetmair Ibid). The term vulnerability encompasses a large range of meanings and implica-tions. While the different definitions have been discussed extensively by other au-thors (i.a. Few 2003; García Acosta 2005; Garschagen 2014; Kelman et al. 2016; Weichselgartner 2016), in this study a specific conceptualisation of vulnerability is of interest. The *Pressure and Release* (PAR) model presented by Blaikie, Cannon, Davis and Wisner is chosen as a valuable conceptual framework for vulnerability. It has been influential in research in disaster risk science since its publication in the book *At Risk* in 1994 (Blaikie et al. 1994). While they present a pseudo-equation defining risk as the product of hazard and vulnerability, their model puts a major focus on the vulnerability side of risk (Wisner et al. 2004: 51). Vulnerability is de-scribed as developing in a three-step process. At the basis of the vulnerability side we find so-called *root-causes* that include factors such as access to power, struc-tures and resources as well as ideologies in a society (Ibid). Root-causes evoke dy-namic pressures which include macro-social factors just as deforestation or specific population dynamics (Ibid). These dynamic pressures result in unsafe conditions on physical, economic, social and organisational levels (Ibid). The model puts empha-sis on the fact that in order to understand the origin of disaster situations, the vul-nerability side and especially the general forces displayed in the root-causes have to be analysed in depth. Wisner and his colleagues, who represent the scientific tradition of political ecology and made use of the concept of adaptive capacity de-veloped by Amartya Sen, present a model that has been discussed broadly in disas-ter risk science (Birkmann 2013: 50ff; Forsyth 2008: 759; Turner et al. 2003). It is one possible entry point into a holistic conceptualisation and offers the possibility to include economic and political processes into the analysis of vulnerability (Can-non & Müller-Mahn 2010: 623). Until today it has been widely used not only by scientists but also by practitioners, e.g. in international organisations for humani-tarian aid (DKKV 2012; IFRC 2006; Morchain & Kelsey 2016; UNOCHA 2013). The concept of vulnerability in today´s academic work on disaster risks has become inseparable from another influential term: *resilience*. It is argued that there exist important complementarities between the two approaches, which "have been kept artificially separate by conceptual constructs, scientific traditions, and lack of inter-action" (Miller et al. 2010: 11). While they are made use of in life sciences just as in social sciences they involve different research foci as well as different meanings (Gallopín 2006: 293). As it is beyond the scope of this study to reconstruct the de-velopment of the term resilience, two major points are made: Firstly, resilience as a concept includes notions of social transformation in risk research. Keck and

Sakdapolrak (2013: 6f) present an overview on the scope of the term resilience, describing it as persistability, as adaptability or as transformability. While *persistability* originated in the description on the stability of ecological systems in situations of disturbance (Holling 1973: 14), *adaptability* includes processes of absorbing disturbance, re-organisation and change (Folke 2006: 259). However, a focus in resilience understanding can also be set on the *transformation* of ecological, economic and social structures (Folke et al. 2010). Mitchell (2010: 344) argues that a resilience approach allows a reflection on the definition of risky and protective practices made by researchers and contrasting them with definitions made by those who perform the practices as part of everyday life. Nevertheless, resilience perspectives pose challenges to disaster risk reduction, as they allow not only for the identification of a desirable state and dynamic within a system, but at the same time might allow the reproduction of undesirable processes (Weichselgartner & Kelman 2015: 253).

While the terms vulnerability and resilience can highlight relevant perspectives towards empirical research on risk, a social constructivist research on risk requires a critical review of its own terms. Taking up the argument presented by Bankoff (2001) for the concept of vulnerability, the importance to deconstruct scientific concepts such as risk, vulnerability or resilience is emphasised. In a critical review Bankoff (2001: 20) highlights the enduring tendency of Western societies to describe other world regions as dangerous places. He locates the discourse of vulnerability in a specific knowledge system that is representative of dominant Western thinking and inevitably contains Western cultural values and principles (Ibid: 29). What academia applies is what Hewitt calls "a *sociocultural construct* reflecting a distinct, institution-centred and ethnocentric view of man and nature" (1983: 8, emphasis in original). The definitions do not only influence science but also policy making and give "further justification for Western interference and intervention in others' affairs for *our* and *their* sake" (Bankoff 2001: 20, emphasis in original). Terms like vulnerability are found in a large range of documents and publications from the Western world, describing adverse economic, social or ecological situations in countries and regions of the Global South (i.a. Antwi-Agyei et al. 2007; Garschagen 2014). Elaborating an ever clearer conceptual outlook of a term like vulnerability and reflecting on the limits of applicability of the term in empirical research is necessary. The subjective attribution and contextual nature of terms like vulnerability and resilience, underlined by Kelman, Gaillard and Mercer (2015: 23) need to be accounted for adequately in disaster and risk research.

In this study the attempt is made not to limit conceptual analysis to the use of scientific concepts from the Western academic. Therefore, the concept of risk used in this study and closely connected concepts are used with two major implications: risk is regarded not as an objective factor but as a concept that is made use of in society in many different ways which reflect different perceptions, intensions, necessities and power relations. Taking into account the conceptual approaches offered by social science, risk is believed to be a concept standing in a larger social, cultural and political context. This context includes negotiations which reflect e.g. in different definitions of risk. These negotiations involve different aspects such as

knowledge, experience, values and necessities which are relevant for decision making (Weichselgartner 2001: 64, 110). In line with geographical risk research and
approaches from social anthropology, it is believed that an appropriate empirical
approach towards risks lies in analysing perspectives and everyday practices of different actors (Müller-Mahn & Everts 2013: 26). Gaining access to socially constructed and shared risk perceptions and analysing the social practices carried out
is a precondition for understanding how humans make sense of and practically deal
with risky or uncertain situations (Bankoff et al. 2015: 2f). A multi- and transdisci-
plinary perspective on risk demands for a historical deconstruction of risk concepts
and the identification of other concepts related to floods. It is emphasised throughout this study that risk concepts have often been used in a manner taken for granted,
although it cannot be assumed that the concept of risk does exist or is of relevance
in all societies around the globe in the same way as it is in the Western world.
Therefore, this research tries to involve non-Western concepts and different social
practices related to them.

4.4 THEORETICAL ACCOUNTS OF SPACE

A third major concept which is of relevance in the realm of this research and which
is connected to social practice theories and risk theories is the concept of space.
Geography builds its main foundations on various concepts of space and spatiality.
The idea of space is fundamentally linked to geography and yet it is a task beyond
serious geographic research to come up with a simple definition of space (Thrift
2009: 85f). The picture of a unified discipline of geography today as well as a clear
historical pathway is an illusion (Heffernan 2009: 3). The different understandings
of space and spatiality reflect a range of different epistemological and ontological
claims (Merriman et al. 2012: 4). These paradigms have developed in different
times but coexist in geography as well as in other disciplines today.

Heffernan (2009: 4) dates the origins of modern geography in Western Europe
to the century after Columbus and links it to a process of transformation in the perception of the world by European travellers (Livingstone 1992 cf Heffernan 2009:
4). Concepts of space however are already explicitly discussed in contributions to
early philosophy and natural sciences (e.g. Platon, Aristoteles) and together with
concepts of time they from the major pillars of philosophical thought. For the discipline of geography, the formation of modern science as well as the political and
religious conflicts in the seventeenth century had important repercussions (Ibid: 5).
Thinkers like Descartes and Newton made major contributions to the concept of
absolute space in the seventeenth and early eighteenth century. Newton described
an absolute space, which "in its own nature, without relation to anything external,
remains similar and immovable" (Newton 1872: 191 cf Werlen 1993: 245). Thereafter it was German philosopher Kant´s account in *Kritik der reinen Vernunft* (1781/
1787), which described space as a representation and argues that this representation
or imagination of space "precedes all external phenomena" (Kant 1922: 19). Even
though the development of the concept of space, spatiality and place and moreover

the also geographically relevant space-time in science and philosophy throughout history has been described at length by a multitude of authors (i.a. Dainton 2014; Disalle 2006; Jammer 1969; Reichenbach 1958; Tuan 1979a), it is the aim of this section to insist on the consequences the multitude of conceptualisations have in geography until today and to delineate those key concepts of spatiality that are relevant for new insight presented in this thesis.

Conceptualisations of space in modern geography have in common that they "abandon the idea of any pre-existing space in which things are passively embedded" (Thrift 2009: 86). Absolute space is replaced by a relational view of space which incorporates agency and relation among human subjects (Thrift 2009: 86). Human geography is not mainly concerned with physical space but with the interactions of humans with the material and the social world. Werlen argues in this line by saying that "social and cultural universes have no fixed spatial existence" (Werlen 1993: 241). Moreover, human geography emphasises the construction of space and spatiality in this social and cultural sense (Merriman et al. 2012: 4). According to Massey (2005: 9) space can be characterised through interrelations at various scales and levels, the coexistence of multiple and heterogeneous entities and as a product under constant construction and change. Putting a focus on the social construction of space however does not oppose accounts that point towards interrelation between the socially constructed space and physical-material aspects of space (Weichhart 2008: 326). Rather, material aspects can be an integral part of the production process. This thought is represented in the "socio-spatial dialectic" presented by Soja (1980 cf Merriman et al. 2012: 8). As Merriman et al. argue, this dialectic understanding is highly relevant for our understandings of *spatialities* or *spatiotemporalities*.

> "They matter materially. They matter in terms of discourses and representations that are mobilized around various spatial concepts. They matter through the ways in which space is performed. And, critically, they matter in terms of the everyday constructions of space that happen in the real world" (Merriman et al. 2012: 8).

Work in the discipline of geography, and with its' characteristic emphasis on spatial categories, has gained broader attention in the social sciences in the early 21st century, mainly with a re-accentuation of spatial categories in these other disciplines (Lossau & Lippuner 2004: 202). The term *spatial turn* however overemphasises the notion of rediscovery of space, as the spatial dimensions of the social had been addressed in geography as well as ethnology long before (Lossau & Lippuner 2004: 202). However, as Merriman and colleagues argue,

> "it is precisely the multiplicitous and heterogeneous nature of space and spatiality – as abstract and concrete, produced and producing, imagined and materialized, structured and lived, relational, relative and absolute – which lends the concept a powerful functionality" (Merriman et al. 2012: 4).

By leaning towards Merriman et al.'s argument, it is not regarded as useful for the sake of this study to list all recent concepts and terms developed in geography in order to analyse space and spatiality. Moreover, as Sheppard (Merriman et al. 2012:

8) discusses, statements on the ontology of space should be modest instead of su-
premely philosophical. It is however necessary to present the core terms and conceptualisations used in this thesis to analyse spatial dimensions and dynamics. In
using different concepts, a dialectic pathway is chosen, which permits to put a spot-
light on the various characteristics of space that are relevant for socio-spatial research. As Soja argues "[p]laces and people are inseparable; we cannot consider the
characteristics of one without considering the characteristics of the other" (Soja
1980 cf Mitchell 2010: 336). Moreover, it is the complex interaction between three
entities, the social, the historical and the spatial which is increasingly dealt with in
research as well as in practice (Soja 1996: 3). Although the relation between the
concept of space and the concept of time is a major point of philosophical and scientific discussion, it is noteworthy to say that "incorporating time adequately into
how we think about spatiality remains a major challenge for geographical theorists"
(Merriman et al. 2012: 9). It is beyond the scope of this chapter to integrate the
discussion of time in relation to space. A reflection on issues of space, time and risk
is added in the discussion of results and conclusion (chapters 9 & 10).

Our entry point into a conceptualisation of space is the specific approach offered by French philosopher and sociologist Henri Lefebvre. In his detailed analysis
of the main arguments and the context of Lefebvre´s work, Schmid (2005) derives
major points of how Lefebvre conceptualises space and the process of production
of space. To begin with, he argues that concepts about space are closely linked to
concepts about societies, a perspective characteristic of social geography (Schmid
2005: 29). Lefebvre's concept of space has to be seen in the context of its development, which is the analysis of urbanisation processes. This argument brings up the
question, how then a theory that has mainly been developed in the context of phenomena of urbanisation could be of relevance and of use for the analysis of spatial
and social phenomena in rural Mexico. Lefebvre´s hypothesis of the complete urbanisation of society (Ibid: 121) at first sight seems to delimitate the areas of conceptual use. However, its main arguments have a value far beyond the analysis of
urbanisation processes today. It is especially the identification of contradictions of
space that has inspired publications on spatial processes of inequality and fragmen-
tations of space which were studied in urban as well as in rural contexts such as
rural Brazil (Lima 2012; Gómez-Soto 2008). Lefebvre makes clear the way in
which he relates humans to space as can be seen in the following quote:

> "They do not merely enjoy a vision, a contemplation, a spectacle – for they act and situate
> themselves in space as active participants. They are accordingly situated in a series of envelop-
> ing levels each of which implies the other, and the sequence of which accounts for social prac-
> tice" (Lefebvre 1991: 294).

This understanding of humans and space is of major importance for this study. Another key message of Lefebvre´s work is that space exists as the result of a production process. The specific interest of Lefebvre however passes beyond the analysis
of space and goes onto the analysis of the production process in which space is
created (Schmid 2005: 203). Lefebvre argues that it is the task of critical theory,
himself belonging to this tradition, to turn away the analytical gaze from products

and towards production (Lefebvre 1991: 26). It is therefore of specific analytical value to look at how he breaks down this process into sub-processes: Production of space is located in the realm of the social, which for Lefebvre includes a material as well as a mental level (Ibid: 71, 415) but passes beyond this dualism (Soja 1996: 11). Lefebvre articulates a critical perspective towards philosophy and describes a major limitation rooted in the primacy it gives to mental processes (Schmid 2005: 200). In contrast to that, he demands a turn "from the abstract to the concrete" (Lefebvre 1991: 415), namely towards "social and spatial practices" (Ibid).

Consequently, describing space as a historical-material product, Lefebvre disbands a dualistic concept of space and describes three different spatial dimensions (Ibid: 39). The first dimension, *espace perçu,* comprises the material production of space, which is built up by spatial practice and forms that aspect of space, which can be perceived (Ibid: 38; Schmid 2005: 208). The description of this dimension by Lefebvre and its interpretation by Schmid bring to a point that social space is formed by social practices which involves an inseparable nexus of bodily-material-social actions. This is one of the main conceptual lines useful for the understanding of space in this study.

Representations of space are at the core of the second dimension, the conceptualised space or *espace conçu* (Lefebvre Ibid; Schmid Ibid). These representations involve certain knowledge producing processes that mainly involve verbal signs and are carried out by scientific and other intellectual actors (Lefebvre 1991: 38f). These processes of knowledge production in turn reflect certain ideologies (Schmid 2005: 218). Representations of space involve language and discourse, but Lefebvre also includes plans, maps and pictures into the realm of relevant products (Lefebvre 1991: 233; Schmid 2005: 216). Although representations are of an abstract nature, they interact with social practice and have an important influence on the production of space (Schmid 2005: 216f). The act of seeing or conceiving space is an act of creating it (Lefebvre 1991: 93f). The *espace conçu* reveals the interrelatedness and simultaneousness of the mental and the material production of space. In pointing out the mutual dependence of mental and material processes, it is possible to identify a core message presented by Lefebvre that can be linked with some precaution to ideas on social practice theory presented by Schatzki (1996, 2001b, 2002). As described in the section on social practices above, they are described as processes which involve mental as well as bodily-material actions. The parallels identified will be discussed further below in this chapter.

Lived space, or *espace vécu* as named in the French original, is the dimension where meaning is produced mainly using images and other non-verbal signs (Lefebvre 1991: 39). Lefebvre describes the process following the production of space as in the following questions: "Dans quelle mesure un espace se lit-il? Se décode-t-il?" (= To what extent can space be read? Can it be decoded?; translated by the author) (Lefebvre 1974: 26). For the process of production of space to be complete, it involves to read and make sense of the representations of space. *Lived space* is a dimension, where social values, traditions and collective experience are represented and enacted (Schmid 2005: 223). Lefebvre (1991: 104) argues that this is the dimension where the languages of art, mainly non-verbal, symbolic expression come into use. At the same time, it is argued that it is the dimension of everyday life, where space is *lived* rather than *conceived* (Schmid 2005: 222). In his analysis of Lefebvre's work, Schmid points out the direct link between the *espace conçu* and the *espace vécu*. The separation between those two dimensions is described by Lefebvre as the result of a historical process of abstraction in which the mind gained a position of dominance over social reality (Ibid: 221). However, social reality is described to involve the permanent interplay between human experience and the creation of concepts (Ibid: 220).

There are several arguments in Lefebvre's concept of space that are believed to purvey a conceptual clarity that is of great value to this study. In opening up a dialectic relation between the mental and the material in the production of space, the primacy of the mental in concepts of space is overcome. The relevance of material aspects of spatiality that are mediated in social practices is given strong emphasis. Schmid contrasts this concept of space to the concept Bourdieu makes, in which social space is an abstract, mental entity in which social positions are represented (Schmid 2005: 208). Bourdieu has a specific relational understanding of social space, in which space represents the social relation, the distance or closeness of actors to each other (Bourdieu 1985: 724). This is expressed e.g. in the description of "social space as an invisible, not presentable reality that shapes practices and perceptions of actors" (Bourdieu 1998: 23, translated by the author). Space in this sense is used solely for the analysis of the social world and its' "social topology" (Lippuner 2007: 265). Bourdieu discusses the relationship between social and physical space more explicitly in his work *Physischer, sozialer und angeeigneter physischer Raum* (Bourdieu 1991). Lossau and Lippuner (2004: 206) discuss as problematic, that in Bourdieu's work, social and physical space overlap and physical space is even described to contain objects which are indicators for the position of a person in the social space. This perspective is argued to present shortcomings for analysis of *the social* and for the analysis of space and spatiality (Ibid). Considering the interest of this research it is assumed here that the identification of material entities in physical space which directly relate to patterns in social space enrich the analytical perspective of this research. The overlaps of spaces and the identification of entities of one "field" or social space in different physical spaces described by Bourdieu (1991: 29) for French cities, provide examples for the complexity of spatiality that serve the analytical set-up of social practice theory for social geographic research.

The concept of space presented by Lefebvre builds a link between mental and material aspects of practices in space. While no direct causal relationship is drawn between material objects and social relations, the interrelatedness of social practices with the mental as well as with the material is highlighted. Edward Soja takes up Lefebvre's triadic conceptualisation and describes the link between mental and material reality as follows: "More concretely specified, each of these abstract existential dimensions comes to life as a social construct which shapes empirical reality and is simultaneously shaped by it" (Soja 1989: 25).

Another aspect of interest in this study is the fact that in Lefebvre´s analysis of the production of space, an important role is given to the body. The body, which according to Lefebvre has been "betrayed", "abandoned" and "denied" by Western philosophy (Lefebvre 1991: 407) is restored and put at the centre of his conceptualisation of the production of space. It is on the level of the *lived space*, that humans imagine and symbolise spatial representations with their bodies (Lefebvre 1978: 281 cf Schmid 2005: 219f). Schmid (2005: 213) takes up these ideas by arguing that the totality of social space originates from the human body. Linked to the before mentioned importance given to materiality and the body in Lefebvre´s work is the general emphasis given to social practices. The historical-material notion of space and the focus on practices in the production of spatiality is what Schmid concludes in saying "[t]here is no space before practice" (Schmid 2005: 204). Working with Lefebvre´s understanding of space, allows an emphasis on the conceptual value of social practices which are used as entry points for the analysis of the relevant phenomena and dynamics of this study, in this case the production of space in a flood context. Imagining the human being in its totality but especially in its bodily existence and expression, brings up a further sub-question for research: How exactly does the human body produce space through social practices and which steps of human action and reflection are involved in it?

Yet another aspect that draws the attention towards the three-fold concept of space given by Lefebvre is the connection between the three levels of space. Lefebvre argues that the different spaces described cannot be thought of in separation but that they entail each other (Lefebvre 1991: 14, 40, 356). The production of space as an interrelated process involving practices of e.g. experience, representation and enacting involves dynamics that cannot be separated strictly into different phases. Analysing processes and the actions that interrelate in a simultaneousness manner is a challenging task however.

A final point made for the use of the threefold analysis of the production of space draws its´ inspiration from Lefebvre´s account on the concept of social space. It describes it as

> "[…] the outcome of a sequence and set of operations, and thus cannot be reduced to the rank of a simple object. At the same time there is nothing imagined, unreal or 'ideal' about it as compared, for example, with science, representations, ideas or dreams. Itself the outcome of past actions, social space is what permits fresh action to occur, while suggesting others and yet prohibiting others" (Lefebvre 1991: 73).

The last sentence of this quote highlights the fact that social space involves past actions. These past actions form part of social practices and these social practices guide future action. It is this detailed description of the "internal order" of social practices but at the same time their dynamic nature that raises a crucial question: Is it possible to link Lefebvre´s thought on the production of space and social practices to accounts on social practices and the *sites* of the social put forward by Schatzki? Where are the parallels, where do the two concepts diverge? What is the additional value gained when linking those two concepts?

In Lefebvre´s line of thought space has a very dominant role in guiding human behaviour and practices. Schmid argues that space contains a variety of orders and
instructions and thereby identifies those activities that can and those that cannot be
carried out (Schmid 2005: 224). Is this dominant role of space in guiding human
behaviour in line or in contrast to Schatzki's argument? We contrast Lefebvre´s
larger concept of social space with Schatzki's account on *The site of the social*
(2002). By quoting passages of text written by the two theorists, the essence of their
conceptual understandings is underlined. Firstly, attention is drawn to a quote by
Lefebvre and the emphasis he makes in the text.

> "'Human beings' do not stand before, or amidst, social space; they do not relate to the space of
> society as they might to a picture, a show or a mirror. They know that they have a space and
> that they are in this space. They do not merely enjoy a vision, a contemplation, a spectacle –
> for they act and situate themselves in space as active participants. They are accordingly situated
> in a series of enveloping levels each of which implies the other, and the sequence of which
> accounts for social practice" (Lefebvre 1991: 294).

In this conceptualisation there is no separation line between a human being and the
social space, both are intermingled. Schatzki makes the same point and expands the
view from a narrow towards a larger scope of the social, which is his new social
ontology expressed as the *site of the social*. He gives the example of banking practices to describe *site* as follows:

> "Rather, banking practices are this third type of site, for requesting reports and other banking
> activities occur as part of banking practices. Requesting and other activities help make up those
> practices, which in turn help constitute them" (Schatzki 2003: 177, emphasis in original).

The *site* of the social in this sense involves certain practices; moreover, it *is* these
practices. Practices contribute to the constitution of a specific site and at the same
time they are part of this site. Grasping the characteristic of practices in this sense
requires a flexible ontological perspective. It can be described as a "flat ontology"
(Schatzki 2011: 14), which objects their ordering on different levels. This perspective assumes that there is "a single plenum of practice[s] [and that] practices and
arrangements form bundles and constellations of smaller or larger spatial-temporal
spread" (Ibid: 17). The essential insight that can be gained from the philosophical
exercises presented by Schatzki is that practices are more than a nexus of *doings
and sayings* but they constitute and are part of a larger process. This is the constitution of society, in it´s perceived, relational and spatial sense. Figure 15 presents a
simple overview of the main analytical levels for social practices and production of
space which are of relevance in this study. It needs to be emphasised that in line
with a flat ontology, the main level that integrates and encompasses the others is
the plenum of social practices. Talking about social practices and about *where* they
take place consequently is a more meaningful analytical question than purely asking
for the physical location of practices and their interaction with the physical space.
It is opening up the analytical arena for an encounter with some of the social, cultural, historical and spatial processes that constitute and dynamically perform and
change social life. In line with this, linking accounts by Schatzki and Lefebvre al-

lows the author to gather some of the missing pieces for a strong account on materiality and the relevance of the body with the description and analysis of broader
plains of social life.

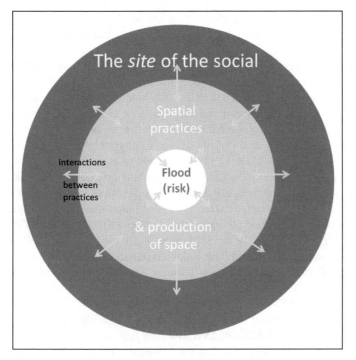

Figure 15: Conceptual perspective on social practices (Schatzki) and the production of space
(Lefebvre) linked to flood (risk) in this study
Source: author's elaboration

4.5 INTEGRATION OF PRACTICES, RISK AND SPACE INTO THE *RISKSCAPES* CONCEPT

Thus far this chapter has shed light on the three major lines of theory relevant in
this thesis. Social practice theory lies the major basis for conceptual analysis and it
is the account made by Theodore Schatzki not only on a more detailed analysis of
social practices but on the nature of the social and a specific social ontology that
are of inspiration for this thesis. The specific view towards risk as a Western concept, as a socially constructed one more generally and an enacted one creates a direct link to social practice theory. Risk is constructed in a material-mental process
that involves social practices. The same can be applied for space. Arguing with
Lefebvre, it is not space that is of interest in the end but it is the production of space.
This process of production involves social practices and it is again in dialectic be-

tween mental and material sub-processes that these practices are performed. Linking social space understood in Lefebvre´s sense with the accounts made by Schatzki on *the site of the social*, promises to reveal how the spatial levels of social practices are part of the larger set of sites of the social. The idea is to identify conceptual nodes which improve our understanding of the production of risk through and in socio-spatial practices. Assumably doing this can contribute to an enhanced understanding of *the social*.

With this perspective it seems possible to overcome what Oliver Smith (2002: 24) calls the trap of geography in describing the "'hazardness' of a place". There are a range of approaches that state that risk is distributed unequally in space (Mitchell 2010: 335). Spatial representations like maps or Geographical Information Systems (GIS) are powerful tools that can convey the message that risk is a characteristic that can be ascribed to a varying degree to physical spatial entities. The case of tropical diseases like Malaria, Dengue Fever or even the Zika virus, shows how health risks are attributed to spatial location while the cause of this risk, a certain biological substance, is repeatedly described as prevalent in certain geographical *zones* (Ibid: 336f). Similarly, hazard zones for hurricanes represent the physical localisation of a risk in quantifiable spatial entities (Kjerfve et al. 1986; Durand, Vernette & Augris 1997). In the fields of disaster risk management (DRM) and development cooperation, the World Risk Index (Weltrisiko-Index) developed by the German based alliance for development *Bündnis Entwicklung Hilft* and experts from UNU-EHS (United Nations University – Institute for Environmental and Human Security), has become a powerful instrument that communicates spatial representations of risk for large parts of the globe. The index "calculates the risk for 171 countries from 28 indicators" (UNU-EHS & Bündnis Entwicklung Hilft 2015: 44) and ranks countries according to factors such as exposure, susceptibility or lack of adaptive capacity in a global comparison. These examples for generalising risk spatially are in clear contrast to approaches of social science and practice that emphasise the differentiation of risks not only between nations but between and even within smaller geographical entities, linked to social, economic, political or cultural characteristics of the people (Curran 2013; IFRC 2014; Lewis 2015; Gaillard 2015). Mitchell (2010: 338) describes the possibility and condition in which "two households are in the same space, but in a different place and thus at different risk" (Ibid). In this description, it is at the level of place that social practices are performed. The social and the spatial construct each other and risk can be identified as a further condition resulting from this process (Ibid: 340). Risk is also constructed at the level of place. It is in line with this relational constructivist perspective that this thesis tries to reconstruct the interrelated ways of constructing risk and constructing space through social practices.

The specific conceptual triad designed for this thesis is chosen not as an end in itself but as a means in order to make sense of empirical phenomena observed on the ground, in the field, in daily life. Selected links between concepts of social practices, risk and space have been presented. In order to emphasise this link, it is the goal of the remainder of this chapter to discuss a specific attempt that links these three concepts in a specific way which is regarded as useful for empirical research.

The concept of *riskscapes* developed by Müller-Mahn (2013) emphasises the relevance to identify and analyse the spatial dimensions of social practices related to risk. The term *riskscapes* had already been used before in environmental health and multi-hazard mapping (Müller-Mahn & Everts 2013: 24) and similar terms such as *hazardscape* and *disasterscape* have been deployed in risk and disaster research (Corson 1999; Mustafa 2005; Kapur 2010). The concept developed in German geography by Müller-Mahn is based on social theory and can be applied in as well as beyond the scientific field of geography. Risk is not foremost regarded as something people are passively exposed to; rather it is social actors and groups who actively construct and take risks (Müller-Mahn & Everts 2013: 28). The main authors who inspired the concept are Arjun Appadurai, Valerie November, and Theodore Schatzki (Ibid 2013: 24ff). Presenting Appadurai´s five -*scapes* of imagined worlds, in which humans locate themselves, the *riskscapes* concept detaches from particular territories and relates specific -*scapes* to practices (Appadurai 1990: 296ff; Müller-Mahn 2013: 24f). Müller-Mahn and Everts (Ibid: 25) emphasise the relevance of Appadurai´s accounts mainly insofar as the concept of landscape is not limited to physical space but as it represents viewpoints and because it gives primacy to the analysis of social rather than individual processes. By stating that *"risk has an impact on the future of the territory, because of its power of re-configuration of the collective"* (November 2004: 274), Valerie November gives important insight into the consequences risk making has on spatial levels. With this she puts forward the argument why social geography is a relevant discipline to analyse risk making. It is the reconfiguration of the social and thereby the reconfiguration of space or spatiality that takes place by risk construction. Schatzki highlights the link between space, time and practice (Müller-Mahn & Everts 2013: 26). Discussing Heidegger´s approaches to spatiality and life Schatzki highlights the notion of "in-volvement" (Schatzki 2007: 43), describing it as follows: "to be in the world is to be involved in it, to be dealing with and proceeding amid (*bei*) the entities that compose it" (Ibid, emphasis in original). He links involvement with the concept of dwelling, thereby connecting social practices with spatial entities (Ibid: 90). Taking up ideas on dwelling presented by Ingold (2000), he argues that landscapes are "the medium for, and outcome of, human practices" (Schatzki Ibid: 91). Ingold emphasises the constitution of landscapes through the social practices carried out in space in the past by arguing that a "landscape is constituted as an enduring record of [...] the lives and works of past generations who have dwelt within it, and in doing so, have left there something of themselves" (Ingold 1993: 152). Schatzki (2007: 50) takes up and discusses Heidegger´s concepts of time and space, highlighting the social dimension of spatiality. Time is described with the metaphor of a journey, which is that time is a "movement into past and future" (Ibid: 58). For this study it is argued that spatial entities such as landscapes and social practices are connected to each other in an inseparable way, as any socio-spatial pattern is constituted by social practices. Taking up the dynamic concepts of time and landscape and linking them to concepts of risk can help to better understand how risks are being perceived or created differently by different actors. As empirical cases show, specific practices aimed towards the reduction of one risk can result in the increase of other risks for local people (Müller-Mahn & Everts 2013: 29). The *riskscapes* concept makes tangible the different processes, subjects and perspectives that are present when risks are being constructed, negotiated or transformed. It also allows the argument that there is a co-production of risk and spatiality, both processes closely linked to each other.

While the *riskscapes* concept addresses major concepts and thoughts which are crucial in this research, the concept in itself provokes a range of questions concerning theoretical and epistemological assumptions. In the following paragraphs these questions are addressed in order to further develop the basis for an analysis of the spatial dimensions of practices related to risk. The term risk is given a dominant role in the *riskscapes* concept and a reflection on the term itself as well as its historical development in Western science is not accomplished in the first publication on the concept from 2013. From a constructivist viewpoint however it is argued that this dominance of the term risk might result in the creation of a bias. In empirical research in flood prone areas, it has shown that a range of expressions is used by different groups of people (e.g. government officials, inhabitants of villages in flood prone areas) in order to describe challenges of dealing with situations in which water overflows a river bank or shore (Rohland et al. 2014: 194). The term risk is not the only term or concept used and is certainly not to be regarded as privileged in the sense of scientific value. Therefore, it is regarded as highly relevant to make use of local concepts in research that takes place outside of the sphere of a researchers´ own socialisation. Consequentially, it is regarded as necessary here to include deconstruction of the different concepts related to risk or uncertainty as well as indepth analysis of those processes that interact with practices of risk making or taking as core parts of the utilisation of a *riskscapes* approach in social geography. Risk is not presented as an absolute social or scientific reality but as one possible result of processes going on in the mental as well as in the material world of social actors. This asks for an integration of specific wordings that are used in different world regions when describing concepts that exist in parallel or in contrast to the terms risk, vulnerability or other terms familiar in disaster risk sciences.

The *riskscapes* concept draws on the descriptions of different *-scapes* by Appadurai, which highlights a concept of space that involves notions of imagination and perception. Moreover, reference to Schatzki´s description of landscapes is given (Müller-Mahn & Everts 2013: 26). In the development of the *riskscapes* concept a clear cut and in-depth development of the understanding of space and spatiality is however missing. It is a task that is open to be accomplished however by scholars using and discussing the concept. Attention is drawn again on the social practice theory presented by Schatzki and on the linkage made to Lefebvre´s accounts on the production of space in the sections above. Integrating this conceptual approach into the *riskscapes* concept can be of major use for the conceptual as well as empirical work on social practices, space and highly relevant phenomena like the social construction of risk. An attempt on this is made in chapter 9.

Whereas Schatzki describes a practice as "a set of doings and sayings" which is organised by four entities that involve the nexus of bodily and mental processes (Schatzki 2001b: 58) (figure 14 above). It needs to be underlined that the descrip-

tions Schatzki makes on the organisation of social practices object a conceptualisation of mind as "a thing or apparatus that causes behaviour" (Ibid.) Rather, social
practices involve performances on a phenomenal level (behaviour) and at the same
time they involve practical intelligibility, i.e. mental processes of knowing when it
makes sense to perform a specific doing or saying. Therefore, bodily-material and
mental processes are interdependent and account for each other in the building and
performance of a social practice. While material aspects are highly relevant in this
understanding, they are described as mediating but not guiding activity (Schatzki
2001a: 20). It can be summarised that neither a mental nor a bodily-material aspect
in itself causes behaviour. It is the complex network of a multitude of bodily-material and mental processes in interrelation that guides social practices.

Recalling the emphasis put on the *site*, where practices take place and which
they are part of, shows that the spatial dimensions of risk are connected to a larger
context, which is not purely spatial but which also includes aspects of time and
teleoaffectivity, namely an "orientation[s] toward ends and [...] how things matter"
(Schatzki 2001b: 55). Including this into the concept of *riskscape*s allows for a connection of risk related practices in space to the larger context in which practices are
produced and carried out in society. Finally, including the three different but interacting levels of production of space described by Lefebvre, the *riskscapes* concept
gains an analytical tool to identify the different steps in which the production of
space and the production of risk related to space are performed.

Figure 16 presents a simplified visualisation of the formation of a *riskscape* of
flood management that involves different socio-spatial practices, which interrelate
through overlapping areas of performance and intelligibility and which all take
place in a larger context described as the *site* of the social and simplified in this
visualisation as a background layer. It is at the bodily and material, at the mental
and moreover at the intellectual and cultural level that spatially relevant risks are
imagined, constructed and performed through social practices. Accounting for the
complex and interwoven character of these processes makes possible the adequate
analysis of the phenomena we encounter and deal with in life. Interacting social
practices linked to flood management – may they be supporting, contradicting or in
an indifferent or other relationship – are at the core interest in this study. Linked to
this interest is the aspiration of the author to improve epistemological and empirical
approaches towards making sense of and dealing with adverse situations.

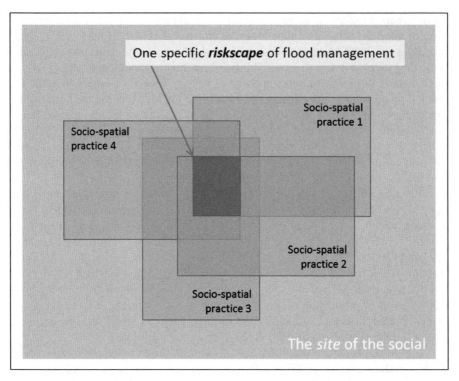

Figure 16: Conceptual setup of a riskscape of flood management in this study
Source: author´s elaboration

5 METHODOLOGY

This chapter presents the methodology of this thesis and thereby it highlights major switch points within the research process. The specific pathway taken in research requires a coherent methodological orientation and it builds a comprehensive research process that demands to perform specific methods at a given time along the way. Methodological decisions have great influence on the research in the short and in the long-term and it is argued here that

> "[t]echnical methodological details matter not so much because they lead to 'successful' research but because they determine the nature of the data being constructed and influence the purposes to which those data can be directed" (Cloke et al. 2004: 163).

The methodological approach is built up on three main pillars of research: (1) grounded theory as a general orientation towards empirical work, (2) *epistemology of the South* as a specific epistemological perspective included in this research and (3) collaborative research as a concrete methodological approach chosen. These approaches do not only address relativity of knowledge but also question individualist processes of empirical research and interpretation. Methodology is on the one hand designed here in order to provide empirical data that help to answer the research questions. The specific methodological approach designed for this thesis allows the development of new conceptual ideas and enables to establish new conceptual links among empirical phenomena. On the other hand, this methodology represents a research practice whose goals lie beyond the generation of empirical data and results. Research in social geography is a social practice and thus has social relevance and consequences. Building up a basis of trust, respect and mutual benefit is not an additional quality of research; it lies at the centre of research and is tangent to methodology. Throughout the research process the researcher and a researcher from Mexico who participated in empirical research in 2015 (called co-researcher in this text) shared knowledge on the methods and the general research methodology with research partners in the field.

In this chapter, the methodological approach is presented in a way that promotes the identification of a specific link between the paradigmatic basis and empirical approach of this study. After an introduction into positionality and considerations of ethics and ownership, the paradigmatic context of research is sketched out by a description of the overall empirical approach and process. This is followed by a description of the specific methods of data generation applied and developed. The last section of this chapter addresses the methods of analysis and interpretation of qualitative data applied for the elaboration of results (see also figure 17 below).

5.1 POSITIONALITY, ETHICAL CONSIDERATIONS
AND QUESTIONS OF OWNERSHIP

Besides conceptual decisions of this thesis also the methodological decisions are based on the specific epistemological fundaments described in chapter 4. The perspective of "social relativity" (Berger & Luckmann 1967: 15) of reality and knowledge among human beings determines and guides the methodology of this thesis. As a contribution to research in social geography, this thesis accesses a range of different forms of knowledge, through concepts, theories and methodologies, through material objects as well as audio and video representations. A range of different sources are used to approach this knowledge, among them the empirical work elaborated within the process as well as a range of different text documents which serve as secondary data. Basing on these sources the thesis creates new knowledge and new sets of information analysed and structured in a specific way. Research and practice carried out here emphasise the need to reflect on the role of the researcher within the research field and the research practice as well as on ethical aspects encountered throughout the research. Authors from the discipline of social anthropology, but also from social geography emphasise the need to reflect in a transparent and self-critical manner about the positionality a researcher takes in the research process (i.a. Browne et al. 2010; Cloke et al. 2004; Merriam et al. 2001; Rose 1997). The research field is understood to be a construct established by the researcher and through the interrelation of different persons throughout the research process (Browne et al. 2010: 586). The phrase "it´s not about them and us, it´s about us" (Ibid) is a specific perspective towards the positionality of a researcher presented by geographer Kath Browne and the activist researchers Leela Bakshi and Arthur Law. The act of positioning addresses aspects of social inequality and vulnerability of social groups that are relevant for the research process and the interpretation of results. At the same time, positionality can overcome a dichotomy of depicting the researched as powerless people and the researcher as the powerful actor (Ibid: 587). Reflection on the positionality allows the identification of the limits of a researcher to gain access to knowledge.

    In the research presented here, a range of persons from the larger case study region in Mexico are involved, e.g. persons from academia, the local village settings, the media and government offices. As the largest part of empirical research was carried out in the local village setting, the reflection on positionality of the researcher in this part of research is emphasised here. Besides aspects of age, gender and ethnicity, the activities of everyday life are relevant in order to reflect not only on differences but on commonalities among all the persons involved in the research at village level. Throughout the research a range of reflexive practices were performed by the people involved. A starting point for in-depth reflection was an exercise carried out during a research seminar held in 2014 at the *Centro de Investigaciones y Estudios Superiores en Antropología Social* (CIESAS) in San Cristobal de Las Casas, Mexico. The exercise of "espejar" (= "mirroring") (personal communication Leyva Solano 07.08.2014, Field notebook 1/2014) allowed reflection on the construction of the specific research question and to reflect on the differences, similarities and synergies encountered between the people involved in the research process. The identification of possible common interests of the researcher and the inhabitants of the case study villages with regard to the research topic and common interests beyond the research topic (e.g. reconstruction of the history of the region, small scale agriculture, health aspects for women) has contributed to the formation of a group of solidarity and exchange of ideas. Practices of reflection formed part of all steps of the research. Basic decisions concerning the epistemological perspective, theoretical concepts and methodologies are not chosen following traditions of scientific disciplines. They are part of an active process of construction by the researcher involving reflections with the co-researcher and other people directly involved in the research. While it is more common practice in academics to highlight the efforts and decisions made by the main researcher or author of a thesis, negating the influence of the others involved in the research is regarded as inappropriate for the case of this research. Taking seriously the notion of participation and the creation of an "us" rather than a "me" and "them" (Browne et al. 2010), demands the reflection on the efforts made by a group of people to engage in the modes of research.

Reflecting on the positionality of a researcher and the ownership of knowledge is not only a question of adequate scientific practice; more importantly it involves key ethical aspects. Qualitative research has a large potential to access knowledge that is not retrievable with methods of standardised enquiry. However, the information obtained and documented by a researcher in a long-term stay in the field provides a researcher with tools that can be misused. Historical examples, e.g. the involvement of ethnographic research in colonial activities, document the use of information in order to perform political control and economic exploitation (Lévi-Strauss 1966b: 126). Awareness and reflection about ethical obligations and the complex process of designing ethically viable participatory research (Brydon-Miller 2008: 203) is regarded a key quality of scientific research. Current developments give a positive outlook on the development of adequate review processes and acknowledgement of ethical aspects in social research (Gelling & Munn-Giddings 2011: 105f). Although the doctoral thesis is an individual project with the name of a single author, the research process and the results gained would not be the same if the researcher had done the research without the co-researcher and the collaborators from the villages. Questions about ownership and the mandate to interpret are not to be answered in a general manner for all data obtained. The level of trust built up in qualitative research allows for access to sensitive aspects of human life. While Wilson and Hodgson (2012: 112) argue that little is known about the specific qualities of trust in social research, Ryen (2008) presents key features of trust in social research with a special emphasis on cross-cultural contexts. As a first step it is regarded necessary for any geographer to be aware and raise awareness of the potential power gap between the researcher and the people researched (Cloke et al. 2004: 164).

A review of the history of the disciplines argues that

> "geographers have acted as if they stand outside the specific historicity and geography of their subjects; this has enabled them to comment on the reality of the subject's view of their own situation, while not allowing the subject's valid version of reality" (Pile 1991: 467).

Involving an interviewed person in a reflection on the expressed view and involving her or him in the conceptualisation and structuration of research arguments is an approach believed to get closer to what is called a theory grounded in the *life-worlds* of human beings. Cloke et al. (2004: 165) opt for the creation of a research alliance between interviewer and interviewed. While the authors of the book *Practising human geography* describe that the understanding of forming an alliance has mainly been used in psychoanalytical work with patients (Ibid), the research alliance built in this research does not include therapeutic or psychoanalytical activities. Rather, the emphasis is put on collaboration in a team of members who all benefit from the activities performed. This is possible through the development of a collective pathway towards addressing various common interests, among them the main research questions of the researcher as well as of the research participants. As the interests that can be addressed in a collective approach are limited, common interests have to be identified in order to make collaboration viable and ethical.

A major challenge in qualitative research is the confidentiality of information. The data obtained, e.g. through narrative or biographic interviews, can contain details of a person's life that make the information attributable to one individual even when names have been changed or anonymised (Mason 2002: 80). The use of visual data can challenge the principle of confidentiality even more (Ibid: 201f). A continuous consideration of ethical principles throughout the research process as well as after the research is indispensable in order to prevent research results from having negative consequences for the participants. This research is based on the ethical principles elaborated in German-speaking social anthropology in the *Frankfurter Erklärung zur Ethik in der Ethnologie* (Hahn, Hornbacher& Schönhuth 2008). The principles comprise general respect, confidentiality of participants, transparency of the research process and the integration of feedback by participants (Ibid 2008: 3). In order to break down these principles into a guideline for research practice, this study bases on the five criteria described by Cloke at al. (Ibid), including specifications on informed consent, privacy, harm, exploitation and sensitivity to cultural difference and gender. In the case of this research, it has been agreed upon with the village members, that the rights of the (audio-)visual material are owned by the village community. However, the permission is granted to the researcher to make use of these data for the purpose of the doctoral thesis, including the analysis, representation and the creation of a backup of all the empirical material. The data obtained through interviews and participant observation, written down or recorded on audio are the intellectual property of the researcher. No personal data is passed on to third parties and names are anonymised in this thesis. As some of the information about the local context and political structures are highly sensitive, it is crucial to protect those people who shared sensitive information with the researcher. In order to assure the confidentiality of people while at the same time giving appropriate credit to the process of collaboration, it is declared that the perspectives presented in this thesis, if not indicated by a quote or another indication towards an external source, represent the scientific perspective and opinion of the author.

5.2 GROUNDED THEORY AND LINK TO TWO "ALTERNATIVE EPISTEMOLOGIES"

This research takes place in proximity to three major methodological approaches that involve certain paradigms and practices. While the first two are described in this sub-chapter, putting a focus on paradigmatic notions, the third one being a more practical approach towards research is presented in the following sub-chapter. The first approach, Grounded Theory is well-known in Western academia, while the other one, *epistemology of the South* (Sousa Santos 2016: 12) is so far mainly known and accepted in Latin America. Grounded Theory is described as "a general methodology for developing theory that is grounded in data systematically gathered and analyzed" (Strauss & Corbin 1994: 273). Breuer (2010: 39f) describes Grounded Theory not as a set of methods limited to the phase of field research but as a general orientation towards doing research and engaging actively with the topics studied and the social groups involved. This approach is directed towards the generation of theory and is called "grounded" because theory is elaborated on the grounds of empirical data. The conceptual outline and research strategy of Grounded Theory was first described by Barney Glaser and Anselm Strauss (Glaser & Strauss 1967). Later elaborations relevant for this research are mainly those by Strauss and Juliet Corbin (1994, 1996). The approach was developed with the conviction that qualitative research methodology "require[s] *redefinition* in order to fit the realities of qualitative research and the complexities of social phenomena" (Corbin & Strauss 1990: 4, emphasis in original). Grounded theory refers to and makes use of principles formulated by pragmatism and symbolic interactionism (Ibid: 5). Putting emphasis on constant change in social life and distancing from determinism of human behaviour (Ibid), the authors propose a methodological procedure that takes these general dynamics into account. Grounded theory approaches presuppose that research questions are or can be adapted throughout field research (Breuer 2010: 54f). Participatory research approaches emphasise that research questions need to be formulated together with research partners from the field or originate in participatory processes (Cornwall & Jewkes 1995: 1668; Grant, Nelson & Mitchell 2008: 593). As empirical research of this thesis was split up into two phases, one in 2014 and one in 2015, the initial research question was revised during and after the first phase in 2014 and scrutinised during field research in 2015.

The orientation towards the basic principles of grounded theory is relevant in a range of ways in this research. During the time when first data was collected, a first analysis of these data was crucial in order to direct the next research steps. Unlike other research processes in qualitative research, grounded theory requires this early analysis as it is believed that first analysis of data can point towards important findings that have to be continuously checked throughout field research itself (Corbin

& Strauss 1990: 6). In the case of this study early analysis resulted in the identification of important local aspects, adaptation of interview questions or it guided the researcher towards specific subjects in a village or organisation for an interview. The early analysis carried out during empirical research involves the formulation of abstract conceptualisations that are labelled in simple and short form (Ibid: 7). In the case of this research, early identified concepts were named with a close orientation to the language used by the people who are part of the research and to the language of the researcher. The identification of sentences and verbal expressions given repeatedly by various discussion and interview partners allowed for the building of a range of locally grounded concepts. These local concepts formulated in Spanish guided empirical research and further conceptualisations. Other early concepts, for which no local verbal expression was identified, were formulated in words by the researcher in English. The building of a range of concepts reveals that some concepts can be grouped into a more abstract category that describes a social phenomenon (Ibid). This process of building categories is helpful in order to discover new relationships within the empirical data and to develop new ideas about concepts and theories on an abstract level. As shown below, in the case of this research, e.g. concepts relating to the phenomenon "flood" had connection to other relevant concepts and were grouped in a specific category. Following grounded theory, the sampling of the people and organisations to include in research is a continual process. Building up on the general research topic and research question, a first sample of individuals, groups and organisations are selected for study (Ibid: 8). However, the identification of new persons and organisations to include in the sample is carried out during research following conceptual considerations (Breuer 2010: 58). Sampling is not done at random but guided by conceptual questions and by looking for enrichment of the conceptual information through a range of "cases, variations and contrasts" (Ibid). Sampling in this study followed the development of concepts through the testing of variations and contrasts, leading to a comprehensive set of empirical data and additional information.

Carrying out grounded theory in the specific research context present in the South of Mexico demanded for an integration of two approaches described shortly in chapter 4. *Epistemology of the South* and *epistemology of the heart* are two approaches that do not only lay a basis to reconsider theoretical accounts on knowledge but they have implications and give direction for empirical research processes. Far from following static rules for practical research, orientation is taken from examples of "how they did it" (Leyva Solano 2015: 27, translated by the author), how researchers who subscribe to the two "alternative epistemologies" do and did empirical research. Two major lessons for the design of empirical research are drawn from the examples of fellow researchers encountered during field research in Mexico and read about in the vast accounts from Latin American research (i.a. Bastian Duarte & Berrío Palomo 2015; Dietz 2012; Puglisi 2014; Sousa Santos 2011). The first lesson is to interact with local people, invite them to become research partners for both data collection and analysis. The second lesson taught by the "alternative epistemologies" is that it is of crucial relevance for the researcher to give attention to, document and reflect on bodily perceptions and impressions the researcher has as well as the co-researcher and research participants express. Co-lombian sociologist Fals Borda describes a research process in which results are won "through the practice, through the empathic development of processes felt within the realities themselves" (Fals Borda 2009: 328, translated by the author). The setup of empirical research in this study has taken up the idea of an "empathic development" (Ibid) within the research situations. Even though, as mentioned before, a separation between cognitive and emotional perceptions is not always possible, this approach attempts to make transparent the information obtained through an emotional involvement. Throughout empirical field research this was accomplished through a continuous documentation and reflection of observations and perceptions by the researcher with the co-researcher and the Mexican tutor. More precisely, this approach was followed by

1. Notetaking of surprising observations in the field notebook during and at the end of each day of field research
2. Notetaking of surprising emotional perceptions and dreams that happened during or in the direct aftermath of a field research phase in the field notebook or in digital memos on the computer
3. Reflection on surprising observations, emotional perceptions and dreams with the co-researcher during and after field research phases and documentation of discussions in the field notebook
4. Reflection on aspects identified as relevant for research with the Mexican tutor in meetings in Palenque and in San Cristobal de las Casas and documentation of discussions in the field notebook or in digital memos and reports.

The continuous reflection and documentation of reflection processes is believed a major task in research based on "alternative epistemologies". The ways of data generation and interpretation under these approaches are not standardised nor generally acknowledged in social research. As Fals Borda (Ibid) argues, opposition towards this approach is expected from social researchers who rely mainly on the traditional instruments of social enquiry developed in academic institutions. In order to underline the specific qualities empirical approaches based on "alternative epistemologies" provide, it is therefore necessary to document a research process in a manner, that allows to recognise the different perceptions included into observation, analysis and interpretation by the researcher.

5.3 A COLLABORATIVE RESEARCH APPROACH

Empirical research in this thesis has been in large parts a collaborative process. The conceptual basis and the methodological design were developed by the researcher. However, in the generation of empirical data and the practical research process the researcher depended on and collaborated with other people. Empirical field research that wants to generate new knowledge on local disaster management strategies and the contexts in which these strategies evolve needs to take into account the perspectives, opinions and concepts of people who live and interact in the case study region. By applying a research paradigm that describes risks as social constructs, the epistemological pathway of this research demands a generation of new knowledge together with or in great proximity to the "research subjects".

Participatory research has regained importance in geographic research in the last decades (Kindon et al. 2007a). While participatory research and collaborative research are often used synonymously, the research carried out as part of this thesis is called collaborative, giving emphasis to the terms used in the research context in Chiapas, Mexico. Collaborative forms of research act on criticism towards dominant forms of knowledge production and governance (e.g. forms which base on a positivist research paradigm), and build on links to critical theory, postcolonial and feminist approaches (Kemmis 2008: 134f; Schurr & Segebart 2012). Inspiration for this research has been taken from researchers who emphasise the relevance of results for the academic as well as for other societal spheres:

> "A good ethnographer is someone willing and able to become a more reflexive and sociable version of him or herself in order to learn something meaningful about other people's lives, and to communicate his or her specific findings, including their wider relevance, to academic and other audiences" (Cloke et al. 2004: 170).

Inspiration and encouragement has been taken from the Mexican tutor of this research, as a person, a scientist and an activist, who presents one of her experiences with collaborative research as follows:

> "Every one of us carried on his or her back a particular history and had very different reasons why to engage in this new form of walking in-net-ted. Specifically, for me it meant the beginning of a seventh attempt in order to find a better way to be in this world and at the same time it represented the possibility to walk autonomously but linked with other members of the net(work), with their communities of origin and organisations" (Leyva Solano 2010: 373; translated by the author).

Collaborative research is a time consuming process. In the beginning goals of the research have to be defined among the participants, throughout, a lot of time has to be invested in the reflection on and documentation of the empirical work of each participant and in the end of a research phase, the results have to be discussed in group again and agreed upon. In this research one challenge was that not all participants of the collaborative process were used to writing down their research questions or findings or had the time to do so. Therefore, verbal expressions had to be written down after discussions in order to document the research process and to comply with the regulations of good scientific practice and citation of primary data that are common in social research.

As this research goes beyond participatory or collaborative research and groups itself among *Participatory Action Research* (PAR), the research does not only follow academic goals but at the same time pursues to empower local people to actively take part in the shaping and reshaping of their livelihood bases. PAR is an "umbrella term" (Kindon et al. 2007b: 1) for different approaches of action-oriented research. Examples from action research show that the approach has the potential to supply tools and ways for social transformation, including the renegotiation of power distributions (Gavin et al. 2007: 66). As Boaventura de Sousa Santos argues that "without epistemic justice there is no social justice" (2005, 2009 cf Leyva Solano 2010: 18), in this research it is argued that if a researcher intends to support social change with her or his research, first of all the researcher has to change how she or he thinks and does research and whom she does and does not involve.

Collaboration was pursued by the active involvement of a co-researcher in the field research phase in 2015, the intensive exchange with the Mexican supervisor in 2014 and 2015, the enrolment of two visiting researchers in a short workshop in one of the communities in 2014, and the long-term involvement, collaboration and regular exchange with the community members of one community that started in 2014 and is ongoing until this day. A short-term collaboration was carried out as part of intensive exchange with two other doctoral researchers in the first weeks of field research in 2014, which allowed for a thorough reflection on the initial concepts and methodological approaches as well as on first experience in the case study villages. The major process of collaboration was performed between the author of this thesis and the co-researcher, a researcher and member of the Maya-Tseltal community from Mexico, in 2015. The two researchers worked together during a time span of around four months (May-August 2015) in the field as well as in a workshop in Palenque (May 2015) and collaborative editing activity in San Cristobal de Las Casas (July 2015). It was decided to work with a co-researcher in the second phase of empirical research, when the focus was put on the generation of audio-visual data. While inviting a different person or local people into a research process is not common in research for a doctoral thesis in geography, it is a strategy discussed in various disciplines that apply qualitative methods including social anthropology, psychology and social work (Hartley & Benington 2000; Heron & Reason 2008; Grant, Nelson & Mitchell 2008; Littlechild, Tanner & Hall 2015). It is seen as an option to gain new insights into a community and to apply new approaches of research which involve the sharing of information and methodological as well as analytical skills. One of the basic ideas is that in fieldwork in a context distinct from the *life-world* of the researcher, some relevant contextual information cannot be understood sufficiently. While the main intension of working with a co-researcher was not to access more data, it was the process of continuous reflection between the two researchers that lead to enhanced options and defined entry points for data analysis. The sociologist was identified as an experienced and motivated research partner by the supervisor from the Mexican host institute CIESAS. Having experience in empirical research in Chiapas and filming local ritual events in the communities of his own ethnic group of Maya-Tseltal people, the co-researcher contributed conceptual as well as methodological ideas. Besides the common goal to work on, each of the two researchers had their own perspectives, interests and ways of doing research. Exchanging upon these differences not only helped to look at the research topic from various angles. At the same time, it put a different light on research itself, on its´ methods, its´ paradigms as well as its´ pitfalls.

Another form of collaboration in this research project was the involvement of village inhabitants in the research and especially in the realisation of audio-visual and visual projects in the village. Members of one of the villages, where empirical field work on floods and local as well as government flood management strategies was carried out, were asked to collaborate in the filming of a video. The topics addressed were the daily life, local customs and floods in the community. While the flood as a topic was set from the beginning, other topics evolved throughout the process of empirical field work. The application of a participatory approach with audio-visual methods was a way of giving community members the possibility to interact with the research questions, to get to know the purposes of this research and to express their opinions about it. At the same time, it was pursued to share with village members the capacity of producing a video. Through the process of involving people from the local context in the process of filming (behind, at the side and in front of a video-camera) and reflecting upon a short part of the video that was presented in the village at the end of the research phase, it was possible to actively involve the capacities, creativity and perspectives of different village members.

5.4 ETHNOGRAPHIC RESEARCH AND
SOCIAL GEOGRAPHICAL METHODOLOGY

The methods used in this study predominantly are those that form part of ethnographic and social geographical research. Ethnography is increasingly acknowledged in geographical research as more than a set of methods but as an approach with significant value for empirical and theoretical processes within research (Scholl, Lahr-Kurten & Redepenning 2014: 58). While the use of ethnography increases in geography, scepticism is still present in the discipline (Herbert 2000: 557). What hinders many geographers from doing ethnographic research is the fact that the structural conditions of research are not always favourable for long-term ethnographic field research (Cloke et al. 2004: 170). Funding patterns and working procedures in the academia result in the fact that ethnographic research is dismissed totally or the time for ethnographic research is limited to only a few months (Hammersley 2006: 5). Notwithstanding the challenges it poses, ethnography is regarded by the author as holding considerable value for geographic research. By reconsidering its´ origins and by committing to its´ epistemological implications as requested by Verne (2012: 185), ethnography is the research practice followed in this study. Due to the possibilities ethnographic research provides to build up a close relationship with local people and their everyday practices, it has been found by the author as offering valuable entry points for answering the research question. Far from believing that ethnographic research gives a more authentic impression of reality, this approach offers an access to valuable information about everyday practices, the larger context of a lived environment as well as the importance of social and material aspects. Lévi-Strauss (1966a: 113) refers to the basic understanding of ethnographic research developed by famous ethnographers Marcel Mauss and Bronislaw Malinowski: "Social facts do not reduce to scattered fragments. They are lived by men, and that subjective consciousness is as much a form of their reality as their objective characteristics" (Lévi-Strauss 1966a: 113).

Criticism on ethnographic research is addressed in this study through a methodology that includes a constant reflective process, the involvement of co-researchers and a thorough documentation of the intertwined process of data generation, analysis and conceptual development. Besides scepticism against ethnography originating in other scientific disciplines, it is the historical background of social anthropology as well as of geography which demand a cautious and reflected use of ethnographic and geographic methods in any research project.

The methods presented in the following section originate in ethnographic research with a focus on social anthropological perspectives. They are designed in a specific way to address the research question of this study which has a specific social geographic focus. The conceptual links between social practice theory and theories of space are specifically addressed by the methods applied. As described before, the specific set of methods did not result from the initial research question alone but it was developed in an integrated manner throughout the research process. While some methods had been chosen from the beginning of research on, others were added or adapted and new methods were developed during the research process. In the following paragraphs, basic information about the methods is presented complemented by a description of their specific design and application in this study. An overview on the methods used for the generation of empirical data and information and on the methods used for analysis is presented in figure 17.

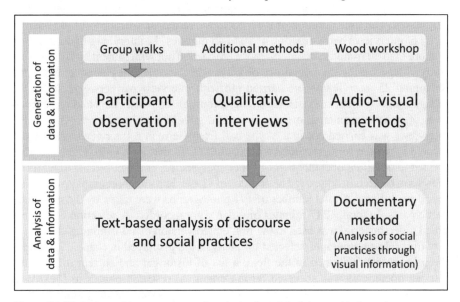

Figure 17: Methods used in generation and analysis of empirical data and information

5.5 DESIGN OF THE EMPIRICAL RESEARCH PROCESS AND METHODS OF DATA GENERATION

The empirical research process in Mexico was performed in two phases which were carried out from July to November 2014 and May to August 2015. During the months May to August in 2015, the co-researcher participated in the practical and empirical research activities. A first exploratory research had been carried out in 2013 in which however no primary data was generated. The empirical research process in 2014 and 2015 involved a variety of different methods. While participant observation and interviews produced a considerable part of data, a range of other methods was used to complement the information. An overview on the methods of empirical research, the format of generated data and methods of analysis is presented in table 4 and the methods are described further below in this section. In the overview it is shown that a variety of types and formats of qualitative data are available for analysis. One distinction of the methods used here can be made between methods that access information through a textual language (e.g. spoken words and written texts) and those that access information through a visual language (e.g. photographs and videos).

Observations, questions, ideas and additional information were written down or drawn as field notes in the field notebooks. Additionally, memos were written in documents by the researcher. A total number of 101 qualitative interviews were carried out. The time duration of the interviews varied significantly. While the majority of interviews had a duration of 20 to 45 minutes, 12 interviews had a duration of more than 60 minutes. 76 interviews were carried out with inhabitants of the field study villages in the municipalities of Playas de Catazajá and Palenque, while the remaining 25 interviews were carried out with interview partners external to the villages, among them activists, chroniclers, journalists, researchers and representatives of non-governmental organisations and government institutions. One of the interviews was carried out in a Skype call from Germany. Interviews were documented in written form, recorded with an audio recorder or on video and were transcribed into written documents for further analysis. An anonymised list of all interviews can be seen in annex 1. The sources of visual data are photographs and audiovisual recordings. A total number of 191 photographs were taken by the workshop participants. In total, 1247 minutes of video were recorded. A selection of the recorded video material was used in the first edited version of a 34-minutes-documentary film. The photographs selected for analysis are displayed in chapter 7. Besides the primary data obtained in the field, a list of historical and current official and internal documents was accessed during empirical research phases in Mexico in 2014 and 2015. Selected information accessed during the exploratory study in 2013 were added to analysis. They are used as sources of secondary data for the analysis of results. Throughout the whole course of research in Mexico, libraries and archives were visited in the following towns: Cuernavaca (Morelos), Emiliano Zapata (Tabasco), Mexico City (Mexico D.F.), Palenque (Chiapas), San Cristobal de las Casas (Chiapas), San Luis Potosí (San Luis Potosí), Tenosique (Tabasco), Tuxtla

Gutiérrez (Chiapas), Villahermosa (Tabasco). The full list of archives, libraries and private collections consulted is presented in annex 2.

Table 4: Overview on methods of empirical research, format of data and methods of analysis

Method	Time period	Involvement of co-researcher	Number/ extent	Format of data	Analysis procedure
Participant observation	2014 – 2015	Partly in 2015	Field notes in four notebooks, 58 memos	Written text, drawings, photographs (field notebook & memos)	Qualitative text analysis (in parts supported by Atlas.ti software)
Qualitative interviews	Both periods (2014 & 2015)	Partly in 2015	101 interviews	Written text (transcript from field notebook, audio audio-visual recording)	Qualitative text analysis supported by Atlas.ti software
Group discussion	2015	yes	One discussion	Written text (transcript from field notebook, audio audio-visual recording)	Qualitative text analysis supported by Atlas.ti software
Photography workshop	2014	no	191 photographs	Photographs	Qualitative analysis of visual information with the documentary method
Video-making	2015 & additional material in 2014	in 2015	13h:06m:50s of video material (2015) and 07h:40m:30s (2014)	Visual recording, edited film, written text (transcript of selected parts)	Qualitative analysis of visual & text information
Screening and reflection of documentary film	2015	yes	Field notes, video recording	Written text (transcript from field notebook)	Qualitative analysis of additional information from text
Group mapping activity	2015	no	One map	Drawing (& documented on photograph)	Qualitative analysis of visual information

Additional methods					
Group walks	2014	No	Field notes	Written text (field notebook)	Qualitative analy- sis of ad- ditional infor- mation from text and visual infor- mation
Wood work- shop	2014	No (support by other re- searchers)	Field notes, photographs of wood fig- ures and of drawings	Drawings (docu- mented on photo- graph) & written text (field note- book)	Qualitative anal- ysis of additional information from text and visual information

In this section, the methods applied and new methods developed are described in an overview, highlighting the specific ways of application and the complementation different methods made in this research. A major method applied in this research is *participant observation*. It is a classical method of ethnographic research which has regained popularity in human geography. A range of scientific contributions from the discipline of geography have been published over the last decades involving participant observation as major research method (Rowles 1978; Kearns 1997; Laurier 2008; Mountz 2010; Walsh 2014; Lorda 2015). As part of a long-term research process, the researcher observes ongoing everyday activities and participates in some of them. Participant observation requires of the researcher high levels of "introspection and reflection" (Kearns 2016: 318). This method involves days or weeks of continuous cohabitation, collaboration in daily activities of a social group and continuous social interaction. Throughout the process, repeatedly or continuously, different "world views, ways of life, self- understandings, relationships, knowledge, politics, ethics, skills, etc. are accidentally or deliberately rubbed up against one another" (Cloke et al. 2004: 170). These situations of confusion or astonishment allow for insights into new views, a process of mirroring the known in the unknown and the unknown in the known. This holds true not only for the researcher but can, if a participatory approach is chosen, be a process of reflection also for and with the other participants of the research process. A range of challenges linked with the method have been experienced by the researcher, e.g. the challenge to memorise relevant information which sometimes can only be noted down after participation ends. However, it is believed here that the main task of a participant observer is not only to remember observations but to engage in a situation of social interaction. While participant observation does not prescribe any fixed activities to carry out throughout a research, the activities develop through the interaction of the researcher and the people of the social group that is part of the research. Depending on the cultural, economic, ecological, political and social context of a village or social group and a range of other aspects (e.g. season of the year, special events, interaction with the researcher), activities vary. This involves a large scope of different activities that are part of various social practices to be identified by the researcher. An overview on the basic activities of participant observation in this research is presented in table 5.

Table 5: Selected activities carried out in case study villages during participant observation

Activity	Empirical data
Walk in the village (together with village inhabitants)	- get to know the village and the main walking routes, - talk with villagers about their perception of the environment - talk with people about the village (e.g. cultural, economic, ecological, social and political aspects)
Accompany village members in carrying out typical livelihood activities in the village e.g. sowing and harvesting on the field, fishing in the lagoon, preparing food in the homes or in the schools, doing handicraft production in the front- or backyards of the houses	- get to know and participate in everyday activities in the village, learn basic skills of importance for local livelihood - get to know material objects used or referred to concerning everyday life and the flood - talk with community members about the everyday life in the community (social, political, economic and cultural aspects) and the daily life during the flood time - talk with community members about challenges (on the personal, family or on community level)
Accompany the boat driver on the river	- get to know the main water routes from and to the village, - talk with boat driver about the village and the river - talk with passengers about the village and the river
Attend other regular activities: School activities Village gatherings (monthly meeting of "ejido" members) Women gatherings (repartition of foodstuff as part of government assistance)	- get to know and participate in everyday activities in the villages - get to know educational system - observe political and social organisation and dynamics in the villages - talk about investigation process - talk with community members about challenges (on the personal, family or on community level)
Attend special activities: Village celebrations (e.g. Señor de Tila; Graduation of primary and secondary school)	- get to know and participate in special activities in the villages - observe social organisation and dynamics in the villages - talk about values, beliefs, traditions

Qualitative interviews are another main pillar of empirical research in this study. Throughout the research process data generated through participant observation and qualitative interviews were compared and contrasted as part of the grounded theory approach. Interviews in qualitative research are an important method in order to get "access to information about events, opinions, and experiences" (Dunn 2016: 150) through direct interaction and exchange of spoken words with people from the research area. An interview is a powerful means of accessing knowledge which cannot be gathered through other methods (Ibid). At the same time, it gives a prominent role to those interviewed or part of the interview situation because of the questions directed specifically to them. This provides the opportunity to value a person and his or her experiences (Ibid). Interviewing involves a series of steps that need to be considered in order to meet the criteria of good scientific practice. Design or preparation of interviews, the performance of the interview itself as well as the transcription and analysis are integral parts of the interview and can take a considerable amount of time (Ibid: 149). Literature on qualitative research presents a range of systematisations of different interview types (Mattissek et al. 2013: 158ff; Hopf 2012; Dunn 2016: 150). In line with Mattissek et al. (2013: 161), it is argued that the common basis of all qualitative interviews is that the interviewed person is asked to express own interpretations and opinions. The degree to which an interview is structured or open depends on the type of qualitative interview. Lamnek (2010: 306ff) distinguishes different types of qualitative interviews regarding criteria such as the types of knowledge addressed, level of standardisation and way of communication (e.g. uni- or bidirectional communication). In this research, the interviews carried out can be distinguished into three major types, which have been adapted from the scientific literature: The short guided interview, the narrative interview and the problem-centred interview. Additionally, the group discussion as a special format of interview carried out in the case study villages is described.

In the first visits in the villages, *short guided interviews* were carried out. As part of a first exploratory process, it was relevant to enquire a range of different people about the flood exposure, flood perceptions, risk management activities and government support to the villages. Personal backgrounds or larger practice patterns were not of main interest in this exploratory phase. Questions from a short pre-formulated guideline (Annex 3) were mainly used as a starting point for conversation and new questions arose from the specific context of each interview. With the topics addressed and new ones identified in these interviews, theoretical sampling took a starting point.

The main interview format applied in this study is the *narrative interview*. This interview format allows entrance into intense research relationship with an interview partner. It was used mainly to access relevant context information around a major research topic and to talk about everyday life, social and biographic information as well as about perceptions and feelings of the interviewed person. The interview type is characterised by openness of the interview process and absence of a predefined guideline. Openness towards the answers given in an interview is a key prerequisite of interpretative social research (Lamnek 2010: 318). Narrative enquiry has been attested the power to "understand a life as it is lived" (Trahar 2011: 5) as opposed to "seeking confirmation for a 'theory'" (Ibid). While narrative interviews are used here as a method to understand practices and their larger context of meaning, the researcher at the same time looks for links and contradictions to existing concepts and theories. Moreover, it is especially the context information revealed by a narrative interview and the openness towards a structuration of a topic and its´ embedding in the context by the interviewed person, which gives priority to this interview format in the study presented here. The specific utilisation of narrative interviews in this research is closely tied to the ontological and epistemological context presented before. Some authors describe narrative interviews as more than an instrument to retrieve stories which at a later time can be analysed. Clandinin et al. (2011: 34) argue that there is the possibility to think about stories as objects, i.e. to analyse them as objects, or to think about stories as living. They present their approach of narrative enquiry as follows:

> "When we begin to engage in narrative inquiry, we need to be attentive to thinking with stories in multiple ways: toward our stories, toward others' stories, toward all the social, institutional, cultural, familial and linguistic narratives in which we are embedded as well as toward what begins to emerge in the sharing of our lived and told stories" (Clandinin et al. 2011: 34).

Engaging in narrative interviews in this research thus means to share stories of life and by this way, to take part in the life of the interview partner for a certain time. The narrative interviews carried out here reveal similarities with the *enviro-biographical interview*, described by Rohland et al. (2014: 190) or *oral environmental histories* presented by Lane (1997 cf Dunn 2016: 161). Narrative interviews in this study were predominantly defined by a unidirectional way of telling stories, which means that the interviewed person told stories about his or her daily life in the present or in the past while the interviewer asked questions. In some cases, where sensitive topics were discussed or when being asked by the interview partners, there was a bi-directional way of interview communication. Here, the researcher answered some of the questions asked by interview partners about the research and background of the researcher, which served to build trust between and provide relevant information to all the persons enrolled in an interview situation. As visual media were an important part of research, it was decided together with interview partners that the discussion of photographs should be included into selected interview situations. In the second half of the research phase in 2015, photographs taken in the respective village and selected photographs of the *life-world* of the researcher in Germany were presented and discussed during various interviews. This allowed for a process of reflection about the economic, ecological and social context in the different *life-worlds*. It showed that discussing about photographs mainly served the purposes of building trust and giving a starting point for discussion, while the following topics discussed and the narratives presented by the interview partners were chosen by both parties involved in the interview situation. The selection of interview partners for narrative interviews was made following the principles of theoretical sampling, as described in grounded theory. While in the beginning of empirical research, the village authorities were approached for narrative interviews, in the course of research and the focus of research on one village, new interview partners were identified due to specific knowledge or experience attested to them by other villagers or due to a specific interest by the researcher and co-researcher. Besides narrative interviews, one *group discussion* was carried out in a village during empirical research. Cameron (2016: 203f) presents a continuum of three different group interview constellations: group interviews, focus groups and in-depth groups. A group discussion pursues different ends than an interview, as the group situation allows a researcher to gain insight into social interactions and opinions in a public or semi-public environment (Mattissek et al. 2013: 183). Although a general discussion topic was introduced by the researcher in this study, group discussion in this research had a mainly narrative character and served the purposes of both gaining access to information on the key topic through enquiry and gaining access to context information and the perceptions and meanings through observation. Beyond the advantages of getting a sum of individual information which can be compared, the group discussion offers additional insights when used as a technique of group observation (Cloke et al. 2004: 160). A group discussion can reveal certain social interactions and patterns in a community, expressed e.g. by the order of speech or by people referring to each other's verbal expressions. Actors engage in building up social ties, e.g. by informing others about their opinion looking for agreement and sympathy, keeping silent and observing or by disagreeing openly to what another person says. The group discussion in this study was carried out in a group of women in the main case study village in 2015. Women who were known by the researcher through participation in a gardening group in 2014 were invited. The original intension of the researcher to carry out various group discussions had to be adapted, as most women mentioned a lack of time to meet in this group repeatedly. The women agreed to carry out individual narrative interviews in their homes, allowing them to go on with their household activities while engaging in the interview.

The third type of interviews carried out was the *problem-centred interview*. This semi-structured interview type is useful when facing time constraints and in formal settings, e.g. when information is asked from government representatives or scientists. A pre-construction of the topic to be interviewed about is common in problem-centred interviews (Mattissek et al. 2013: 167). Pre-defined questions are included in an otherwise relatively open interview situation (Ibid). In this research, the problem-centred interview was mainly used as method to engage people identified as relevant regional subjects other than the village inhabitants and scientists knowledgeable in a certain thematic or professional field of interest in this study. Other scientific accounts denominate the type of interview as expert interview (Bogner et al. 2014: 11). It has to be pointed out however, that the status of an expert is given according to the specific topic or context of a research (Ibid). As in this study the different practices involved in flood risk management are of major empirical relevance, the definition of expert versus lay people is problematic. The author refrains from dividing the persons interviewed into a group of experts and a group of lay-people. In the starting phase of empirical research in 2014, the identification of interview partners had been prepared through internet research on official actors of importance in the field of flood risk management in Chiapas and scientists from Mexico, who had published relevant literature. Following the first interviews in 2014, other interview partners were identified through a snowball sampling process (Berg 2009: 51). Sampling was complemented by conceptual considerations or specific questions from the qualitative research in the case study villages.

5.6 AUDIO-VISUAL METHODS

As part of the methodology of this study, a range of audio-visual methods was applied. The relevance of the visual in the discipline of geography is emphasised by various authors within and beyond the scientific discipline (Pink 2013: 26ff; Rose 2003; Thornes 2004). It is especially the reflective power set free by visual methods which is of high additional value to a research process that involves conceptualisations of space and spatiality, materiality and embodiment. Notwithstanding its great methodological and conceptual strengths, the application and development of elaborate visual methods as a dominant part of empirical research is still a novel approach and underrepresented in geographic research. In order to highlight the relevance of visual media in and for geography, a short introduction into visual media is presented here before the specific methods applied in the empirical process are described.

Visual media are present almost everywhere in the discipline of geography. The visual is a sense that in geography has been of high importance from the very beginning of the discipline. Tuan (1979b: 418) argues that geography as a discipline developed as a visual approach towards landscapes. Visual media bring about certain arguments with their own language and ways to convince (Miggelbrink & Schlottmann 2009: 185). In the history of the discipline, the visual has mainly been used with the intension to represent data in an objective manner, in forms of topographic and thematic maps or models. Photographs were mainly used as illustrations for written text and videos were used as documentary footage. Both, photographs and videos were used with naïve assumptions on the possibilities to represent reality, lacking critical reflection on the constructed nature of these media. Images contribute to the construction of realities, based in time and space (Ibid: 193). Moreover, images play an important role in the functioning of society, in the actions and practices of members of society and in the construction of meaning. At the same time, it must be considered, that geographical representations are "active, constitutive elements in shaping social and spatial practices and the environments we occupy" (Cosgrove 2008: 15). Visual media are part of discourse just as other forms of reason, like spoken or written words. Visual media are never objective, never separate from the context in which they are produced, never independent from the person who generates them. Therefore, it is of crucial importance to reflect on processes of visualisation and the contexts in which these are made.

Different options exist, to make use of the characteristics of visual and audio-visual media for empirical research. Audio-visual productions such as fiction and documentary films are of interest in cultural and social geography (Aitken 2007; Curti 2008; Escher & Zimmermann 2006; Lukinbeal & Zimmermann 2008; Roberts 2013; Schlottmann & Miggelbrink 2015). While one group of visual geographers has so far mainly focused on the analysis of audio-visual products created by other people, other researchers increasingly argue for and actively engage in the production of videos and photographs (Crang 2010; Garrett 2011; Hunt 2014; Mistry et al. 2016; Oldrup & Carstensen 2012; Pink 2013). Moreover, participatory and action research approaches linked to different disciplines like social anthropology, media studies, geography and others increasingly involve audio-visual methods. Visual or audio-visual material produced within empirical research can be used as part of different methodological and academic contexts. As a first mode, Garrett (2011: 524) highlights attempts to publish videos as part of the general publication process of scientific results. The format of a digital video still faces a range of obstacles upon the encounter with the general publication procedures, as they are designed i.a. for peer-reviewed journals (Ibid). Among the exceptions of journals that integrate videographic publication formats are the interdisciplinary journals *Visual Studies* (formerly Visual Sociology), *Liminalities* and *Geography Compass* (Garrett 2011: 524; Pink 2013: 18). A second mode, used frequently in geographic research is the integration of photographs and video recordings as data and documentations of empirical research (Garrett 2011: 525). While video recordings made in early anthropological research were regarded as objective representations, Garrett (Ibid: 527) argues that "[e]ven if objective representations were possible, objective footage could not be objectively consumed". Another mode described in literature is "reflexive filmmaking" (Ibid). As this study includes photography, drawings and video making, the third mode is named reflexive audio-visual production. The specific qualities of film are described by David MacDougall: "[T]he film is a conceptual space within a triangle formed by the subject, film-maker, and audience and represents an encounter of all three" (MacDougall 1978: 422 cf Garrett 2011: 528). Similar descriptions can be given for photography in geographic research. The fourth mode described by Garrett (Ibid) is "participatory video", adapted here to participatory audio-visual production. This method allows different people to work together in the production of photographs or a video. It provides the opportunity for the participants to express opinions, ideas and impressions in their own ways and languages (Ibid). In a publication on experiences with participatory video making in Chiapas presented by Jiménez Pérez and Köhler (2012), specific characteristics of the medium and the process are described by a participant:

> "Video strengthens our oral and visual traditions [...]. Video allows us to form our knowledges in new mediums of communication; at the same time, it presents a new technology which currently is more adequate for our people than written language." (Jiménez Pérez & Köhler 2012: 334f, translated by the author)

The specific skills required for the elaboration of audio-visual material and for the analysis and critical reflection reveal, that visual methods cannot arbitrarily be included in empirical research. There needs to be awareness on the side of the researcher that in a participatory and reflexive approach he or she can also become a possible phenomenon under study (Flyvbjerg 2002: 132 cf Garrett 2011: 528), be recorded on photographs or video and discussed about in the research reflection process. Thorough consideration of the opportunities and the challenges in a research process is required to decide whether visual methods are of considerable value to a study.

Among the most used methods in geography that involve the active production of visual empirical material are methods of mapping and modelling. Representing

physical locations together with other relevant information has been a major practice of geographic work since early production of cartographic maps until most recent formats of complex Geographic Information Systems (GIS). Among the participatory methods are the drawing and sketching of places and objects on paper or
on computer, the designing of three-dimensional models of the environment and
other format of spatial representation (Pain 2004: 653ff; Pfeffer et al. 2013; Wood
2005). These participatory methods allow the expression of "relations with and accounts of space, place and environment" (Pain 2004: 653) through visual media.
Mapping methods applied in field research cannot follow a predefined pathway but
need to be adapted according to the requirements and ideas of the participants and
research situation (Roth 2009: 213). These methods are of specific use throughout
the research process not because of the empirical material elaborated alone, but especially because they stimulate an active reflection about conceptualisations of
space and spatiality (Roth 2009: 211; Wood 2005: 159). Moreover, participatory
mapping involves "performances of the past, the present, and often divergent, imagined futures" (Sletto 2009: 445), opening a floor for reflection about experience
and aspirations linked to spatial entities. Mapping is regarded as a highly valuable
method due to its double quality, firstly as a method of production of representations as empirical data and secondly as an entry point into conceptualisation during
the research process.

In this thesis, different entry points into the realm of the visual and the audio-
visual are taken. Visual empirical data has been generated throughout the phases of
field research in 2014 and 2015. Three major visual methods used in this study can
be distinguished: A three-day *participatory photography workshop* was carried out
in August and September 2014 that involved photography as a visual method of
empirical research. The workshop was held with 15 students from the age 15 to 18
of the local secondary school of one village, their teacher and the researcher. The
students were divided into groups of two or three by their teacher. Each group received a one-way camera with 27 photographs and was given the task to take photographs of places and objects in their community which they liked most and those
which are regularly affected by the floods. The students were given three hours of
time to think about which photographs to take, to go to the places and take the
photos and to return the cameras to the researcher. After the photographs had been
developed in Palenque, the researcher returned the photographs to the students in
the second day of the workshop, which was held around three weeks after the first
workshop day. A group discussion was held in the classroom and each group of
students selected 3-5 photographs for an exhibition in the community hall. In addition, two questions were raised by the researcher which had to be answered by the
student group in a written form. The questions were:

1. What do I/ do we like about our community?
2. How do I/ do we imagine my/ our community to be in five years?

The students´ answers were discussed the following day and written down in short form on the paperboard below the photographs. A day in the following week was selected for the exhibition in the community hall and all village members were invited throughout the week. From each group, one student was selected to present the photographs and explain the written text during the exhibition.

During the second phase of empirical research carried out in 2015, the researcher and co-researcher initiated a *participatory video-making process* in the main case study village. The research team stimulated a process of representation and reflection about the community, similar to processes described in scientific literature as "collaborative representation" (Banks 1995: paragraph 1) or "ongoing process of creating community" (Sandercock & Attili 2010: 37). The aim was to produce a video and reflect on it as a team. The researchers involved the community members, especially a group of women that the German researcher had worked with in 2014, in the process of doing a film about the village. One major topic was set by the researcher to be the flood in the village, but the process was open to include those topics relevant to the group. The idea of telling "one´s own story" or a story of the village was put forward in a meeting with the women´s group. It proved difficult to involve the women as a group into the video making process due to time and other constraints faced by women. Therefore, a more individualised approach was followed. Two women actively participated in the recording of videos throughout the research phase in 2015. They were given an introduction into the handling of the camera and were provided with continuous support by the co-researcher and researcher. They filmed other women in interview situations, during everyday activities as well as during a village celebration. Moreover, a range of men and children, who showed interest in the participatory activity were included in the group of participants. Different from the photography workshop, video-making was used primarily in order to stimulate interaction with research participants and conversation about the research topic. The camera was used as a tool for collaboration and expression in a visual language to enable a reflection about life and everyday practices from a new perspective. Besides the filming process and reflection about this activity, a first edited version of a documentary film, edited by the co-researcher and supported by the researcher, was shown in the village. A screening of the documentary film was carried out in the village in August 2015, embedded in a discussion round lead by the researcher and co-researcher.

In 2015, a participatory mapping exercise was carried out with youths of the major case study village. In a two-hour workshop held by the researcher in the community hall during school holidays, around ten youths and children of ages ranging from 10 to 17 years participated. The task presented by the researcher was to drawn a map of the village on a white sheet of paper in format DIN-A3. The specification of the drawing was that it should represent the main places that are at risk during or that are affected by the yearly floods. While the map in itself is a possible source of relevant information for this thesis, the major value of the mapping activity was the reflection about life in the village and the social organisation of life. At the same time, the mapping activity provided exclusive time for interaction with young research participants, who did not speak up during meetings with adults. While the original map remained in the village, a photograph of it serves the purpose of doc-umentation for the researcher. The photograph of the map is not used for visual interpretation, but relevant information obtained from observation during the activity is integrated into analysis.

5.7 INTEGRATION AND DEVELOPMENT OF ADDITIONAL METHODS

Besides the core methods of this research, participant observation, qualitative interviews and different audio-visual methods as described above, a range of additional methods was carried out in the field. These methods were chosen and developed mainly in order to stimulate interaction between the researcher and the village inhabitants and to complement participant observation. In 2014 two *group walks* were carried out with the women group in one village. The idea for group walks was brought up by a group of women participating in a community garden project. Walking as a research practice, as part of methodology and even as a part of epistemological perspective is discussed in various accounts by social researchers, with primacy in accounts of a phenomenological tradition and with research based in urban settings (Funke-Wienecke 2008; Lee & Ingold 2006; Pink 2013). Prominent examples are presented in the publication *Ways of Walking: Ethnography and Practice on Foot* edited by Tim Ingold and Jo Vergunst (Ingold & Vergunst 2008). The specific interest in social practices in this research puts the focus on specific empirical methods that give access to a multi-dimensional understanding and analysis of practices that expand beyond the aural and visual levels, including haptic, tactile and other sensual levels. Together with the researcher it was decided to do a group walk after the ending of a day´s work in the community garden. During a first 45-minutes group walk in September 2014, the women walked along those areas in the village, that usually get most affected by floods. During this time, the water level in the river was still moderate, not affecting the settlement area of the village. Through indication of specific locations in the physical space and through conversations and story-telling about the flood, additional information was accessed by the researcher. This activity involved passing by foot along backyards, fields, gardens, grazing land and other public as well as private areas of the village. In October 2014, a second group walk was carried out. This was at a time, when the water level was rising and affected the road access to the village and some of the fields. The activities of this group walk included walking to those locations that were affected by the rising water in the river and the lagoon, passing through the water by foot, measuring the depth with wooden sticks, catching fish and collecting ripe corn cobs from flood-affected fields. The activities, observations and conversations of the group walks were documented through field notes, photographs and short video recordings. Besides the information accessed through verbal communication and the visual impression of the physical environment, the act of walking together with the women was a method that involved mental as well as bodily processes. The main value of this group walk method can be seen in the temporal formation of a social group that walked together, exchanged ideas, experiences and shared the same path for a limited time. The social bonds created between the researcher and the women during these group walks built a basis of trust, which added to a general acceptance of the researcher among the women in that village. Another activity carried out as part of the larger methodology was a two-day *wood workshop* in the primary school in the major case study village. In the social sciences, various approaches are discussed to involve the creative arts as part of research methodology (i.a. Hernández 2008; Quiroga 2015; Scribano 2013). Creativity is understood as a platform for performance which gives access to the expressions of research partners about everyday life (Scribano 2013: 33). Creativity and the creative arts are used in this thesis believing that deeper insight into social practices may be gained. The wood workshop involved the activity of designing and constructing dolls and vehicles made from wood and local materials. The main objective of the workshop was to reflect on ideas of nature and everyday life of the children by working with local organic material from the community. The activities were realised in collaboration with two co-researchers who have been working in co-research teams and with children in local communities of Chiapas. David Arijas, Mexican artist and facilitator of youth workshops, and Noemi Dulibe, researcher and activist at CIESAS Sureste and the Transnational Network of the Other Knowledges (RETOS), were invited to carry out the workshop and other research activities for a period of two weeks.

5.8 METHODS APPLIED IN ANALYSIS AND INTERPRETATION

The analysis and interpretation of data obtained during empirical research follows the major objectives and research question of this study. The overall objective is to understand social phenomena, in this case everyday practices of flood risk management, and reflect on the meanings they hold for those involved in them in a participatory manner. These steps influence the results of this study in as much as the methodological steps described above and the broader epistemological and ontological assumptions that lay the basis of this thesis. It is therefore crucial to present the steps of analysis and interpretation in a transparent manner and to link them to the overall research process. Making sense of phenomena through the analysis of written text and visual documents involves steps of description and interpretation. The identification and analysis of social practices which is of major interest in this study was carried out in parts supported by discourse analysis. As emphasis is also put on the performance of social practices on bodily and material levels, the analysis of other data than those generated through spoken and written words became relevant, pursued here mainly through an analysis of visual material. Van Maanen (1990:7 cf DeLyser 2010: 346) argues that the relationship "between words and worlds is anything but easy or transparent". The specific ways in which in this thesis analysis and interpretation is made through the use of different formats in which knowledge is reflected are presented in the following section.

5.8.1 Discourse-oriented analysis of text and interpretation

Text analysis is carried out in this study in order to gain access to new knowledge on the research topic and especially in order to identify entry points for answering the research question. Different types of text are accessed, among them transcripts of direct verbal communication (interviews), transcripts of observations by the researcher and co-researcher, as well as documents and graphs that include written text accessed from a range of different groups and organisations that were identified as interacting with groups and subjects in the case study region. The information of interest in the texts analysed can be grouped into four different categories: (1) new information and knowledge, (2) opinions and beliefs, (3) local concepts describing social practices, (4) specific discourses. While information on the first two categories are accessible mainly through direct extraction of text passages, information of categories three and four involve a more complex process of analysis and interpretation. The decision to undertake discourse-oriented text analysis was taken in the second phase of empirical field research in 2015, when a large amount of interview and visual data had already been gathered and important steps in concept identification, following the procedures of grounded theory, had been accomplished. The early identification of local expressions and concepts in the spoken language accessed and documented in empirical research by the researcher indicates towards the relevance of dominant discourses in the social practices of flood risk management. This empirical relevance recommends discourse analysis as one method of choice for qualitative data analysis in this study.

In human geography, discourse analysis has become a prominent tool for research in the last 25 years (Dittmer 2010: 274). Influential works of discourse analysis in geography are Edward Said's work on "imagined geographies" (1987 cf Dittmer Ibid) analysis of discourses of development (Crush 1995), of environment (Norton 1996), on geopolitics (Dodds 1996) and, more recently discourse analysis in urban research (Lees 2004). Michel Foucault as a prominent author of poststructuralist discourse theory puts emphasis on general notions of power and avoids class-based interpretations (Dittmer 2010: 277). Following Foucault, different discourses construct different "regimes of truth" (Foucault 1980 cf Lees 2004: 103). A discourse-oriented analysis of the different texts available in this study is one approach by which the social construction of risks is approached. Two key authors of discourse theory, Ernesto Laclau and Chantal Mouffe (1985 cf Dittmer 2010: 278) emphasise that different ideologies exist and compete in society. Categories like ideology or politics in their understanding are no objective entities but are created as "sediment of discourse" (Glasze & Mattissek 2009: 155, translated by the author). Some versions of discourse theory place social practices in the field of discourse (Laclau 1995 cf Dittmer 2010: 278). This thesis however looks at the relation between both from a different angle which conceptualises discourse as part of social practices. The difference is significant insofar as from the specific understanding of discourse and social practices held in this thesis, social practices include a variety of processes that might not be primarily part of a discourse but part of a practice. It is believed for example that beliefs, bodily performances, and material aspects that are part of social practices cannot be identified through an analysis of language and discourse alone. This thesis follows an understanding of discourse as the language-based parts of social practice (Fairclough 1992: 63, 66). Fairclough (Ibid: 64) argues that "discourse is a practice not just of representing the world, but of signifying the world, constituting and constructing the world in meaning". He presents a three-step approach of analysis, which is followed here: *Textual analysis* as first step reviews single words, text patterns and grammar. In a second step the analysis of *discursive practice* focuses on expressions and statements that can be linked to policy and other relevant debates (Fairclough 1992: 78f). The analysis of *social practice* is carried out as the third and last step through review and conceptualisation of the "general ideological context within which the discourses have taken place" (Lees 2004: 104).

A diverse set of different original formats of data is translated into transcripts in written format for the purpose of a discourse-oriented text analysis, including orally expressed information from interviews, field notes and memos. The transcripts elaborated from the original data are analysed through a process of interpretative-hermeneutic classification and bundling of text passages (Glasze et al. 2009: 294). It is accomplished through coding as well as building and merging of categories supported by the software Atlas.ti (version 7.5.10). The software is used only for the analysis of data transcribed into format of written text, while the interpretation of visual data is carried out in a separate process. The crucial step in analysis of the transcripts is the building of codes. In this thesis an inductive process of coding is followed, which does not allow for a predefinition of codes by the researcher but requires that codes are built during the process of review of the data as abstractions of the content of the text or implicit concepts identified in it. In this approach, categories and codes are developed in the process of looking through all the empirical material (Glasze et al. 2009: 296). It is an iterative process as the process of coding and building of categories becomes refined during the process. Some of the codes built here are kept in the original language, while others are translated into or described in the English language. The thorough and repeated revision of interview texts allows the identification of regularities in the expressions of different persons. These regularities in the text can provide hints on regularities in the discursive constitution of meaning (Ibid: 294). Discourse-oriented text analysis rather than following a steady and pre-defined order is an "artisanal" process (Hoggart et al. 2002 cf Dittmer 2010: 279). It involves openness and flexibility towards the empirical data while at the same time the initial conceptual ideas developed need to be constantly revised or enhanced in order to adopt or dismiss them.

5.8.2 Qualitative analysis and interpretation of visual data

When working with visual media, the basis of production and analysis needs to be that every image has a polysemic character. As the famous and compelling image of the "Duck-rabbit-head" used by Ludwig Wittgenstein in his *Philosophische Untersuchungen* (2001 [1953]: 1025ff) exemplifies, in one and the same drawing a spectator can see very different things. Together with this example, Wittgenstein presents the argument of *aspect-dawning* widely used in the philosophy of perception until today (Ibid: 1028ff). Noticing a new aspect in a known image fundamentally and irretrievably changes the way a person looks at the image and perceives it. When analysing visual data one needs to consider carefully the polysemy of an image and the different possible "levels of meaning" (Gombrich 1972: 2). The features of an image cannot be described exhaustively by text, because "no verbal description can ever be as particularized as a picture must be" (Ibid: 3). The meaning of an image is not solely determined by content and structure, but it is embedded in social norms, which provide meaning to them (Miggelbrink & Schlottmann 2009: 186). Miggelbrink and Schlottmann (2009: 194) argue that images can be used as entry points into the understanding of past and contemporary societal spatial relations as well as of human-nature-relationships (Ibid). Thereby, the authors point towards the specific interest in this study and underline the relevance of the visual to empirical research and conceptualisation in human geography. It is believed here that images create a space for reflection about and representation of general world views and more specific views on space and society. Following the general theoretic assumptions made in this study, it inextricably includes the everyday performance of these views and aspirations through social practices. Visual media here is analysed in order to gain insight into the internal representation of the *Lebenswelt* and practices. The photographs taken throughout the photography workshop in 2014 are the major source of visual media for analysis. Added to that, a selection of video sequences identified as fundamentally relevant were selected for analysis from the vast amount of video recordings. Both the photographs and the video recordings are analysed with an analytical tool: the documentary method.

The refinement of methodological techniques in the social sciences in the last 25 years has mainly promoted text-based interpretations, leaving behind interpretation of images (Bohnsack 2009: 953). It is the documentary method, developed in the tradition of Karl Mannheim, that is regarded here as an adequate approach to get closer to the meaning of images. The method is grounded in an ethnomethodology approach and has in the past mainly been applied in the disciplines of social and educational sciences (Bohnsack 2013: 9). While attention of researchers from other scientific fields has grown (Ibid: 18), the application of the method in geography is still limited. The documentary method was developed taking into account the shortcomings which rose from the dichotomy reiterated by objectivist perspectives in quantitative social research on the one hand and phenomenological approaches with subjectivist perspectives on the other hand (Ibid: 10f). In a praxeologic tradition developed by Mannheim in sociology of knowledge during the 1920s a change of the perspective from a "what" (What is actors´ perspective of reality) to a "how" (how is reality produced in a practical manner) (Ibid: 13) is presented. Garfinkel (1967: 77f) describes the method presented by Mannheim (1923/1957) naming it the documentary method of interpretation. The documentary method allows for the interpretation of an image in a two-step process, referring to the work of art historians Erwin Panofsky and Max Imdahl. In a first step, called formulating interpretation, it is described what can be seen on a picture (pre-iconographic and iconographic analysis following Panofsky). This is followed by an interpretation on an iconographic level, the description of visible actions (Bohnsack 2009: 960). This description of actions is a crucial step because the contextual knowledge of the person interpreting the picture has to be largely suspended. It is only common-sense knowledge about an activity displayed on a picture that should be included in this step of interpretation. This can be the knowledge about the "history of types" (Ibid, translated by the author) displayed on a picture (e.g. the generalised knowledge about a scene displayed on a picture as representing a family meal) or the "history of styles" (Ibid, translated by the author) to be interpreted (e.g. the generalised knowledge that certain clothing of persons in a picture is typical of hip-hop culture). In the second step, the reflecting interpretation, results of step one are reflected on by the researcher and documented in a transparent way (iconic interpretation and iconological-iconic interpretation following Imdahl) (Bohnsack et al. 2013: 15). The structured process of formal interpretation carried out in this method achieves a distance between the researcher and the data she or he has produced or accessed in empirical research. It is by this way of alienation from the data and subsequent reflective approximation, that the process of interpretation is enhanced and improved. Referring to the art historian Max Imdahl, who had proposed a reconstruction of the formal elements of a picture mainly with a concern about aesthetic aspects, Bohnsack (2009: 956) adapts the approach for the purposes of social sciences.

In this research, photographs as well as videos are included in the analysis. The general challenges that arise in the interpretation of video (video-supported observation) from the specific perspective of the spectator cannot be overcome with this method. However, it allows for a very detailed reconstruction of verbal and nonverbal elements, especially to bodily-performative practices (Bohnsack et al. 2013: 20). These bodily performances of activities form part of the relevant social practices in this research. The method is regarded as especially valuable for the interpretation of ways in which people construct spatial entities and how they interact with space, both regarded in this study as key social practices. The analysis of videos with the documentary method requires an in-depth analysis of single video frames. While an analysis of each frame of a video is beyond the scope and intension of this study a selection of frames is chosen for in-depth analysis. Frames are selected from those video sequences which give access to information on relevant practices linked to flood risk management or the larger context of everyday practices in the case study region. As each frame is rich in visual information, the analysis provides insight into some of the key social practices from a novel perspective. This provides additional information in the process of understanding social practices in their complexity.

6 SOCIAL PRACTICES OF LIVING WITH FLOODS PERFORMED BY VILLAGE INHABITANTS

The previous chapters have presented the research interest as well as the conceptual and methodological approaches of this study. As the research paradigms and approaches suggest, key access to new knowledge and scientific insight can be gained from in-depth empirical field work. As such, the core of this study lies in the empirical process carried out and the empirical results originating from this process. It is the "material" obtained in the field which is of major interest, while material is meant here as a metaphor, which comprises data recorded in a digital and in a physical sense just as information accessed and elaborated by the researcher, co-researcher and research partners in case study villages. The cluster of chapters 6 to 9 presents empirical results structured by the main conceptual strands identified. Local concepts identified together with research partners are the entry point for this enterprise. As the core interest of this study lies in the social practices that are related to flood management in the Usumacinta River region of Catazajá and Palenque municipalities, the outline of these chapters is also structured by the major social practices. This chapter presents the core local concepts which were identified in the field, analysing the ways in which conceptualisations are made by local carriers of practices concerning real life phenomena such as floods and other social and ecologic dynamics of relevance. A myriad of social practices are identified and are presented here through a selection of examples, thereby giving an in-depth insight into what is summarised as *the social practices of living with the flood*. Another focus is set in the following chapter (7) which presents a specific type of social practices related to flood management, which are labelled *the social practices of anticipation*. Thereafter (chapter 8), a look is taken into the specific *interrelations between social practices* that involve different groups of people (as human practice carriers) and produce specific practice patterns in the case study region. In the last chapter of this cluster (chapter 9), results of the previous chapters are combined in an analytical approach derived from a combination of practice theory and the *riskscapes* concept, focusing on the spatial dimensions of flood related practices.

6.1 LOCALLY IDENTIFIED TOPICS AND CONCEPTS

The analysis and interpretation of qualitative data generated in empirical research of this study are fundamentally based on the identification and use of concepts. As Merriman et al. (2012: 5) argue, a concept can be imagined as "a reaching out with the hand, a way of dealing with the most abstract ideas as if they were a collection of things". As part of the grounded theory approach, the generation of conceptual ideas during empirical research is a key process for the creation of new knowledge. The empirical data originating in processes like observation, participation and reflection contains information that is used for conceptualisation. At the same time, it is relevant for the researcher (in this case including the co-researcher) to be susceptible towards conceptualisations used by the people of the case study region.

In an early stage of empirical research in 2014, a list of topics was noted down, which deemed relevant in the empirical research up to that point. Figure 18 presents the preliminary topics and concepts identified in 2014. Presenting this figure is chosen in order to make the processual nature of concept building throughout the whole process of research transparent, starting with concepts retrieved from local expressions identified during an early phase of empirical work.

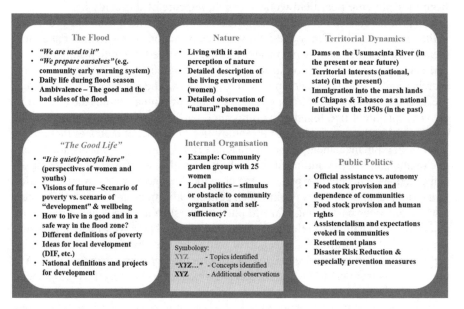

Figure 18: Preliminary topics, concepts and observations made during empirical research in 2014
Source: author's elaboration

The main topics identified as relevant concerning the general research topic of flood risk management and in line with the research question were the following: *the flood, "the good life", nature, internal organisation, territorial dynamics* and *public politics*. Among the first local concepts identified in the field were three phrases that village members expressed repeatedly when asked about how they prepared for the flood, how they experienced the flood and how they experienced life in the villages. These phrases were documented during interviews and participant observation. The first concept is called *"we are already used to it"* (="ya estamos acostumbrados"/"ya somos acostumbrados" in Spanish original). This expression is discussed further below in more depth, as it has become a key concept of this study. The second local concept identified in the early stage of research can be summarised in the expression "*we prepare ourselves*" (="nos preparamos"). Throughout research it became obvious that different activities in the villages form part of what was called preparation and that different types of information about related activities and practices could be accessed through interviews, observation and visual methods. The third local expression identified in an early phase of research is the concept of "*the good, quiet and peaceful life*". It is summarised from a range of different expressions such as "*here you live well*" or "*here you breathe fresh air*". These local expressions were linked to the concept *La vida buena* often referred to as a counter-discourse towards development discourse (i.a. Fornet-Betancourt 2010; Lopez Intzín 2015; Macleod 2008; Mejido Costoya 2013).

6.2 IDENTIFICATION OF KEY SOCIAL PRACTICES

Empirical field research in the case study region shows that flood management is a broad term that subsumes a large set of social practices. In this chapter it is relevant to take a detailed look into those social practices that are carried out in the case study villages and that link directly or indirectly to the flood. Practices of flood management are entangled in a complex network of larger social practices that form social life. One way of getting access to these social practices is through the spoken word of people. Interview and observation transcripts are used here to analyse the patterns of practices and to identify key practices which guide decision and bodily-material action concerning floods. The text-based information that has been elaborated during field research is analysed here in a discourse-oriented text analysis (see also chapter 5). The procedure used to structure analysis is the forming of codes and categories from the empirical material. This procedure is supported by the software Atlas.ti, which allows the attribution of codes to text passages, to summarise different codes into a category and which gives an overview on the different codes and categories built. The comprehensive analysis of all transcripts after the ending of empirical research complements the identification of concepts and the preliminary identification and building of codes carried out during the field research phases.

The repeated review of transcripts led to the consolidation of the list of codes and to the building of more abstract categories. In total 80 codes were built and assigned to the transcripts gradually. Many text passages carry multiple coding, as the textual analysis showed that some verbal expressions hinted towards several relevant aspects concerning the research topics or that verbal expressions provided information on various levels of interest. Through a revision and amendment of codes, some codes that contained almost identical information were merged and at the same time other, more precise codes were distinguished where necessary. After finalisation of coding, the codes were ordered in more abstract categories.

Table 6: Categories ordered by numerical prevalence in the text analysis

Category		Nr. of codes grouped	Number of total grounding	Category		Nr. of codes grouped	Number of total grounding
1	Flood	12	448	13	Weather_Climate	1	40
2	Village	9	263	14	Risk	3	39
3	Experience with government	7	223	15	Biography	1	35
4	Livelihood	2	118	16	Politics	3	34
5	Materiality	3	105	17	Germany	1	25
6	Civil Protecion	7	97	18	Resettlement	1	21
7	Territory	2	90	19	Women	1	20
8	Practices	5	79	20	Organisations	3	15
9	History	5	77	21	Culture	1	14
10	Investigation	3	62	22	Health	1	8
11	Migration	2	42	23	Other	5	103
12	Development	2	40				

Table 6 presents 23 categories formed after the finalisation of coding. The categories are ordered by their numerical prevalence in the text analysis. The numerical value is not used as a means for quantitative analysis of textual data. Rather, it is pointed out here that the building of codes and categories are interpretative steps of analysis, which follow purposes different than those performed in quantitative analysis. The building of codes is grounded in the categories and codes encountered during empirical research in the ethnographic field research. Strong prevalence of some categories in the text represents which topics were most talked about during interviews and most observed or discussed during participant observation. Categories like "flood", "village" or "experience with government" were formed according to the high relevance these topics revealed in the general research process. Although some of the codes which are regarded as having major importance for the identification of conceptual insights have high numerical prevalence in the texts, other codes regarded as crucial are not represented dominantly in the texts. In order to

make use of the codes and categories in a qualitative analysis they need to be trian-
gulated with the concepts identified in the field research phases by the co-research
team and supported by research participants from the village. Of specific interest in
this regard is to link concepts identified in the field with those codes that were
formed based on local expressions. The representation not only in the empirical data
but in the expressions in original language allows insight into local ways of conceptualisations of the *life-worlds* and relevant ecological, political and social dynamics in the case study region. Table 7 presents those codes that were formed on
the basis of local expressions amended by a translation of the words and phrases
into English. These local expressions or words were chosen due to the dominant
role they played in the interviews and participant observations in the case study
villages. They are of major relevance for the discourse-oriented analysis of the text,
as it is believed that important information can be drawn from them for the analysis
of social practices and corresponding conceptualisations.

Table 7: Codes formed using local expressions from the empirical data

Code in original language	Translation
Autosuficiencia; "somos autosuficientes"	Self-sufficiency; "We are self-sufficient"
Mentira	Lie
Narcos	Drug dealers; drug traffickers
"No es lo mismo (cuando viene la creciente)"	"It is not the same (when the flood comes)"
Olvido del gobierno; "Somos olvidados del gobierno"	Oblivion by government; "We are forgotten about by government"
"Somos conformistas"	"We are conformists"
Sufrir; "Sufrimos"	Suffer; "We suffer"
Tierra	Land, Earth
"Ya estamos acostumbrados"	"We are already used to it"

The following sections present the main concepts identified in the analysis and
which subsume the social practices of living with the flood in the case study regions.
Six main concepts have been identified, which represent a comprehensive network
of social practices. The examples underline how practices express as nexuses of
bodily-material and mental performances. While in some examples the bodily-material aspects dominate and are separated from the mental aspects for purposes of
analysis, in other cases the mental aspects are highlighted. It is however emphasised
that the linkage between the bodily-material and the mental is suspended for analytical reasons only while the general linkage is emphasised. The six concepts and
related social practices are analysed using a considerable amount of original interview material, transcribed and translated into English. Moreover, a small number
of photographs taken by the author in the field are selected to represent the material
aspects of selected social practices. Presenting detailed accounts of verbally expressed and materially performed social practices of flood management in the case

study villages provides the basis for understanding how social practices are built up, how they interlink and how they merge into larger patterns of social dynamics.

6.3 FLOOD, FLOOD RISK OR LIVING WITH FLOODS

This section presents an analysis of major empirical material gained in the field concerning the research topic of the flood. Qualitative analysis of interviews and memos has resulted in the generation of twelve different codes that can be sub-sumed in the category of flood. These codes give a first impression on the facets in which flood is of relevance in the case study region and are therefore presented in table 8. However, the building of codes is only one step in a thorough analysis and interpretation. In the following paragraphs, different aspects identified in the text are analysed in depth and used for a comprehensive interpretation of the social prac-tices related to floods in the case study villages.

Table 8: Codes related to the category "flood" (built throughout qualitative analysis with Atlas.ti) presented in alphabetic order

Code	Grounding in the text	Code	Grounding in the text
discourse_floodpractices	108	flood_negative	55
discourse_prepareforflood	28	flood_positive	16
flood_bodilyexperience	5	Flood_protectionmeasures	32
flood_care_animals	29	no es lo mismo_flood ("it is not the same_flood")	7
flood_exposure	129	prevention	1
flood_fear	15	ya estamos acostumbrados ("we are already used to it")	23

In the case study region flood is a phenomenon that involves the rising of the water level in the river Usumacinta and the adjacent lagoons and other small water bodies surrounding the settlement areas or agricultural land. The most common verbal ex-pression for the flood in the language of local people is "la creciente" that derives from the verb "crecer" (="to grow"/"to increase"). Inhabitants of the case study villages use this term more often than the word "inundación", which is the direct translation of the word flood. Another local expression used often when referring to the flood is "vamos a pique", which can be translated as "we go to the peak". In order to get access to local understandings of the flood, it is relevant to describe and analyse local expressions and discourses around the flood, as well as non-verbal expressions as parts of flood related social practices.

Interview 2014_C_7: Lines 98-101[1]
[P1]: Aqui estamos acostumbrados. Y el rio lo tenemos medir. [D1]: El ri...? [P1]: El rio, el rio. Lo tenemos medido. Por ejemplo el tiempo de la inundación, ya sabemos hasta donde va llegar.
Translation into English language
[P1]: Here we are used to it. And the river, we have it measured. [D1]: The ri...? [P1]: The river, the river. We have it measured. For example, in the time of flood, we already know until where it will reach.

The quote from a male village inhabitant expresses that the village members have developed specific ways in which they measure the water level in the river and which allow them to predict the extent of flood. The water level of the river or in other water bodies like the Catazajá lagoon or the Nueva Esperanza lagoon is under constant observation. Participant observation carried out in the self-organised public transport system of one village, has allowed access to one aspect of a flood related practice linked to *knowledge about the flood*: In the times of increasing rainfall and water level rising in the months between July and October, constant observation of and discussion about the water level takes place in the vehicles of privately organised "public" transport. People sit on two wooden banks in the back of a transporter or stand in the middle between the two banks. The driver sits in a front cabin with one seat for additional passengers. The drive from the village to the nearest town takes around 15 minutes in which the transporter passes over an untarred road with agricultural fields and the river on one side and other fields and the lagoon to the other side. The water level in the river and in the lagoon can be observed from the transporter. The transporter is a key place of observation and conversation about the water level because of the importance of road access for the village inhabitants and as a key situation for social interaction and information exchange with village members as well as outsiders such as visitors or teachers. When the water level in the river reaches a certain height, the road to and from the village becomes flooded making access by road impossible and demanding for access by boat. During participant observation in September 2014, conversation about the water level included sequences of estimating how many days it would take until the road would be impossible to pass due to the flooding. Various passengers and the driver estimated that if rainfall continued in the same way, the road would become inaccessible in the following two to three days (Field notebook 2/2014 30.09.2014). Another facet of *knowledge about the flood* is linked to sources of information about rainfall and water levels in upstream regions of the river in Chiapas, Tabasco and Guatemala. While no integrated water level information system of the river Usumacinta is accessible by the village members, informal communication channels complement

---

1   Remark of the author: For the purpose of readability, the subsequent interview sequences are presented in their English translation, while the Spanish original transcripts of these sequences together with the translations are added in annex 4

information about water levels in upstream regions. A system for radio-communi-cation was installed in the house of one village member through which communi-cation exists with the regional centre of civil protection in Palenque (I_2014_C_8). However, the exchange of information shared through this medium is comple-mented informally by the village members. Several villagers mention that during the rainy season, they keep informed about water levels in Tenosique through phone calls and mobile text messages with family members who live in that town. If they receive the information on a certain level of water in Tenosique they estimate the time when rise of the water level will affect them (Field notebook 2/2014 21.10.2014).

A crucial part of *knowledge about the flood* exists as bodily knowledge, derived from past experience and from social practices performed for a period of many years. Several expressions reveal this bodily aspect of flood knowledge. During interviews and conversations people in the villages used parts of their body to pre-sent how they measure the height of a flood. The hand is an important measuring device, as the following quote of a male village member reveals: "There is missing "one quarter" until there is no passage with the local transport anymore." (Field notebook 2/2014 21.10.2014). "Un cuarto" (="one quarter") is the width of one hand, is a common unit for measuring and describing the water level. In an inter-view carried out with a villager in his field at the side of the river, the interviewed villager indicated the level of water on his body. He recalled a past flood presenting that in parts of the field, the water reached until the location of his belt and in an-other part, he was immersed in water until his throat (I_2015_C_22: 410–412). An-other interview repeats the notion of the water level measured by the own body: "Because the water reaches until the waist. Or deeper in parts" (I_2015_C_7: 70–71). Besides the bodily experience of the water on a tactile level, the flood also has specific acoustic features which are described by village members.

Interview 2015_C_17: Lines 45–50
[P1]: [...] Ah, there is a stream that when the water enters, it sounds like a motor or something, like "Hiiiii". When the water is low, and it flows, this one. Like an, this, which falls from large height, how is it called? A cascade. Yes, this is how the water sounds.

The *spatial extension of the flood* in the villages can vary strongly from one flood season to another. The main areas that were mentioned by the villagers to be differ-entiated concerning flood affection are the lands, where they grow food or rare cat-tle on the one hand and where the settlement is located on the other hand. Agricul-tural land in general lies in lower parts of the villages and the settlement areas are located in higher areas. Agricultural land thus is affected first when floods occur. However, the settlement areas can also be divided into those areas that are located directly at the side of the river or the side of the lagoon and those that are located on small mounds. Conversation partners mention that in the year 2011 a flood hap-pened with the most severe effects of all floods they remembered.

Interview 2015_C_7: Lines 93–97
[P2]: Yes. Yes. It is about four years ago, there was a large flood that water entered here, into the kitchen five centimetres. Into the house ten, but the house is lower. Here we had never gone into the water. But that time yes.

Villagers describe that the seasonality of floods has changed in the last few decades resulting in a reduced reliability on the seasonality of wet and dry seasons. One interview partner mentions that floods have become stronger since the year 2007 and that they rise higher than average since then (I_2014_C_16: 1–2). In 2007 the lowland areas of Chiapas and Tabasco experienced severe flooding, which gained international and media attention due to the strong repercussions in the city of Villahermosa (Paz Ojeda 2009). For the year 2014 villagers describe that while in total three floods occurred, the first flood came unexpected in the month of June, destroying considerable parts of the agricultural crops on farmers' fields (I_2015_C_7: 142–145; I_2015_C_22: 388–389).

The analysis of expressions that concern the *perception local people have about the flood and how they evaluate* flood events at first appear as a series of contradictions. Whereas flood is described as something negative, at the same time the positive effects of the flood are underlined by the villagers. The negative aspects mentioned by interview partners mainly concern hygienic conditions during floods, the inaccessibility of the village and the loss of harvest and animals. Moreover, women express that they fear the flood mainly for three reasons: They are afraid to travel on the boat as many of them do not know how to swim, they are concerned about their children because they could drown in the waters and women fear the wild animals, among them crocodiles, spiders and snakes, that approach or enter the houses during flood times. In one interview, the female interview partner presents her perception and evaluation of the flood and compares it to the perception and evaluation of other people in the village. As highlighted in the text below, she repeatedly switches between positive and negative accounts:

Interview 2015_C_14: 8–20; 32–51
[P1]: Well, for me, the flood is something, well, not very pleasant because you imagine that everything rises. Wherever we move, it has to be in the boat because you cannot travel in another way. And although for many well, they benefit, others don´t because for those who work, has to stop his work for many months if the flood is a big one. Who works, for example who sows corn. Aha, and in this time they don´t have work, those who do, who sow corn. They leave to work outside the village. Mhm, they leave to work outside the village. […] And I, because of that I tell you yeeees, I am afraid of the flood, for, because there are snakes. Although others have benefit because they say: Now come the food supplies [laughter] […] Yes, because of the food supplies. And as it is flood time, who knows from where all these people come out but many people go out for the food supplies. Aha, and I, well on the other side it is beautiful because a lot of fish enters there into the lagoons. When the water level gets lower people fish a lot of fish just as I told you. A lot of fish. But also for the people it is very, very like that, how should I tell you? Like. They have problems because of the animals, those who have cattle they have to bring them to up, to

> higher grounds. And those who don't, well, have to rent higher lands as well, to bring them there. And on one side it does good but on the other side it does bad as well the flood.

Representing this sequence enables to highlight a range of patterns that are relevant for the analysis of flood perception. The switch between positive and negative aspects of the flood shows an ambivalent evaluation. While the interview partner presents more negative evaluations, that concern her personal experience, she also perceives parts of the flood effects as positive for herself. This is the situation when there is an abundance of fish, at the ending of a flood. Two times she presents an expression of "they benefit", referring to other village members, mainly those who benefit from the food supply offered by government during flood times. Three activities are presented which form part of flood related practices in the village. One concerns the movement in and outside the village, which is organised on small boats. The other two concern the economic activity of people during flood times. Those people, who normally work on their fields, need to temporally migrate and get employment outside the village and those who have cattle need to organise the transport onto higher lands or make financial effort to rent lands on higher grounds. Moreover, the interviewee presents a range of different aspects which concern the daily life of people and the specific challenges during flood times. A group of nouns that hints towards everyday life, changes during flood times and relevant action taken by village members can be highlighted: *The boat* as the only means of transport during flood times; *corn* as the main crop and staple food in the village, *food supplies* as the main support given during flood season, *snakes* as dangerous animals that appear during floods, *fish* as animals that serve as highly valued food in the village, *cattle* as animals that have high economic value for villagers, are mainly sold to large cattle owners outside the village and are only eaten on special occasions. Besides clearly positive or negative evaluations, expressions about the flood by villagers also involve humour and wordplays. This is exemplified by the denomination one inhabitant uses for his village by calling it the "Venice of Chiapas" (I_2014_C_22: 1).

 In different local expressions and discussion topics raised by village members one observation is significant for this research: the *lack of the term and concept of risk* in local discourse on floods. While theoretical and empirical interest in risk phenomena was one starting point for this research, the empirical material from the villages did not confirm that risk was a major topic related to the flood. As discussed above, flood is related to positive and negative effects at the same time and while emotions of fear and sadness are evoked by specific phenomena related to flood occurrence, there are also positive emotions involved. From the empirical material gathered in the field, only two interview partners explicitly used the term risk when referring to the flood.

> Interview 2015_C_7: Lines 93–97
>
> [P1]: Because now we lose. If we are going to sow now we take the risk that the flood comes and seizes us and then when we prepared the tamal, the corn, and we lose it.

The male interview partner describes the activity of farmers who sow corn as risky, because the harvest will be lost if a heavy flood occurs. The risk of losing corn has substantial nutritive and economic disadvantages, as people produce their main food from corn, which they have to buy at expensive prices in town if not produced on the fields. Moreover, another disadvantage is the effect of losing corn. In the quote, the interviewee refers to the time in the year, when people in the village prepare *tamal* which is a food made out of corn that is traditionally prepared all over Mexico in the month of November. Preparing *tamales* and offering them to the spirits of dead ancestors around the festivities of *Día de los muertos* on 1 of November, is one of the most important cultural traditions in Mexico. Not being able to give this offering is a possible detrimental effect of the flood. The analysis of the structure and tenses used in the quote shows a relevant pattern: While the quote starts with an expression in the present tense "now we lose", the interview partner continues in the future tense to express the plans to sow corn "If we are going to sow now". This tense is followed by the expression of risk-taking in the present tense "we take the risk that the flood comes and seizes us". The preparation of *tamal* is expressed in the past tense "then when we prepared the *tamal*, the corn," while the effect of the flood is again presented in the present tense "and we lose it". The assumption is made that referring to *tamal* in the past tense, underlines the remembering of past experiences with processes performed in the present and future. The other interviewee who mentions the concept of risk gives distinct evaluation of it:

Interview 2015_C_7: Lines 398–399, 402–403
[P1]: For us, the risk doesn´t do anything to us, the risk, because it is like always, well when the rains start, it begins to grow [...]. Here no, here no, here we see it coming and we protect ourselves. To us, it doesn´t do anything.

The description of the text passages in the above paragraphs allows the identification of a specific discursive practice involved in expressions about the flood. Verbal expressions of contradiction or of ambivalence stand out. Five pairs of ambivalent expressions have been identified (table 9).

Table 9: Analysis of ambivalence in discursive practices concerning perception and evaluation of the flood

Negative	concessive	Positive
We suffer	But	We are already used to it
It harms us	But	It benefits us
We are afraid	But	We see it as something natural
It does bad	But	It does good
We lose	But	It is beautiful

While fishermen emphasise the economic benefit they have at the beginning of the flood when an abundance of fish enters the lagoon, mothers emphasise the fear for their children to get swept away by the current. Evaluations contrast not only between different people, as even the evaluation by one and the same person includes positive and negative aspects. Farmers for example highlight that the loss of harvest causes great damage to them; however, they emphasise that the sediments that accumulate on their fields after a flood contribute to soil fertility. It can be summarised that there are multiple layers of flood perception and evaluation as part of dealing with floods. Besides perceptions and evaluations, the social practices of living with the flood entail a range of other aspects which are analysed in the following paragraphs.

The expression "we are already used to it", introduced above as one major local concept identified, is a crucial empirical representative of social practices of living with the flood. The three following interview sequences allow for a more in-depth analysis of what these practices involve.

Interview 2015_C_22: Lines 359–365
[P23]: About what we have lived with the floods, well we already see it with naturalness because we have already got used to this, to what is nature, the, the floods because that is what it is. The floods it is year after year. And well this, we are already so used to it that we already, already they do not scare us.

In the above sequence, the female interview partner introduces her sentence by saying "what we have lived with the floods", thereby connecting the flood not only to her life but to the lives of her family and village members as ongoing processes. She presents the flood as part of nature and that therefore people in the village regard it "with naturalness" and "have already got used to it". The knowledge about the cyclical occurrence of floods "year after year" allows village members to adapt accordingly.

Interview 2015_C_20: Lines 8–18
[P1]: Well the flood, it is, sometimes it surprises us. But we are already used to it and year after year, we expect it. For that when the rains start. Surely that the hurricanes. And all the water that is brought with it, to say the wind, there in the direction of the forest. It goes down again the water of this river abounding in water, that is called river Usumacinta. And here we are prepared, when the water continues. Until up the levels. Well we are also preparing ourselves for the case that the level rises.

While the male village member in the above interview sequence highlights that he prepares for the time when the water level rises, the female interview partners in the sequence presented below focus on the fact that they are used to the flood and do not need to prepare:

Interview 2014_C_7: Lines 129–135
[P5]: Prepare, no. Here we are used to it, to the date that it comes, we are up to date, we already know the date that…
[P1]: Already, already.
[P4]: We are used to what happens…to what happens.
[P3]: We are already up to date, already.
[P1]: Like now we are already up to date, the river we already see that already is, it is rising, already.
[P5]: It doesn´t scare us. Because we know that year after year this happens.

This sequence stands out through its´ patterns of repetitions. Expressions like "we are up to date", "already" and "we know" are used by different women in the interview. They highlight that they know what to do and when to do it, as they have a year–long experience with the dynamics of the flood. Living with the flood is a social practice that involves bodily-material and mental processes including emotional aspects of preparation. A specific materiality can be identified here to form part of the practice. Other than through an analysis of flood discourses through spoken words, a range of material aspects were identified through a visual approach by the researcher. The extended walks through the village in group as well as by the researcher alone or accompanied by the co-researcher, allowed for an identification of objects, structures and material patterns that may be attributed to the flood. In several yards of houses that are located at a distance of over 100 metres from the river bed, the so-called cayucos (small fishing boats) can be found (figure 19, colour plate p. 143). These small boats indicated that although the flood was absent at the time of empirical research, material patterns of flood practices were present in physical (and social) space.

Another example is presented in figure 20 (see colour plate p. 143): In a house, that is located at around 20 metres from the river bed, the fridge in the kitchen is placed on a wooden structure that raises it around 10 centimetres from the ground. While the water was absent at the time the photograph was taken, the fridge stood on this wooden structure, while the other furniture in the house stood on the ground. Initially the researcher understood these material objects as witnesses of a past flood; however, the idea of the witness was subsequently expanded and opened up a scene for reflections on the temporal dimensions of flood events: The material objects can be both regarded as witnesses of the past and as indicators for the future. One interview partner mentions in this regard: "we already have certain inheritance to have this, at the door of our house a cayuco (small fishing boat)" (I_2015_C_17: 165–166). At the same time, the emphasis of village inhabitants that the flood is something that happens "year after year" (I_2015_C_20: 10) reveals the perception of a cyclical temporal character of floods. An important material aspect of living with the flood in the past involved the type of houses people used to live in in the river area in earlier decades. Houses called tapanco are two-storey houses made out of wood. During dry season, people used to live in the lower level of the house while during the flood period, people used to move up their goods and smaller animals and lived together with these on the second floor. While in the case study villages this type of architectural structure is not present, it is mentioned that other villages along the river Usumacinta still have two-storey houses of the *tapanco* style (Field notebook 2/2014 2.09.2014). While the *tapanco* style houses are not present in the villages, a different material structure called *tapesco* serves the care of animals: Before a flood occurs, some village members construct wooden platforms on stilts as small stables for their animals, such as chicken, geese, turkey and pigs. Another model of *tapesco* is not built on stilts but floating, with plastic bottles or large containers filled with air placed below the wooden stable (I_2014_C_7: 25–28).

Another facet of the social practice of living with the flood is the temporal mobility of people. In flood times, a few village residents who can afford, move to their relatives in town. Others move to houses of family members in the villages that are not affected by the rising water level and those who are severely affected move to the secondary school which serves as temporary shelter (I_2015_C_14: 93–97). Others have to or decide to stay in their houses, although it means to live inside the water. For one male interview partner, the main reason to stay in the house is to protect the possessions the family has from getting stolen (I_2014_C_6: 48–56). One girl at primary school age presents her experience of living with the flood in the following way:

Interview 2015_C_13: Lines 22–40
[P1]: Those who live at the river bank, everything fills up with water.
[D2]: Everything goes into the water? And what do they do in these times, during?
[P1]: We need to put up with the water inside the house.
[D2]: What do the children do inside the house when there is flood?
[P1]: Well [she laughs]. They play.
[D2]: And there the water entered into your house?
[P1]: Yes. From there where there is the nona tree and it filled.
[D2]: And how do you feel?
[P1]: Well sometimes it can fill the house.
[D2]: But how do you feel, are you happy or sad? What do you feel?
[P1]: Happy, because there are those small fishies.
[D2]: How do you catch them?
[P1]: With a plate [laughs].

During floods, children don´t have school and can only play inside the house, even if it is flooded (I_2015_C_14: 202–203). For parents the care for children is a major concern as many fall sick with fever and there is general worry that they could drown. Two groups of people perform specific social practices related to the flood. One is the group of fishermen, who catch fish in the lagoon or in the river and benefit from an abundance of fish and high economic activity during the beginning of the flood. The other group are the lancheros in one of the villages, the drivers of the motorboats who transport people on the river. In the case study village that is directly located at the side of the river, the motorboat drivers are all members of one family.

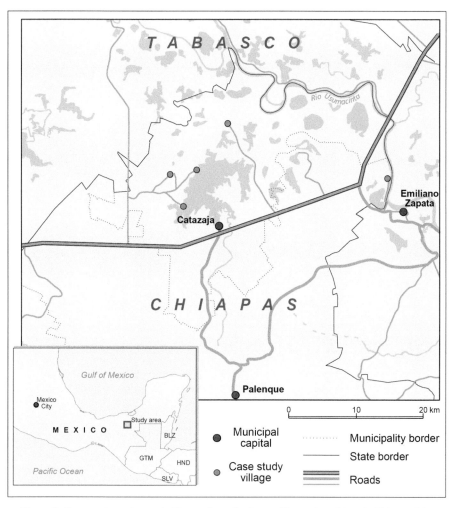

Figure 2: Case study region and villages along the Lower Usumacinta River in Chiapas, Mexico
Source: Stephan 2018

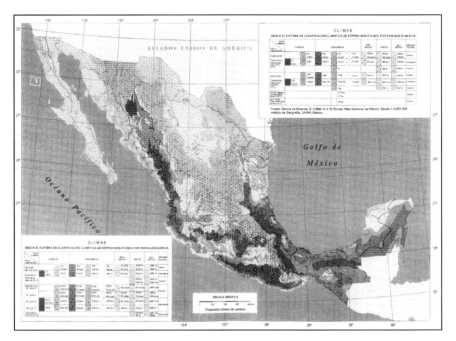

Figure 4: Climates of Mexico
Source: García 1988

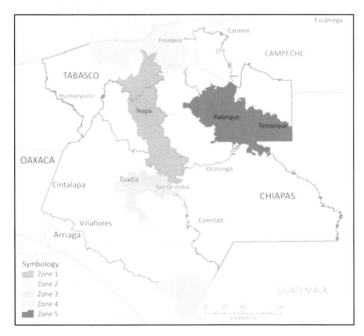

Figure 6: Priority regions for intervention proposed by BID
Source: adated from BID 2014 (legend translated by the author)

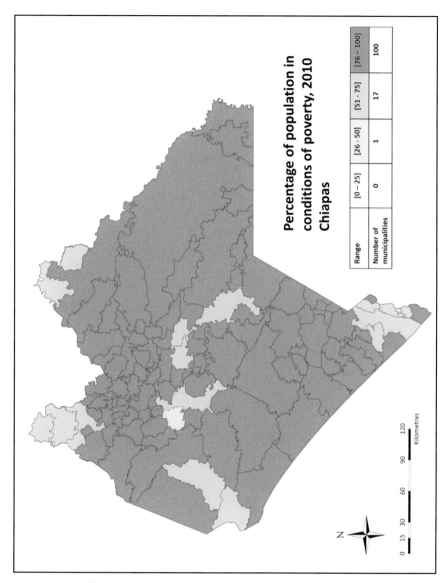

Figure 9: Percentage of population in conditions of poverty in Chiapas in 2010
as defined by CONEVAL
Source: adapted from CONEVAL (2010) using ArcGIS (version 10.1/2012)

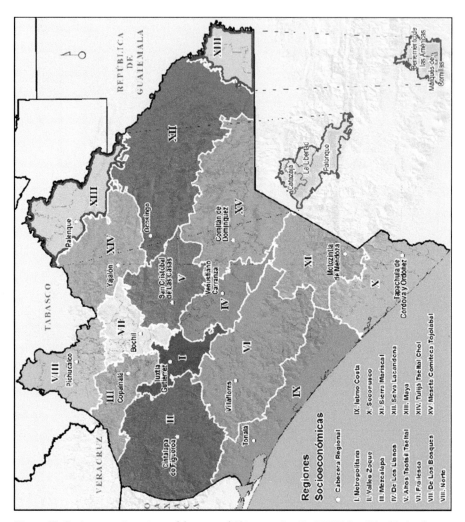

Figure 10: Socioeconomic regions of the state of Chiapas – Región XIII "Maya" highlighted
Source: adapted from CEIEG (2015)

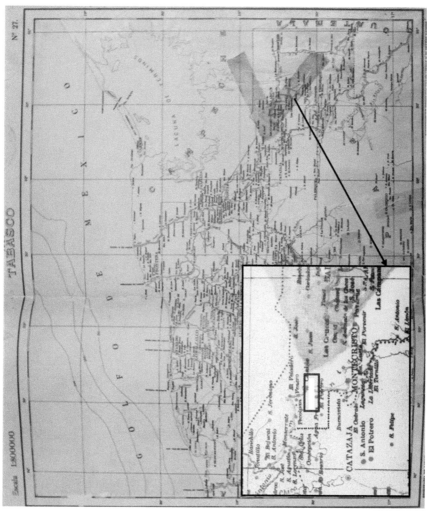

Figure 12: Historical map of the state of Tabasco from 1921 (Location of case study village indicated in red in the detail map on the left)
Source: adapted and anonymised from Gobierno del Estado de Tabasco (1982)

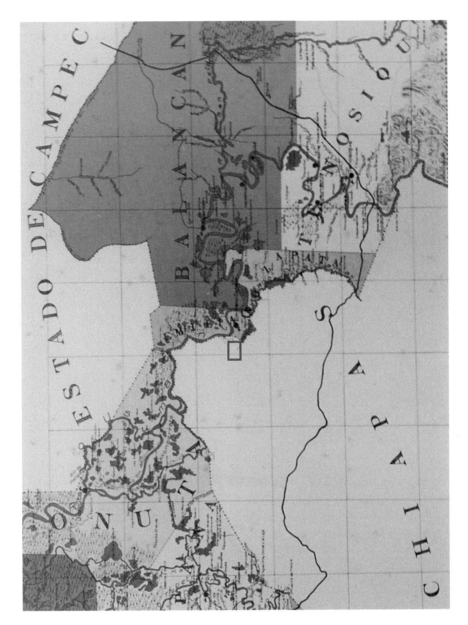

Figure 13: Detail from a historical map of the state of Tabasco from 1952
(Location of one case study village indicated in red)
Source: adapted from Gobierno del Estado de Tabasco (1982)

Figure 19: Photograph of a wooden boat (cayuco) in the front of a village house
Source: C. Stephan (2015)

Figure 20: Fridge on wooden platform in the kitchen of a village house
Source: C. Stephan (2014)

Figure 21: Clay figure and stones found by villagers in one case study village
Source: C. Stephan (2015)

Figure 22: Photograph 1 selected from photography workshop
Source: Student group 1 from case study village

Figure 23: Drawings in photograph 1 as part of visual analysis with the documentary method
Source: author's elaboration on basis of photograph 1

Figure 24: Photograph 2 selected from photography workshop
Source: Student group 2 from case study village

Figure 25: Photograph 3 selected from photography workshop
Source: Student group 3 from case study village

Figure 26: Photograph 4 selected from photography workshop
Source: Student group 4 from case study village

Figure 27: Photograph 5 selected from photography workshop
Source: Student group 5 from case study village

Figure 28: Photograph 6 selected from photography workshop
Source: Student group 6 from case study village

Figure 29: Photograph 7 selected from photography workshop
Source: Student group 7 from case study village

Figure 30: Snapshots from video sequence 1
Source: Documentary film 2015 (produced by members from one case study village, co-researcher and researcher)

Figure 31: Snapshots from video sequence 2
Source: Documentary film (2015)

Figure 32: Snapshots from video sequence 3
Source: Documentary film 2015

Pronóstico de lluvias (SMN)	Nivel de alerta	Tipo de lluvia
De 0.1 a 25 mm De 25 a 50 mm	Azul	Lluvias Lluvias Fuertes
De 50 a 75 mm	Verde	Lluvias Muy Fuertes
De 75 a 150 mm	Amarilla	Lluvias Intensas
De 150 a 250 mm	Naranja	Lluvias Torrenciales
Superiores a 250	Roja	Lluvias Extraordinarias

Figure 35: Classification of alert levels according to amount of precipitation
Source: Insituto de Protección Civil n.Y.

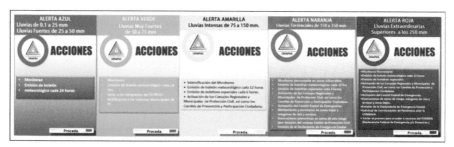

Figure 36: Different actions ordered according to rain alert levels
Source: Insituto de Protección Civil n.Y.

Figure 37: Photograph of a scale for flood level measurement in one case study village
Source: C. Stephan (2014)

Figure 38: House donated by government after major flood in one case study village
Source: C. Stephan (2014)

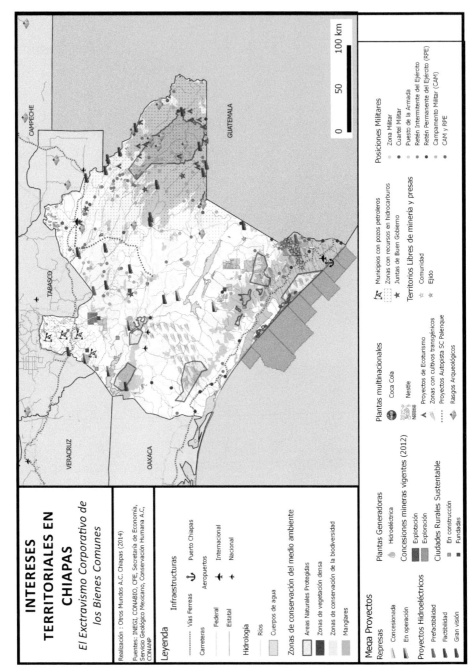

Figure 39: Map entitled "Territorial interests in Chiapas"
Source: OMC (2014)

They transport people and goods throughout the whole year; however, when the flood occurs, they are the only providers of transport to and from the village. Their daily routine changes fundamentally during flood times: Each passenger has to be picked up at the door of his house while during normal days, passengers gather at the side of the river at a specific time. The passage to the nearest town Emiliano Zapata takes around one hour and if the current, rains or winds are strong, the journey can be dangerous. One account of the activities of a lanchero are presented in the following quote:

Interview 2014_C_7: Lines 52–64
[P1]: In the motor boat you arrive until here, until the door of the house.
[D2]: Yes?
[P1]: Yes.
[D2]: Is it not very, very low? With the motor, no?
[P2]: Ah no, but you use a paddle.
[P1]: Mhm.
[P2]: That´s it. Here we call it paddle or canalete. And you use it to paddle when the water level is low. The motor doesn´t work here. There outside yes because it is deeper, but here no.
[D2]: Mhm.
[P2]: And you go around and around. You enter, you leave.

While some activities stay the same during the times of flood, like the preparation of food or the care for animals and children, most other activities change, stop or are replaced by different activities to satisfy daily needs. It becomes obvious that social practices related to the flood are intermingled with larger social practices in the villages. In the following sections further practices are identified which are part of the larger set of the social practices of living with the flood.

6.4 LIVELIHOOD, THE "GOOD LIFE" AND DEVELOPMENT

People in the case study villages perform a range of different practices in order to satisfy their basic needs. One of these needs is providing nutrition. Village inhabitants repeatedly express that they grow the basic food for their daily supplies on their milpa, which is the local name for the agricultural production of various agricultural plants in one field. Additionally, they keep small animals (e.g. chicken and turkey) in their yards. People with more economic resources keep cattle on grazing lands and sell these animals to buyers from around the country. Moreover, fish is caught in the lagoons and, to a lesser extent, in the river for consumption and sales. Besides livelihood practices performed inside the village others are carried out outside the village. One interview partner presents livelihood provision in his village in the following ways: "We are self-sufficient" (Memo_I_150603, Annex 7) "and what is here from the region is consumed in the region" (I_2015_C_4: 88–89). Discourse about self–sufficiency is found in many interviews and observations from the case study villages. This includes various accounts on the benefits of eating locally produced food. While increasingly people buy convenience food from outside the village, discourse about food is dominated by an appreciation of what is described as natural and healthy food from the villages.

Interview 2015_C_7: Lines 60–67
[P1]: Here you are going to eat a fresh *mojarra* (local fish variety). You are going to gut a fresh chicken. [D1]: Mhm. [P1]: And if you have it in the fridge, it would be for one or two days. Not like they have them in SuperChe and all, where they change the label and it stays. [D1]: Yes. [P1]: Yes, it is very different.

Two different ways of life are presented by interview partners from the village. While many inhabitants grew up in the same village, some have spent part of their lives living in towns. These interview partners compare "the good life" in the village with life in towns. Moreover, temporal dynamics are described in the way that life in the village during the youth of interview partners is compared with life in the village today. Local discourse presents life in the village in a specific way, which highlights the quietness, the peacefulness and the good environment and way of life one can experience.

Interview 2015_C_1: Lines 114–122
[P4]: Many people say that [...] (anonymised) is quiet. That they like it. Because living in a town, life is very fast. Or everything costs something. And here at least, we keep it going. Because we have a lime tree, one of orange, or we grow corn, for the tortilla. We sow beans. Pumpkin. [D1]: Mhm. [P4]: And well in a town it is different.

Another female interview partner who had lived part of her life in large towns, emphasises that "[h]ere it is quiet. We still breathe fresh air" (I_2014_C_18: 10). The following quote presents further sequences of this discourse of "the good and quiet life" in the village:

Interview 2015_C_7: Lines 10–16
[P1]: [...] (anonymised) is quiet. [...] (anonymised) is a, it is a community, well, where people are dedicated to working. Sowing millo, others with their little work in, in a shop. Others dedicate to the picante, to the chiva, to the broom stick. And until today it is a quiet place. Sure that many enter in order to make their, their things here but everything quiet.

The word used most frequently to describe one of the villages is the word "tranquilo". While in the sequence presented above "tranquilo" is translated as "quiet", the meaning of the word can also be "peaceful". It is in the last sentence of the above quote, that the notion of "peaceful" shows relevant. The female interview partner mentions that people from outside sometimes come into the village "to do their things", but that this does not affect the village. The word "rollo", here translated as "things", may in the context of the sentence have a negative connotation. One problem repeatedly brought to the village from outside is drug trafficking. The untarred road that leads through the village is used by drug traffickers and members of gangs as a secret path to cross the border between the states of Chiapas and Tabasco (I_2015_C_33: 17–19).

While storage of large amounts of food for the flood times is not common in the villages, the importance of corn harvested from the fields is emphasised by one interviewee saying "even if we have no money, we have corn (I_2014_C_7: 114). While people mainly rely on their income and family support during flood times, in recent years the supply of food stuff by government has steadily increased in the villages. The need for the food supply called *despensa*, is discussed controversially in the villages. A major farmer in one of the villages describes the provision of *despensas* critically and emphasises the autonomy of villagers in the provision with food during flood times:

Interview 2015_C_32: Lines 463–467 & 470–473
[D1]: But what do you think? During this time of the flood, they come to bring despensa for example, but you say that this is not necessary. [P1]: Well the truth, we take it because they bring it. [D1]: Yes, sure. [P1]: But in reality it is not a thing that we would lack a lot, no? […] because we gather it for us as I tell you, basically how we achieve to take out all we can of our corn harvests, we have for the posole (drink based on corn) and the tortilla (corn bread) and with my daughter we leave to fish with a fishing–hook, with the net, as you can and we have food. There is the chance to go and kill a wild game, we kill it and have food, exactly.

In contrast, the wife of the interview partner argues that those people who don´t grow corn and normally buy *tortillas* are highly dependent on the food supplies:

Interview 2015_C_32: Lines 478–484
[P2]: But what I say, I, for me with what comes it is more than enough, because he what he says there comes no soap, there comes nothing, for example, for cleaning the bath, this is what he says, but there come many things that indeed is lacking. He says that not much is lacking, but there are people here who truly they need it, and yes the despensa does good to them because there are many people as he says who harvest corn, we have to make tortillas all the days, but there are many people here who don´t sow corn and they buy tortilla.

Changing practices in production and consumption, leaving behind self-sufficiency and relying on food, convenience products or even fertiliser and technologies from outside the village can be linked to discursive practices of "modern" ways of life and "development" in the village.

The idea to move towards places outside the village is for most interview partners not a realistic option. People do not intend to live in any other place as the options to gain access to new land for living and for agricultural production are limited. One aspect highlighted by interview partners is their being enrooted in the villages: "Practically here there is already, here are their, here are their roots [...] Here we feel well" (I_2014_C_7: 135–137). While most interview partners emphasise their interest to stay in the village, younger interview partners express ideas to move to places outside the village. In most cases, those village members who pursue advanced education or need to work outside of the village do not intend to return to the village as permanent residents. Temporal mobility of both male and female village members is of great importance for the economic basis of most families in the villages (I_2015_C_11: 234–236). Male inhabitants strive for employment on palm oil plantations in neighbouring municipalities or states which requires daily commuting from and to the village (I_2014_C_18: 36). Another option is employment on offshore platforms of the oil company PEMEX in the Gulf of Mexico which results in absence from the villages for a period of three months followed by two to four weeks of vacation (I_2014_PR_12: 27; I_2015_C_37: 3–4). Women who work outside of the village predominantly work in hotels in the main tourist destinations of the Yucatán peninsula which often results in permanent migration. Other women find employment in restaurants or supermarkets in nearby towns, which results in daily commuting from and to the village (I_2014_C_4: 8).

Discussions with students from secondary school about the futures they imagined for their village, identified two different perspectives: One can be described as the optimist development perspective, that highlights the installation of infrastructures and services as tarred roads, hospitals, libraries and churches and emphasises that the positive characteristics are already present in the village, such as unity, honesty and hard-working people, and will become even stronger in the future. The other perspective presents a more pessimistic perspective towards the future, highlighting poverty, caused by the loss of animals and vegetation and sales of property to people from outside. The large contrast between the two perspectives may indicate towards current changes in the practices of livelihood and "the good life" in the village, which are discussed at more length in chapter 9.

6.5 CREATING AND PERFORMING COLLECTIVE IDENTITIES

In this section those aspects of discourse are analysed that give insight into the formation of the social relations inside the village and the building of identity through social practices. Various connections exist between the perception of floods, sense of community and identity. These connections are established and reemphasised as parts of dynamic social practices. In the case study villages, a series of myths and stories exist about the founding of each settlement and about early traditions. Some of the myths are widely spread in the region or even all over Mexico. But there are myths specific to the villages or municipality. In the case of the villages around the Catazajá lagoon, there is a legend about *El hombre pez*, the fish man (Annex 6).

The main figure of the legend, the creature half man half fish, is depicted as a powerful agent regulating the access to fish and to other natural resources like precious
trees in the region. The offering of a female virgin is presented as the only way for
the Maya to gain favour of the creature. The myth can be seen as part of a set of
rituals of human sacrifice presented for problem solving in diverse cultures from all
around the world (Davies 1981). The murder of the creature by the people, the bal-
ance of power is interrupted which results in changes in the natural environment.
One of the major points of interest for this study is the fact that the occurrence of a
flood season and a dry season is described in this legend as a result of a divine
punishment cast by the gods to the Mayan people, who are the original inhabitants
of the lagoon and the Palenque region. As todays´ inhabitants of the Catazajá lagoon
perform a repetition of the legend and pass it to coming generations the assumption
can be made that people regard the punishment to their ancestors as part of their
history and as part of their identity.

In the case study village that is located by the side of the river Usumacinta, a
legend exists that recounts the founding of the village and the origin of its name.
The elder people in the village tell, that there was a pirate or a noble man with the
last name […] (name anonymised), who had fled from Spain on a ship. Upon the
arrival in the Gulf of Mexico, he navigated the ship upstream the river Usumacinta
until he reached the place where today the village is located. Upon unloading the
ship, a treasure fell into the river and could never be found again (I_2014_C_10:
18–38). Some village members mention that the treasure must still be inside the
waters of the river as nobody has found it yet. The narration tells that after the death
of the nobleman his children decided to return to their country of origin but that
since those days the place was attributed this family name (I_2015_C_17: 540–
541). Today dispute exists inside the village about the validity of the legend and
other origins of the village name are claimed (I_2014_C_14: 323–330).

In the case of the first legend a direct link can be assumed from the former
inhabitants of the lagoon and the people who live there today. The Mayan tradition
and language is emphasised in this legend including the importance to respect the
natural environment and divinities. This contributes to an identity of Mayan descent. In contrast, the legend of the nobleman is a story of travel and discovery by
a European who passes his name and thereby part of his identity to the village. Of
specific interest is the idea of the treasure which in its discursive practice attributes
specific value to the village. Moreover, a direct material representation of the treasure can be identified. People in the village frequently find objects made from clay
and even pieces of obsidian and jade, who originate in Mayan culture. Some of the
pieces found by the villagers still remain in the village, as exemplified in figure 21
(see colour plate p. 144), while many have been sold to people outside of the village.

The story of the nobleman can be identified as a representation of the collective
history of many people in the region. It is the history of conquest and the mixing of
different ethnic groups which at later times continued through marriage between
local people and Europeans. In the following interview sequence, the history of the
village and the personal family history of one of the elder inhabitants of the village
is recounted:

Interview 2014_C_10: Lines 57–72, 75–85

[P1]: The duke had the name José María [...] (anonmyised) and here he lived. Those from Zapata who did this with the land, gave it the name [...].

[D2]: José María...

[P1]: José María [...]. It was his last name, [...]. And as here lived this, this family from Zapata who were the owners of these lands, and they gave the name [...]. This is what is stayed, [...]. Until the day.

[D2]: And this history came to you, they told you? This story was told to you by your father, or by your grandfather or by, how?

[P1]: A man who lived in the house by the side, a driver carried the story, the legend. And travelling to Tuxtla he showed it to me. And when I started to read, I [...]

[P1]: There it was where I read a small part because it is a large legend.

[D1]: Aha.

[D2]: And you said that your grandparents were Spaniards? From the side of your father or of your mother?

[P1]: From my mother. The fathers of the father of my mother were, they came from Spanish. So the father of my mother was quite white, still. He didn´t speak very well, he didn´t speak very well. And here they got married. Already with people from here.

Another aspect relevant for the formation of a collective identity in the villages is the close relationship village members have to non-human beings. The non-human category created by the researcher comprises the flora and the fauna as well as other visible and invisible entities. Animals play an important role in the daily activities and in the larger livelihood system. As an important economic factor for village members they are the focus of everyday activities for a large share of village inhabitants. While women mainly are responsible for those animals in the backyards of their houses, it is predominantly men who take care of cattle. Besides these animals, a large variety of other animals is present in the everyday lives of people in the villages. Among the wild animals are iguanas that live at the river banks and in the trees, crocodiles in the river and lagoons and a large variety of snakes, fish, insects and birds that have their typical habitat in the wetlands of the humid tropics. The close relationship village members have to animals has repercussions in flood related social practices. In the coding of the interviews it is visible that one of the major concerns local people face in preparation for the flood and during flood events is to protect their animals (code "flood_care_animals"). Detailed description of the practices of caring for poultry during floods and the focus on their suffering, shows that people feel pity for animals that have to experience a flood (I_2015_C_7: 98–101). As the interview sequence below highlights, people emphasise that they suffer with the animals when they get lost in the current and die.

Interview 2015_C_19: Lines 57–64

[P1]: But it also brings disadvantage, because there we have to alleviate, to suffer with the animals. Be it a cow, the horses, even with the birds.

[D1]: They die?
[P1]: Yes. Sometimes a tapesco is made, they climb up but sometimes they fall asleep, they fall, the current takes them and they drown, they die.

In a poetic form, a primary school student from one case study village presents her impression of her village and of the flood. While this poem gives entry points into the analysis of a wide range of different aspects of flood perception, at this point it is aimed to drawing the readers´ attention onto the specific ways in which identity of the village is created and how the poem represents the relationships between village members and nonhuman beings. A translation of the poem is presented below, while the poem in its´ original form as an interview sequence in Spanish language is added in annex 4.

> "All of these beautiful twelve years that I am now,
> have served me to get to know my village.
> A village small and full of love
> where it´s people is very hard-working and humble of heart.
> You can hear the birds at the dawn of each day,
> people welcome you with happiness.
> But what a sadness when everything starts to change
> because autumn comes and the birds are not to be heard any longer.
> The river grows and the work of the farmer is lost,
> all of which was happiness unfortunately becomes sad.
> When the rains start the river only grows and grows,
> some birds fly away and are not seen again."
> (I_2015_C_11:1–13)

The flood is described as something that changes life in the village causing people to become sad. Two non-human beings can be identified to play a major role in the poem. The first is the river. It is described as the river that grows. Negative consequences like the loss of the harvest follow. The second type of non-human beings are birds. In the description of the normal life in the village, birds are those beings that indicate the beginning of each new day. When the situation changes in autumn, the first thing noticed by the author of the poem is that birds are not heard any longer. It is especially the last sentence of the poem, in which the birds play the major role. It is described that when the flood starts, "birds fly away and are not seen again". The author of the poem puts emphasis on the fact that some birds disappear due to the flood and do not return. The analysis of the poem underlines the connection between lives of humans and lives of animals. The poetic figure of the bird can also be understood as forming part of collective identity of people in the village.

The occurrence of the flood is regarded as a process closely linked to nature. As the quote presented above highlights, people in the villages regard the flood "with naturalness" and they count it to "nature, the, the floods because that is what it is" (I_2015_C_22: 359–363). People make sense of the flood event as a natural phenomenon and at the same time connect them to experience made as part of their belief or religion. In case of the following interview sequence, the suffering that people in the village experience is linked to nature. While suffering is regarded as a normal feature of life on earth, it is expected to end only when someone dies and comes to heaven. It is referred to a passage in the Bible, in which Jesus Christ calls people to come to him (Matthew 11: 28). This quote is referred to in the interview as a hope for the ending of the suffering, including the ending of the experience made during the floods.

Interview 2015_C_17: Lines 488–497
[P1]: Yes, well it is a sacrifice, we go, but it is nature. Nature and it has to be endured.
[D1]: Mhm.
[P1]: Yes. Even Jesus says, come here. And we go to heaven. [Laughter] Until then we suffer. Yes. That´s how it is.
[D1]: And what do you think, why are there these changes of climate, of water, of rain?
[P1]: It is nature, nature.

An empirical example which mainly supports the assumption of a close connection between natural and cultural phenomena is a religious tradition celebrated each year in one of the case study villages. It is the celebration of the *Señor de Tila*. A painting that is described to depict Jesus Christ is kept in the Catholic Church in the village and worshiped on the Thursday of the feast of Corpus Christi. It is called the "Lord of Tila" (="Señor de Tila") because narratives exist that in the village Tila in Chiapas a figure of Jesus appeared. A dark coloured wooden figure of the crucified Jesus is worshipped until the present day. Scientific literature presents a link between the *Señor de Tila* and other figures of a black Christ which are common in other Mexican states like Oaxaca and in a major location of catholic pilgrimage in Esquipulas, Guatemala (Navarrete Cáceres 2013). Some people from the case study village pilgrimage to Tila, which is located at a distance of approximately 150 kilometres, to spend the main festive day (I_2015_C_4: 40–45). In the accounts made by one male interview partner from the village, the importance of the festivities is highlighted:

Interview 2015_C_4: Lines 4-28
[P1]: Well, what happens. Here, what is the festivity, we are used to that in these dates, which is the day of the "Señor de Tila", which is the patron of the community, we are used to making a celebration. The mass, after the mass, finishing the mass, the redistribution of the food. In some occasions some fellows have given a cow. For the food. And other fellows bought tortilla, the stews, the cups, plates. The way to live together. The day of tomorrow, which is the day of the Señor. There where we gather in the church and for, in this way we celebrate the Tila. The celebration of us here in the community.
[D2]: One can see that you take your time for that. Because today the preparations are the whole day, isn´t it?
[P1]: Yes, today it is practically worked the whole day. And tomorrow, it is the whole day. In fact, the church, we have to stay from two in the night until then, the father arrives to start the

mass and leaves. The food and at two, three o´clock in the afternoon we already go home to relax in the house. This is one of the important things that exists, it is a custom of many years already.

The interview partner describes in detail the order of the festivities. It is a tradition that has been installed in the village around forty years ago, in the 1970s with the buying of the picture of what is also called *The Saint* in San Cristobal de las Casas (I_2015_C_4: 33; Field notebook/1/2015, 04.06.2015). The celebrations are perceived by the village members as "a custom of many years already" (I_2015_C_4: 28). In the interview sequence presented and throughout the whole interview the notion of sharing is a dominant discursive pattern linked to the celebrations of this religious festivity. It is explained in detail, how villagers gather before the festive season begins in order to discuss about the financial investments made by the different village members, to distribute specific tasks and to discuss the timetable. It is emphasised that everything happens "in harmony" (I_2015_C_4: 48). The event reproduces practices of sharing in the case study village and supports notions of collective identity. Considering the time in the year, when the celebration takes place allows the assumption, that the date of this religious festivity few weeks before the starting of the rainy season also connects with the occurrence of floods.

A range of local expressions obtained from the interviews in the main case study village allow an in-depth insight into the type of collective identity created and performed through discourse. The expressions are self-descriptions by the local villagers, referred to here as the *We are-phrases*. A list of sentences identified is presented in table 10. A considerable amount of expressions linked to the sentences mentioned in table 10 is presented in the first person plural. This underlines the assumption that identity is in large parts a collective process reproduced through social practices. This is specifically relevant for flood related practices. Preparation for a flood, prevention measures, experience during the flood and recovery after a flood are all processes expressed in the interviews as collective ones. The question "How do you prepare for the flood?", which was asked repeatedly by the researcher many times was answered with the expressions "We have to be ready", "We are already used to it" or "We prepare ourselves". In the light of the flood experience, general parts of the collective identity become visible. The expression "we are already used to it" is discussed as a main feature of flood related social practices above but here it is referred to again to assume that it expresses a more general perspective towards life. In the interviews, the expression is also used in order to describe other processes or conditions which people are used to. This is e.g. being used to the bad condition of the road and the cumbersome journey to places outside of the village or even to the journey in a motor boat in times of flood (I_2014_C_1: 22). People also express that they are so used to living in their village, that although they grew up in other villages and states and have lived outside the village for a long period of time, they are more at home in the village than elsewhere (I_2015_C_1: 307-310). The expression "we are forgotten about" used by different interview partners, refers to oblivion of government actors towards the village. It is linked by village member to flood situations in which there is hardly any assistance given to them by regional authorities but it is linked to other situations in which support from government is necessary.

Table 10: Expressions of collective identity identified in the empirical material – "We are"-phrases

Sentence	Code attributed or general topic	Source
We are self-sufficient	Autosuficiencia	Memo_I_150603
We are a family	[NameOfVillage]_descriptions	Interview 2015_C_11
We are equal	[NameOfVillage]_descriptions	Interview 2015_C_11
We suffer	Sufrir_WeSuffer	e.g. Interviews 2015_C_6, 2015_C_17, 2015_C_32
We are already used to it	Ya_estamos_acostumbrados	e.g. Interviews 2014_LB_1, 2014_C_7, 2015_C_1, 2015_C_20
We prepare ourselves	Discourse_PrepareForFlood; River_Usumacinta	2015_C_20
We have to be ready	Discourse_PrepareForFlood	Interview 2015_C_3
We are forgotten about (by government); We have the stain of obliv- ion	Olvido_gobierno	Interviews 2014_C_16, 2014_C_5, 2014_C_7
We are conformists	Somos_conformistas	Interview 2014_PR_4

As it is emphasised in the interview sequence below, people from one case study site especially emphasise that all houses built in the village have been built from the efforts of the village members, without support by government:

Interview 2014_C_5: Lines 69–78
[P1]: But we are forgotten here, we are forgotten about.
[D1]: By whom? Forgotten by government?
[P1]: By the government. He who has his house over here. Although we worked here, sowing millo, sowing chile and from this we got through because. And this of the harvest give to us, we invest it in 100, 100 blocks. And in the other year like that we go on buying another thing until we come to a certain point. And it goes.
[D1]: Yes.
[P1]: Like that? But from government? There is no help at all. It is not that this fellow, this fellow would have got his house because of the government. No, no.

As a flipside of the notion of oblivion from government, village members empha-sise the collective work they do in order to improve the living conditions. This con-trast between external oblivion and commitment of village members builds a strong pillar of collective identity. The notion of suffering that is highlighted at various points throughout the analysis in this chapter is another major feature of collective identity. It refers to experiences of suffering that date back to the oppression Mayan ancestors experienced (Field notebook 2/2015 5.07.2015) as well as to current suffering related to the flood but also to other dynamics (I_2015_C_17: 101–102; I_2015_C_1: 270). The argument that suffering is in fact part of the collective identity is underlined by the reference a female interview partner makes to a song which is commonly known by elder people in the region. Linked to the remembrance of the song is an explanation of the interview partner that the village members stand together and support each other when a misfortune happens:

Interview 2015_C_1: Lines 276–291
[P5]: I would not change. No, no.
[D2]: But how is the song? Somebody knows it so that you can sing it?
[P5]: No, it seems that it is like that: This is my people, which I would not leave for anything even if I suffered the same.
[P6]: Ah, yes.
[P5]: Even though sometimes, well, although sometimes we ruin and among us, right? But we have this that if to someone a misfortune happens. Well, me, my person, I cry for them. We can have such or such difference, eh, whatever it may be, but a misfortune that happens to a fellow, I feel it.
[P3]: It is cooperated.
[P5]: Yeeees.

The song referred to in this interview sequence is the song "Sin Fortuna" by Gerardo Reyes, a well-known Mexican singer (Annex 6). A strong notion of identity is expressed in the phrase "this is my people", which interview partners in the case study village refer to frequently. Finally, the expression "suffer the same" from the song is expressed by the village members as a regular experience during times of floods, which is part of their lives.

6.6 RETELLING AND RE-PERFORMING HISTORY

Building up on the legends and history of the case study villages mentioned in the previous section, in this section remembering and remaking of historical processes as part of social practices are analysed. History and histories are understood here as a means by which meaning is attributed to current processes and through which everyday life is interpreted. At the same time, a reflection about the past inherent in parts of the empirical material provides a basis for the analysis of different temporal dimensions involved in social practices of flood management. A selection of key historical processes in the villages and in the case study region is presented here. This allows the analysis of how history is being told by village members through certain discursive practice. Far from an attempt to reconstruct one "true" version of history or a chronological account, different pieces and patterns of the histories made and retold are presented.

Not much is known by the inhabitants of the case study villages about Mayan history in their region. However, the pieces found in the villages, like dolls made of clay and precious stones and the fact that some villagers have visited large archaeological sites stimulate reflection about Mayan people and history. The Mayan history of the case study villages is exemplified in an interview sequence presented below. As the co-researcher, who is a member of the Maya-Tseltal population participated in the interview, the presence of Mayan people in Mexico today was openly discussed about and actively informed by the co-researcher.

Interview 2015_C_10: Lines 146–184
[P1]: Yes. I could not tell if it was Mayan population or not. […] (anonymised) was, maybe they came like, well like now, here there is a Mayan ranch, and another Maya one, and another one. But maybe they came, well because as they left dolls. […]: Like all these people. Because they were stronger than us. Here in the ruins of Palenque. In the Palace of the laws, as they call it, they are so tall, every step and there they stepped up with loads, with stones to work there. They were more than we are. We one small stone, we cannot stand it. This people was very tall. When they discovered, what they worked in 49, they discovered many things. There are tombs, skeletons of the people. They were of this size. It was called cali, cali. This people was tall.   [D2]: Mhm, mhm.   [P1]: They were beautiful. But this is from another era yet, from another era. It is very old. I don´t know why this people disappeared, the Mayas. Why? Before the diluvium?   [D2]: Well here we remain.   [P1]: What?   [D2]: Here we remain the Mayas.   [P1]: Ahhh?   [D2]: Yes, there are the Ch´ol, the Tseltal, the Tsotsil, those from Tabasco there they are, I don´t remember but it´s the same, there are the peninsular Mayas. So, we remain. No, well there was a population before, which dispersed.   [P1]: How did they end?   [D2]: Well, they also migrated to other places.

A story is told of the travelling of Mayan people along the river Usumacinta until the Atlantic Ocean. The interviewee compares the Mayan people with today's inhabitants by saying "they were stronger than us" (I_2015_C_10: 157–158). Referring to the diluvium, the interview partner introduces a biblical event into the discussion and tries to locate the disappearance of the Mayan high culture along a chronology of events that are described in the bible. The co-researcher argues that Mayan people continue to live in the region and presents himself as one of them, thereby bringing in a new perspective, the one of the continuation of history today. The linkage of different systems of reference, the early cultures of the American continents and biblical event, can be seen as an attempt of local people to locate themselves in a historical line of ancestry. In another sequence of the interview, the interviewee links the question of his own origin with a fundamental question of human kind: "And here is the doubt. Are we from the monkey or are we from the soil?" (I_2015_C_10: 307–310). In this sequence the interviewee contrasts the theory of evolution with the theory of creation of man by god from soil. Furthermore, it represents the contrast between scientific explanations and predictions with other explanations and beliefs not only for questions of ancestry but also for current problems, including floods and other detrimental events. Another part of history that is repeatedly referred to in the interviews is the origin of the case study villages. Some accounts describe processes of exploitation that took place in the region during and after colonial times. One of these processes involved the extraction of the *tinto* tree in which parents or grandparents of today's residents were involved.

Interview 2015_C_10: Lines 97–112
[P1]: Well, among what my father told me that he still lived part of that. He cut a lot of tinto and took it out. It is used for a paint. Tinto gives a very beautiful paint. Who knows how many more they took out. Yes, they took out a lot of wood. [...] Further ahead I don't know which other woods. This tinto, it's what they cut into pieces over there. Who knows which other things, because they know how to work, they knew how to work. How many other things. They worked, they just knew how to work. It is some while ago.

The interview sequence presents the process of cutting the *tinto* tree and the qualities of the liquid of the tree as a dying agent. Linked to the work in the timber extraction is the expression of hard work mentioned repeatedly in the sequence, which points towards the experience the ancestors of today's residents made in this land. The history of the village members is also referred to in explicit references made to the last names of village members:

Interview 2014_C_14: Lines 419–438
[P1]: This, the family name came from Belize, they brought Mister [Name], they brought José as was my fathers' name, they brought him for the cutting of timber. [D1]: Yes. [P1]: That there was a lot, lot of forest here, everything was forest so in order to cut it he was like very able, who directed the workers and, and they brought him, he came with his wife [Name],

Irishwoman, of Irish origin. [...] And the girl was raised and, and later she found a man and so my father was born. But they didn't take the last name of the father but of the grandmother, my grandmother and so stayed my last name [Name] because if it wasn't, if the father had accepted, the last name [Name] would have ended there, [Name] wouldn't exist here. It is very close to but he was coloured the man he already was from the Antilles, there fro, from the Caribean, he already was black, he yet didn't come white [laughs]. He yet didn't come white that is why all we are like this, half, half black.

The sequence exemplifies how village members recount and create history referring to population dynamics during colonial times. This includes populations from Europe, from the Caribbean, which includes African ancestry as a consequence of slave trade, and from the local area of what today is Southern Mexico. Emphasising that ancestry can be an important aspect of identity, the passing on of last names that originate from colonial dynamics continue to influence performances of identity in the case study villages. In parts the last names are presented as a remainder of oppression and exploitation, while other village members express pride to have European ancestry.

Before becoming an *ejido*, the land where the village is located today comprised a cattle ranch and grazing land and was owned by people of the town Emiliano Zapata in the state of Tabasco. Different interview partners recount the history of the foundation of the village as a story of struggle for land of a few men standing together against large-scale land owning families (I_2014_C_14; I_2015_C_32; I_2015_C_33). In the interview sequence presented below, the former land owners are mentioned. The struggle for land took place in the time period of the Mexican Revolution, as direct reference is made to Emiliano Zapata, one of the main historical figures of the revolution and to Porfirio Díaz, who was president of the Republic of Mexico when the revolution started in 1910. The interview partner presents the foundation of the village as a history of organisation, fear and fight for land that his grandfather and father were involved in.

Interview 2014_C_14: Lines 45–73

[P1]: And the owner was the sister of [Name], [Name] but she got sick and upon dying he was the one who came, no? [...] Well he started to exploit it while he was looking for ways to improve it. Well, he had good politics this mister. He was federal senator of the Republic and there [...] Later, well, it started to come some people to open here for their workers. There were some who were called [Names] various people. Later there came others, other people the [Name] and yet those are family of us. Father of my woman. [Name] and finally, others and my father came as well, [Name] with my grandfather. Finally, the community already was quite big. And yes they gave themselves the idea that, that it had to be fought. So, this was Emiliano Zapata with Porfirio Díaz with this, yes, Emiliano Zapata. He, who procured the lands, fighting the lands for the farmers. And here there was a Municipal president of Zapata who was called [Name], he supported the farmers a lot. And they said to, to fight, to fight and everything. And they go to assure it, they win here, but it costed a grand case, the mister as he had money, he brought soldiers, and the people was afraid. Of course they were afraid. The mister said that they should leave <because if

not, I will put bullet.> <Well put it then, put it then.> And the president supported them. Very good and papers and office, before no, it is that they were farmers, no? Proletariat they called the farmers. [D1]: Proletariat. [P1]: to the worker like to us now. Proletariat. After that came the agrario. [D1]: It came the …? [P1]: The agrario, the agrario.

The interview partner mentions that a range of men (names left out in the English translation for reasons of anonymity) took up ideas of the Mexican revolution and decided to fight for land. A different interview partner mentions that "all those who started to struggle the lands were his workers" (I_2015_C_32: 15), while he refers directly to the former owner of the land, Don Ovidio Jasso whose family owned lands in Tabasco as well as in Chiapas. The conditions in which the workers of the large estate owners lived and worked were very rigid and violent (I_2015_C_32: 26–34). In the account of historic events, the language used by the interviewee switches from the past tense to the present tense and from the third person form to the first person form. It presents the words that might have been spoken by the main protagonists of the fight for land. This includes the landowner, who is said to have expressed a threat against those men fighting for land with the sentence "because if not, I will put bullet". This form of presenting the conflict through a verbal argu-mentation with the protagonists talking to each other represents a dramaturgical means which is part of a specific discursive practice performed by the interview partner. The people fighting for land in the early twentieth century are denominated by the interview partner as proletariat, while in the following sentence he mentions that workers like today's population in the village are also called proletariat. By attributing himself and the present generations living in the village with the same name of proletariat, the continuation of a class struggle is suggested. The inter-viewee makes reference to the agrarian reform that followed the Mexican Revolu-tion, by using the term *agrario*.

Interview 2014_C_14: Lines 85–90
[P1]: Troop of fighters, workers, roughnecks for the land and everything. Until then came, there was an engineer to measure. They measured all the lands of what was […] (anonymised). It was 600 hectares. [D1]: 600 hectares. [P1]: And they gave half, 300, to […] to the group and the other 300 they gave to this mister. To Don [Name].

The ending of the revolution is presented in the interview sequence above by the arrival of an engineer. As part of the agrarian reform, land surface was measured, thereby formalising land ownership along a metric system. Empirical research in the archives of the states of Chiapas and Tabasco as well as in the Office of Agrarian Administration in Palenque documents that the process of village formation and enlargement involved various steps involving the declaration of *ejido*, claims for extensions by the villagers and formal declaration for the extension of lands. Far from being a singular historical process, land dynamics through sales and redistributions are ongoing processes until the present day.

Retelling history has been identified as a valuable entry point into the different time scales encompassed in social practices. The remembering and the re-enactment of historical events and processes analysed in some examples in one of the case study villages allows for the assumption that discursive practice that involves history discourses is also an important part of social practices. Discourse about history involves the experiences of ancestors and the lessons learned from them. More fundamentally, history can involve the question of one´s origin or descent, whether in a geographic sense (Which place am I from? Which continent am I from?), in an ethnic sense (Who are my ancestors? Am I coloured or white, European or American?) and even in a religious or metaphysical sense (Why do I live? Am I a product of evolution or a creation by god?), just to name some questions identified in the analysis. For the case of social practices of flood management, the link to local histories of a region, a village and it´s people is found valuable, as it promotes the identification of key patterns of discourse about the past that influence both the perception of processes in the present and the continuation of practices performed in the past in the current contexts. Given the case that one of the villages was formed only in the 1950s, with a population that predominantly did not live at the side of the river Usumacinta and did not exist as a village community before, allows for a revised look at the flood management practices identified so far. Moreover, in the light of flood dynamics and increasing difficulty to predict the time and severity of flood exposure, the importance the land has to the people who have fought for it in the past has to be regarded again. The key historical process that continues as a part of everyday practices and that is directly linked with practices of flood risk management is the struggle for land.

6.7 FIGHTING FOR LAND AND THE IMPORTANCE OF "LA TIERRA"

Land and access to land is a major topic related to social practices of flood management in the case study region. As such it was one of the major aspects discussed by research partners from the field. "La Tierra", which is the Spanish expression for land, earth or soil is the primary resource needed for settlement and for economic activity in the village. Dispute for land is not only a historical process but an ongoing concern in the present and one of the major challenges for the village. The growing interest in land and the scarcity of land expresses in rising costs of land in the case study villages. While no exact price for a piece of land is given by interview partners it is mentioned that land is available only at a high price (I_2015_C_9: 93–94). Different attempts have been made by villagers to get access to new lands that are not affected by floods. However, as the lands offered or supplied by government did not meet the requirements of the villagers or processes of land acquisition were hindered by external actors, many villagers decided to stay. The following interview

sequence from one case study village emphasises the fact that long distances between the village centre, where the houses of people are located and agricultural
lands are evaluated as a negative factor in daily life.

Interview 2015_C_17: Lines 331–340
[P1]: That is why we stayed here. That is why some people say that we suffer because we want to. Because we had gone there. But what gave a change from one to the other. It was not for more than houses that one. It is very beautiful, because there stayed this family [name], there they have their land. Those did not sell it, they stayed there. And it is very beautiful the land. But they have to travel in order to see their animals. They have cattle there. They are always travelling.

Many people who had been offered land on higher grounds sold this land in order
to live close to their agricultural land. The interview partner presents the motivation
to continue to live in the village, which is the proximity to one's land and animals.
In a later part of the interview, he mentions what he did with the money from the
sale of the land: "I bought a *cayuco* like this, a motor as well. Mhm. In order to go
[…] I took advantage of my money" (I_2015_C_17: 361–364). The phrase "in order
to go" refers to the possibility, the small fishing boat and motor offers to be independent and travel on the water during the times of flood.

Different interview partners from the case study village hint towards a period
of time in the 1980s, when additional land was donated to them. As one interview
partner mentions, people regard the land donation as a "souvenir from this Juan
Sabines" (I_2015_C_17: 365–366). Juan Sabines Gutiérrez was a Mexican politi-
cian and native of Chiapas and had the position as governor of Chiapas between
1979 and 1982 (Gobierno del Estado de Chiapas 1982; INAFED 2010). The interview partner shares a memory how he and two male village fellows undertook action during a large flood in the region to ask the governor for help:

Interview 2015_C_17: Lines 368–402
[P1]: We are flooded and in this time they almost didn't help in nothing. For as much as we asked for food supply, no, no, no. They didn't help us. So we went to Tuxtla, me with [name], other mister whose name is [name], the three of us went. We were there like five days, asking for an audience. And they didn't give us the, they didn't give us the audience. We stayed there, day and night, this mister has a lot of work. And he went to take a rest, or maybe, because he asked to, or well […] Until I say to my fellow: <When this group enters, you go with them>. Ah well, and when a group came that had an audience, he entered and after they had talked to the governor, with this mister Juan Sabines, who in peace may rest, he talked to him. And understand that he was a little, well, his vocabulary was like with rudeness. <What happens, why don't you tell me that you are flooded? How humble you are there. And how long have you been here?> <We have been here for three days.> <Tomorrow I will see if I can come with you.> And I say, he said: <Yes, we are waiting.> And he came here. He was here with white trousers. And I gave him a cayuco so that he could get until the school building, the one that is destroyed. That was a part where we didn't want that he fell. <No, here I go.> And he went in until up. And he was here with us. It was that he, we asked him to give us this land. But he sent the engineer and said that

> it wasn´t seen, that we wanted it not for agriculture but to live. Yet he gave us this terrain that I
> say.

In the interview sequence, the interview partner switches to the present tense and to direct speech, when retelling a sequence of conversation from the past. In the conversation that is presented to have happened in the state capital, the governor´s words are expressed in the first person singular and in the present tense: "why didn´t you tell me that you are flooded?" (I_2015_C_17: 387-388). At the same time, the interview partner presents his own words in first person singular and in present tense, like in the sentence "Yes, we are waiting" (I_2015_C_17: 392). The text elements and grammatical features identified here are part of a discursive practice identified in another interview sequence above. The effect of direct speech and presentation of a conversation in a way similar to a piece of theatre, the impression is given that the interview partner remembers the event in the history of the village and as part of his biography as if it happened in the present. In the text it is emphasised that only through perseverance, patience and courageous action in the right moment, it had been possible to get through with their plea to the head of government, which resulted in the donation of lands to the village.

Current interactions with government officials also involve the topic of land. Various interview partners and conversation partners from participant observation mention that the mother of the current governor of Chiapas has recently bought lands in direct vicinity of one case study village. This land incorporates a *finca* and large lands for cattle grazing. While most conversation partners name the mother of the governor as the owner, one interview partner argues that the land is owned by the governor of Chiapas Manuel Velasco Coello but officially registered under the name of his mother to prevent conflict of interest with his political office (I_2014_C_15). Another interview partner reports that around one hundred villagers had formed a group and argued for their interest in the acquisition of this land and *finca*, but that due to a "fraude" (=fraud) they had not been given the chance to acquire the lands (I_2014_C_5: 85-100). The question asked by the researcher, why this land had been acquired by the family of the governor of Chiapas, resulted in scarce answers but it is emphasised that the *finca* was of high economic value and well-known for the high quality cattle that is reared there (I_2015_C_17: 568-569). Moreover, conversation partners argue that large parts of the adjacent land are not affected by floods and that former owners of the *finca* had rented lands to the village members to put their animals on during times of floods but that since the owners changed, they have not been given this opportunity (I_2015_C_6: 115-116). In the interview sequence presented below, the land acquisitions and subsequent changes in the natural environment are described:

> Interview 2015_C_17: Lines 565-569 & 581-591
>
> [P1]: The governor is who they say is the owner of this finca. Finca Nueva Esperanza, very famous. Because they raised here, it was purely the finest cattle. Only beautiful cattle. As they were rich people, who raised only finest cattle here. [...] But the problem that was there that the

> governor already knows, maybe he is not aware but he already knows. This, there was a pajarral.
> A pajarall where so many birds of a certain class slept. And they fell these trees. So the birds went
> to flee. And I don't know who indicted them but they chased these birds. These wild birds entered.
> There is somebody who takes care of them, as well.

In the interview sequence a strong emphasis is put on the birds and the fact that they
fled from the region after the felling of the trees. The interview partner repeats that
"it was them", referring to the new owners of the *finca*. The destruction of the natural habitat is expected by village members to stimulate negative changes in the
lagoon and river ecosystem. Land as a major resource fought for in the past and in
the present, is still the main basis for livelihood and freedom in the case study region. The struggle for land and the close observation of changes made concerning
flora and fauna, reveals the importance those practices have in life of the villagers
in general and specifically in flood seasons. As interview data presented above indicates, the social practices of flood management in the villages involve a range of
activities that protect and care for animals. These activities are partly linked to access of dry land:

> Interview 2015_C_6: Lines 107–122
>
> [D1]: And are there no other ways how you can prepare?
>
> [P2]: We, we have a communal terrain as we call it us here, where the ejidatarios can come to
> pasture our cattle. But we cannot put more than ten heads of animals per person and what remains,
> well, you have to rent a ranch over there then. Before those from there they rented, which is now
> from the government. Yes, they gave us pasture there. But today, well, it is owned by a man who
> [...].

Ejido members own communal land which serves as the place of rescue for cattle.
The provision of land as a strategic resource can be identified as a long-term practice formally installed and practically organised in group by all *ejido* members in
the village. The organisation of communal land and the collective organisation of
the village as an *ejido* are two strategies that foster a flexible flood management and
at the same time protect legitimate access to land. The identification of a direct
interconnection between flood management and access to contested land is relevant
for challenges on the local as well as on other levels. Struggles for land which are
part of the history of the region and of people's identity continue in the present and
the threat of losing land is tangible; thus, there is a rejection of ideas to leave the
villages through formal resettlement schemes to places where the flood does not
have effects. Ideas to resettle the people from the case study villages have been
promoted repeatedly by government officials. While in the case study villages in
the Catazajá municipality, resettlements have been carried out at a small scale for
few families, the case study village in Palenque municipality has not experienced
resettlement. Resettlement as a measure to prevent people from getting flooded is
an idea promoted increasingly by government authorities and civil protection in the
case study region in recent years. The discourses of different actor groups related

to resettlement ideas are analysed in chapter 8. In the empirical material gathered in the village, resettlement is a controversial topic. Some interview partners emphasise that the donation of land following three years of severe floods in the 1980s would have been an option for resettlement of the village (I_2015_C_32: 513). At the same time however villagers emphasise mistrust in government (I_2014_C_5), negative experiences with people from the villages surrounding the donated lands (I_2015_C_32: 542) or the general neglect of the idea to live in any other place then the own village (I_2015_C_1: 253–254). Familiar links to the place and the fact that people are used to certain conditions of living prevent openness towards living in other places. According to one interview partner, people would miss many things if they were resettled, they would even miss the flood (I_2014_C_14: 300–305). Similar expressions can be found in the other case study villages (I_2014_PR_1: 26; I_2014_LB_1: 9; I_2014_PTA_1: 5–6).

Besides formal schemes or informal processes of resettlement that are discussed for and in the case study regions, a range of other current processes create new challenges and possible territorial conflicts in the future. One of these processes are current plans of the Mexican government, to construct a hydro-energy system for the production of electricity on the river Usumacinta. While possible resettlement schemes for the hydro-energy system would predominantly concern settlements upstream of the location Boca del Cerro, other direct consequences of the construction of dams or small hydro-energy systems could affect downstream settlements. The possible effects and personal opinions people from the case study villages express towards the ideas of the construction of a dam are presented in various interview sequences. In general, the topic of a possible hydro-energy project to be built on the river was never brought up by the interview or conversation partners during field research. As it had been an initial research interest of the researcher, questions about whether people had heard about plans and what was their opinion about it were asked repeatedly in the first phase of empirical research. However, the hydro-energy project never evolved as a major topic brought up by the research partners from the case study villages. In the interviews and observations from 2014, people mention that they have heard about the plans to construct a dam on the river Usumacinta but add that they have not heard any new information since two years or longer (I_2014_C_4: 14) or evaluate the information they received as "only rumours" (I_2014_C_8: 5). Repeatedly, events that occurred on the river Grijalva in 2007 are named as a possible scenario for the Usumacinta River in case of a dam construction. One interview partner mentions in this regard that a dam "means danger, because [we] saw the case from 2007 in Villahermosa" (I_2014_C_5: 13–14). Another interview partner evaluates a dam as a disadvantage for his village and describes a clear scenario that could involve the "danger of flood, when it rains much and the doors of the dam open fast" (I_2014_C_8: 6–7). In 2015 information about current activities towards the start of construction work in Boca del Cerro were mentioned by an interview partner village member who argues that "it will harm us. In the agriculture. Fisheries as well" (I_2015_C_36: 7).

6.8 INTERACTING WITH GOVERNMENT

Among the empirical examples analysed above various accounts point towards another key topic relevant in flood management and in social life of people in the case
study region. Interaction with government is a topic through which specific ways
of production and perpetuation of conditions of inequality can be identified. In general, interaction with government takes place in a wide range of ways and results in
specific forms of discourse and social practice. There is variation in the personal
experience individuals have with government. Notwithstanding, the empirical material obtained also points towards a collective experience with government. This
collective experience is linked to what is described above as a collective identity.
The empirical material presented here permits to establish a link between experience with government and practices of flood management.

Interview sequences show that the relationship with government actors including political candidates is characterised as ambivalent. While promises are believed
and sought after, at the same time people remember false promises and denial experienced in the past. Key areas of interaction with government and an evaluation
of this interaction by village members are displayed in table 11.

Table 11: Analysis of verbal expressions concerning government interaction

Topic and quote	Source
Land extensions "But yes, we had **options** that they told us through the agrarian reform."	Interview 2015_C_9: 31–32
Donation of land "But we had this **souvenir** from this Juan Sabines."	Interview 2015_C_17: 365f
Support by DIF regional "And yes, **she has not left us alone**. And here we go, living, we will see what else comes with the government that supports us with the road."	Interview 2015_C_17: 633ff
Support with the construction of the road by municipal and state govern- ment "But candidates who have come here to say that yes. That **they will sup- port**. They think they can do it. It´s that, they were in a committee, that if they win, they will go with us, to talk with the governor, that this place which is close to his finca and he **doesn´t support** with the road. Be- cause the presidency cannot, with this work. It is much money. Now the state government, this one yes."	Interview 2015_C_17: 637ff
Promises made by president of DIF Chiapas "It was the mother of the governor. Ah, **she promised many things** there. She wanted to support with the road, these things. **But the prob- lem** [...] and since then **she has become a little distant**, distant."	Interview 2015_C_17: 579ff
Construction of houses "Because here you **cannot find any house** which would have been given to us by government." "**Government does not give**."	Interview 2015_C_32: 776– 777

Different support programmes by government "And the luck that **government, well, it supports us**. With 67 and older, they give us 1150 Pesos, with this **we persevere**. [...] And this with the cattle, which is called Progran, one is Procampo and another one is Progran."	Interview 2015_C_17: 435ff
Specific types of interaction with government during floods	
Distribution of aid by government actors "Sometimes even the **municipal president has come**. And they them- selves come. I have never taken care of this because you feel **unsatis- fied**. They want that even the people who live in Zapata, but those who are from here, what is given to them? [...] Well, the people who don´t suffer nothing, who are there, **it is unjust** that they are given."	Interview 2015_C_17: 670– 673 & 676–677
Local civil protection committee "But in the hour yet when the flood comes, imagine nothing else, **they don´t look for the civil protection committee**. [...] For anything they do or leave they look for the Comisario Ejidal or the Agente Municipal, to those from civil protection well **they come to see as third**.	Interview 2015_C_32: 188– 190 & 185–186

Among the topics referring to interaction are questions of land extension and the construction of physical infrastructure such as the main road and houses. As these topics are crucial for the village members, evaluation of interaction with government in these affairs is highly ambivalent and emotional. Less critical is the topic of government programmes for the support of specific groups of population, which is evaluated positively. However, the interaction with government in the case of flood management again is a highly controversial topic. As can be seen from one interview sequence, the installation of a local committee for civil protection has not resulted in the sharing of power in the village in times of floods. In fact, the committee stands behind more traditional formal authorities in the village. The distribution of food supply during floods is identified to be a critical activity as it has created conflict in the past. Therefore, the process is formalised as much as possible and carried out by the formal staff of civil protection, not by village members or authorities (I_2015_C_17: 670–671). The perceptions of government hierarchy and the importance of interaction with actors on the different levels is highlighted in the following interview sequence:

Interview 2015_C_6: Lines 130–144
[P2]: I think of this, the president does not support us. In the president is everything. He told me here once in the government secretariat, that he would go to Tuxtla. That all this would go with the president. If the president does not support it, there is nothing. [...] [P1]: Voice, voice and vote to give faith that, that, that, that what is asked for, would be reality. [P2]: Yes, exactly. [P1]: They, they are the indicated for ... He is like a judge who says: I liberate you. And why? Because I consider that there is no guilt. The same are these municipal presidents. All of those who are bosses. They say: This one goes, this one not.

The interview partner describes that without the support of the municipal president, the demands made by villages to the political leaders of the state are not heard. The interview partner compares the role of a municipal president to the role of a judge. The sentence "I liberate you" can be seen as the representation of what the interview partner expects from a good municipal president. However, the fact that this person has the power to decide about freedom or captivity ("this one goes, this one not") is not approved for by the interview partner. The ambivalence towards government action and emotional parts of evaluation can be identified more clearly in the following interview sequence:

Interview 2015_C_6: Lines 50–54 & 64–72
[P1]: To tell the truth, what the government does, that one day it would look to, it would put its, the gaze towards here. With a dam and a street like it is in Jobál, we would not go to "Pique". [...] There come the, these, there come the councils and with the hope that they would support us and they don´t give us anything, the least. They don´t give us anything. What do they come to bring? A small food supply equivalent to 100 pesos or 150 pesos. Already they say: "[...] (anonymised) was benefitted with how many thousands of pesos". And this is a lie. This is a lie.

In the first part of the sequence, the notion of hope for attention by government is dominant. By the expression "one day" and "put its gaze", the interview partner makes use of imprecise formulations. The expressions represent notions of hope but at the same time notions of improbability. What people hope for is expressed by the interview partner in his demand for a wall structure that would act as a dam for the river and prevent the road and the fields from getting flooded. By making reference to another village, he points towards the fact that other villages receive support. In this case, Jobál is named as a neighbouring village on the other side of the river Usumacinta, which belongs to the state of Tabasco. The assumption identified here is that there is an unequal support to flood-exposed villages in Chiapas and in Tabasco.

Besides reception of external aid, interaction with government takes place in a range of other ways. Some village members emphasise their individual initiatives to approach government in order to receive basic services and support in times of crisis. This active demand for support is described as a key activity in the historical formation and development of one of the case study villages (I_2014_C_10: 222-225). Membership in political parties is described repeatedly as a pattern that produces ruptures in the social cohesion of villages. Conflicts arising from different political opinions are felt most strongly during times of political campaigns. Observations from the field in 2015 reveal that both women and men are active in the campaigns of political candidates. The decision for a political party divides groups of village members and provokes fierce conflicts within families and larger social groups (Field notebook 1/2015 21.05.2015; I_2015_C_34: 13–14). As can be seen in the following interview sequence, some village members interact with government representatives in relationships of friendship:

> Interview 2014_C_10: Lines 226–229, 234–237 & 246–248
>
> [P1]: Because one looks there for a friend, a political friend of the good ones. Isn´t it? And there I had a friend who worked in, in which office did this mister work? Ah! He was delegate of programme and budget. I went to Tuxtla looking for a credit to buy some land […]. He was a friend of ours. We went there every now and then, yet, yet we knew him a lot [laughs] […] But well, it didn´t proceed to more. It didn´t proceed to more. […] I have, I have political friends there, which I would like to visit. I think one of the closest to government for candidate of the municipal government of Palenque is called […] (anonymised). A good friend, a good friend.

The interview partner distinguishes between good and bad politicians. In going to the centres of political power, in this case the state capital Tuxtla Guttiérez, he has in the past been successful in making friends in government positions. By saying "it didn´t proceed to more" he argues that in some cases a friendship does not provide permanent support. While people in the village actively seek to gain support by government through approaching people at high levels, through participation in political campaigns and through making friendships, the outcome and success of this interaction for the village or for personal aspirations is unclear.

The analysis of a larger sequence of sentences in its context allows insight into the links village members establish towards different political and government actors from outside. Furthermore, a set of explanations for current situations of violence and inequality in the villages are identified, which link to larger political conditions in Mexico. The following interview sequence is analysed in detail to identify key patterns in discursive practice and the representations of social practices.

> Interview 2015_C_10: Lines 372–416
>
> [D2]: Because something happened or something occurred?
>
> [P1]: Because of what one can see.
>
> [D2]: Mhm.
>
> [P1]: One can see. Everything that you can see. You can see and hear so many things here. It´s like three years ago, two years, the shooting there down of […] (anonymised). There like at two or three kilometres. A gang of narcos and a gang of government. The shooting happened.
>
> [D1]: Mhm.
>
> [P1]: But strong performance. Why? For what we are talking about, still of this brigade. Well-armed. They fired bullets. Many people died.
>
> [D1]: This is how long ago?
>
> [P1]: It´s like four years ago.
>
> [D1]: Four years.
>
> [P1]: Aha. The soldiers passed by, the government. And they clash there with the bad guys. Them, they were well, well-armed. And this was a shooting. And now, this government, it is like one week ago, one week [laughter], two parties. There they come, the last house from here. The Green with the PRI. I was very disturbed. If I had […] they would have killed someone. They cut him here [indicated on his body], here, to one from the Green. And up there, in the entrance to the highway there were the people like this. A PRI-ista. He was waiting for the Green that they all were killing each other.

[D1]: Why?

[P1]: This is where I go. This is where I go. But apart from this personal, this people from, this people from, foreigners. Bad people as I said. They were there, giving five thousand pesos to each of them, they arm them, and let's do this work. This is where I go. There is the pickup, from the Green.

[D2]: Ah, they are still there?

[P1]: Hee?

[D2]: They are still here with the pickup from...

[P1]: Yes, from PRI or from the Green. The PRI, they were about to take everything from them. This is already type of war.

[D2]: Yes.

[P1]: Mexico is in war. Mexico is in war. A confrontation like this, and it is already hard, it is already hard. This is war, we are in war.

The main topic of the interview sequence is conflict brought into the village from outside by external actors or taking place in near vicinity of the case study villages. The interview partner emphasises that over the years he has closely observed many incidences of violent conflict brought from outside into the village. In the description of a shooting that happened near the village some years ago, the group of *narcos*, *narco* being the local word for drug trafficker, are called a "gang" ("banda" in the original Spanish text) and the group of government is called a "gang" as well. Two actor groups which formally are very different are here both described by the same term, which has a negative connotation. Moreover, both groups are described as "well-armed" and violent. A more recent violent conflict in the vicinity of the village involved two political parties. Members of the *Partido Revolucionario Institucional* (PRI) had a confrontation with members of the *Partido Verde*, the Mexican Green Party. As a consequence, village members were interrogated by police, while the interview partner emphasises that "to all the village members from here they put question and question" (I_2015_C_10: 433–435). Another group of people is described as "this people from, this people from, foreigners". While it is not directly said, where the people he describes are from, the fact that they are described as foreigners makes it possible to assume that the interview partner refers to people from Central American states like El Salvador, Honduras and Guatemala. These countries are known for the high amounts of illegal migrants passing the border to Mexico through three different main routes, one of them passing by Tenosique and Palenque (Isacson, Meyer & Morales 2014: 13). While some of the migrants engage in drug and human trafficking in Mexico, a large part continues the route of migration heading towards the USA. Drug traffickers exert strong influence on the political and economic conditions in the case study region and repeatedly come into violent conflicts with law enforcement agents. However, they also interact and build up social as well as economic relations with people in town and villages in the region (Field notebook 1/2015 25.05.2015). The interview partner embeds local conditions of violence and conflict brought to the village by external actors in a more general condition to be found in the Mexican republic. This link is made in a discursive pattern trough the description of various local cases of conflict, calling them "war" and then mentioning that "Mexico is in war" and "we are in war".

As the analysis of empirical material shows, interaction with government is a highly ambiguous topic in the case study villages. It includes descriptions of inequality in interaction patterns as well as a range of emotional expressions of disapproval towards government action or government neglect. At the same time, people engage actively with individual government representatives and hope for future support. More generally, interaction with government is described to take place in conditions of war, whereas the country as whole as well as the region are described as involved in this war. The social practices of flood management have to be reviewed in light of this perceived condition. Conditions of social inequality, violent conflict and ambiguous interests of government individuals and political parties are relevant contexts for flood management in the case study region.

This chapter has revealed a large range of social practices of flood management performed by members of the case study villages. Discourse-oriented analysis of interviews and other texts has allowed the identification of some of the key social practices of flood management. Moreover, some social practices have been identified that are closely linked and interact with flood management practices in various ways. Among them are practices of performing collective identities, retelling and re-performing history, and social practices of a continuous fight for land. It has been possible to identify some of the bodily-material notions of social practices through the analysis of empirical material, which supports the conceptual outline of the social practice theory applied in this study. In order to emphasise the importance of the bodily-material aspects of flood management and the specific nexus of bodily-performed and mental activities involved in social practices, a selection of specific social practices related to flood management is analysed in the following chapter. This allows a description of some of the key dynamics of preparation and anticipation of floods and how they are performed in and reproduce space.

7 BODILY-MATERIAL NOTIONS IN FLOOD RELATED PRACTICES OF VILLAGE INHABITANTS

Living with flood events in villages of the lower Usumacinta River involves a large range of social practices performed by different groups of people at different points throughout a year. In the expressions and performances identified in the field it shows that local people understand and conceptualise floods not as singular events but as phenomena within a continuum of processes. Taking a detailed look into the performances, the bodily performed and mentally mediated *doings and sayings*, enhances our understanding of floods and the larger social dynamics in which they are embedded. Among the social practices some *doings and sayings* stand out in their material relevance and in their bodily aspects observable in physical space. This chapter highlights the bodily-material notions of selected social practices of flood management identified in the case study villages. By looking into a set of visual data gained in a participatory manner during empirical research, some bodily-material aspects of social practices can be identified and analysed. This empirical approach both enriches and transcends discourse-oriented analysis of social practices. A novel approach towards the material levels of social practices is chosen, which includes different perspectives of *life-worlds* and the epistemological bases on which to grasp these perspectives. The documentary method described in chapter 5 is applied in the analysis of visual data and allows gaining insight into specific practices of living with the flood, especially practices in the anticipation of and preparation for floods. This chapter presents various photographs taken by village members and selected sets of frames from video sequences. The analysis of the practices identified supports the empirical and conceptual basis of describing a complex pattern of social and spatial dynamics. Discussing these dynamics is regarded an indispensable step within a critical reflection on flood management in the case study region.

7.1 MATERIAL REALITIES OF LIVING WITH FLOODS

The social practices of living with the flood involve a large range of *doings and sayings* that reflect on the material levels of the *life-worlds* of local people in the case study region. In the analysis of material and bodily-material entities that form part of social practices specific patterns can be identified that provide access to a detailed understanding of the social practices of living with the flood. Key topics and concepts which reflect in the visual empirical material are identified and discussed. They are summarised in table 13 in the following section. Among others, topics and concepts identified as key features on the material level of living with the flood comprise changing social and spatial proximities, social practices of anticipation, creation, protection and reconfiguration of space through practices of self-determination and water spaces vs. land spaces. Besides the identification of these topics, the process of identification of material notions of the practices allows the resketching of those processes in which social practices are formed. Moreover, it is assumed here that it is possible to gain insight into the processes by which social practices endure over space and time and how they can change. A central result of analysis is the proposition that processes of social transformation in the area of flood management in the past can be traced in material objects and bodily-material performances. As discussed further below, it may even be possible to anticipate and identify initial steps of social transformation in ongoing patterns of *doings and sayings*.

In this chapter, selected photographs are presented which were produced in a three-day photography workshop carried out with fifteen students in sixth grade of secondary school in one of the case study villages in 2014. The composition of the workshop group and context within the research design is described in in chapter 5. From all photographs produced in the workshop (191), it was decided to select seven photographs – one from each group of students – for interpretation. The interpretation of photographs with the documentary method comprises a structured and detailed process involving various steps of description and interpretation (chapter 5). These steps are divided into two larger processes, the *formulating interpretation* and the *reflecting interpretation* (Bohnsack 2009: 960). The *formulating interpretation* follows a highly structured procedure of formal analysis. This step of analysis provides a description of pictures in form of text which is needed in order to carry out the second step of analysis. The second part of interpretation process is called *reflecting interpretation*. This process is split into various predefined steps, while each step of interpretation is documented and carried out in a transparent manner. In the following paragraphs, the seven photographs are analysed and interpreted. As a complete presentation of the comprehensive analysis following the two steps of the documentary method would conflict with readability and overview in this chapter, the documentation of the process of analysis is added in the annex. In this chapter, the complete process of the documentary method is exemplified for one photograph. For the remaining six photographs, only the interpretation with reference to social practices resulting from the analysis is presented and the full analysis is added in annex 5.

7.1.1 Changes of distance and proximity

Table 12: Analysis with documentary method (photograph 1)

1. Formulating interpretation (following Panofsky) WHAT can be seen on the photo?
1.1 Pre-iconographic interpretation   A muddy pathway surrounded by green grass and vegetation. The mud path is very wet, has some small holes filled with water. The path leads towards a house made out of stone in the upper part of the picture/in the background of the picture. Besides the house there is a palm tree.   **1.2 Iconographic interpretation**   The picture was taken in a tropical area. This can be recognised in the dense vegetation and the palm trees. The one-storey house made out of bricks (concrete) and with an aluminium roof gives hints that the people who live here do not have a lot of money. However, they are also not poor people, because these would not be able to afford aluminium roof and concrete, but would live in a house made out of wood or bamboo.
2. Reflecting interpretation
2.1 Iconic interpretation   **2.1.1 Planimetric composition**   The main line in the picture goes from the right front of the picture to the centre and then turns right. It points towards the house in the upper middle part of the picture. The centres of attention are the house and the water holes in the mud.   **2.1.2 Perspective projection**   Long-shot. Picture taken from a standing position. The view is directed slightly to the ground. The focus of the picture lies on the water hole in the pathway in the centre of the photography.   **2.1.3 Scenic choreography**   The picture is parted by the pathway into a left part of the picture and a right part. Both parts are characterised dominantly by green vegetation. In the left part there is more grass in the foreground of the picture, on the right side there is more grass in the centre of the picture with some shrubs towards the right side of the picture. The house is in the background of the picture but located at the middle of the upper part of the picture. A palm tree stands at the left side of the house. Other trees are found to the left and the right of the house, whose tops are beyond the picture. Over the house the sky can be seen and it is largely covered with clouds.   **2.1.4 Relation of sharpness/blurriness**   The foreground of the picture is sharp, especially the green grass on the lower left part of the photography and the earth holes filled with water in the right front and in the centre of the picture. The left and the right parts of the picture are blurry. The house and the palm tree and the sky are blurry too. The brightest parts of the picture are the sky and the house as well as the water holes,

in which the white of the clouds reflects. The upper right and upper left part of the picture are in dark green, due to tree vegetation and the shadow it casts on the grass. The foreground and centre grass is in brighter green.

2.2 Iconological-iconic interpretation (Identification of habitus/meaning)
The picture is dark as the sky is covered with clouds. The mud and the water are very important topics in the picture. The house is an important topic in the picture, too. One imagines that the producer of the picture shows the activity of walking towards the house and having to pass the mud. The house is quite far away from the producer of the picture, he/she has to walk the muddy path, but he/she wants to reach that house It is a distant position and at the same time a challenging position, as the way clearly leads towards the house but it is not an easy way to walk.

As a central topic of the photograph (see figure 22 and figure 23, colour plate pp. 144–145) it is possible to identify the distance towards a destination that is sought after. The way of production of the picture allows the assumption that the muddy pathway, which is a general asset to be found in the case study region, creates an obstacle on the way towards the house. Walking along muddy pathways is however part of everyday action in times when rainy season starts. People in the case study region have to travel large distances by foot or in vehicles in order to reach their villages. In times of the rainy season, felt distances get larger because the conditions and ways of transportation change in so far that it takes more time, more money and more preparation in order to get to and from the villages. Even the way to houses inside the village, one´s own home or the house of relatives or friends, becomes more difficult and is felt more distant than during times of dry season.

A question the photograph and empirical research in general raises is if during times of the rainy season and the flood, social ties become stronger and more stable or if they become more lose. Data obtained from interviews assumes that while the relationships in the larger social groups of a village become more lose, people focus strongly on mutual assistance in the core families. Various interview partners emphasise that people don´t support each other during times of flood and that everybody takes care of his or her own family (I_2014_C_3: 44, I_2014_C_5: 12). However, some examples of collaboration in the care for animals and compassion beyond family boundaries are given (I_2015_C_7: 97–98, I_2015_C_1: 286). The photograph demands reflecting on the links between the social construction of space which is guided partly by perceptions of distance and proximity on the social level and material aspects which are made use of in the construction of space. The producer of the photograph creates an impression of distance and obstacles in the material world, represented in the pathway filled with water. Distance however it is argued here does not involve the material world alone but also the social world. Social practices in times of the rainy season are performed in a way that core social relations in families grow in importance. This includes the provision of food for family members, visits to others´ houses organised through transport on horses that can pass muddy pathways and on small fishing boats in times of severe flooding. Moreover, family members whose houses get flooded move to relatives in unaf-

fected parts of the villages. It can be argued thus that in times of flood, social proximity is performed an underpinned increasingly through spatial proximity. Reconfiguration of social practices includes therefore a reconfiguration of spatiality. This
reconfiguration is in general a slow process which does not occur from one day to
another but which develops at the pace of the rising of the water level.

7.1.2 Changing dynamics on the river and anticipation of floods

In figure 24 (see colour plate p. 145), the main topic is the arrival in a small motor
boat on the river Usumacinta to the village. As it can be seen in the lower right part
of the water body, the river Usumacinta, some shrubs stand inside the water. It indicates that the photograph is taken in a period of the beginning flood season. The
main focus in this picture is put on the driver of the boat who looks into the water.
His action represents a key social practice in the case study villages: The close observation of the water. As empirical research in this case study village as well as in
other villages has shown, observation of water levels and climatic conditions are
performed and discussed about constantly throughout the year. In times of floods
however, observation and exchange of information within and beyond villages becomes more frequent. It is not only the visual appearance of the level of the water
but also the colour, movement and it´s sound, which is observed by local people.
Not everyone in the village is knowledgeable in the same way about the characteristics of the water. It is mainly the fishermen, the farmers and the drivers of the
motor boats, who can recognise the characteristics of the water most easily. However, constant communication about these characteristics within the village gives
access to updated information on the stage of a flood or the development of weather
to almost all village members. In the photograph, the water level is far from unusual,
which is reflected in the relaxed bodily expressions of the people on the photograph,
especially the women. They sit and wait for guidance by the male boat driver. The
photograph is taken from a rather distant position from the people. Another aspect
of relevance in this photograph is that the shady ground which lies in the front of
the photograph makes almost for one third of the picture. At first sight it looks like
empty, unused space. No details can be recognised, only few spots of light are shed
on the ground, which might indicate a pathway towards the water line. It can be
assumed that this ground is however not empty space but that it represents a part of
the village territory which normally gets flooded first in times of rising water levels.
By representing this part of space, the producer of this photograph anticipates a
situation of flood which is waited for by the people in the village. The flood interacts with the material lifeworld of the village in a way that a specific spatiality is
created. This spatiality is created however not by the water and the land alone but
in direct interaction with the people. While in this photograph the boat is relatively
distant, in times of floods, which is anticipated here, the boat is a central object to
be seen and to be used by all villagers. In this process of taking a photograph the
future situation of the flood is anticipated. This activity goes in line with the general

social practices of anticipation to be found in the villages. These can be identified e.g. in discursive practices analysed above (chapter 6). People express that they are used to the flood and that they wait for it every year (I_2015_C_20: 9–10). The preparation that takes place on a mental level has been identified as a crucial part of social practice. However, social practices of anticipation are also represented in a range of other activities which involve the material world, as identified in the example of the fridge and the *cayuco* discussed in chapter 6 and in the example of the water level and transportation on the boat in the photograph discussed here. As a key pattern within the social practices of flood management in the case study villages, the social practices of anticipation are discussed in the description of the *riskscapes* of flood management in chapter 9.

7.1.3 Protecting livelihood

The main topic of the photograph (figure 25, colour plate p. 146) is cow raising. It is represented in the picture as a main social practice for the generation of livelihood in the villages. Cows are specifically precious and can only be afforded by people with comparably high financial resources. The reproduction of cattle takes place in the village and is important in order to provide more livelihood options in the future. Slaughtering of a cow and eating of the meat only takes place at special occasions in the villages, but cows are raised and sold to large ranches and cattle owners from Chiapas and Tabasco. Cows can be identified as a type of insurance or saving in the case study region. They are sold in times of crisis, when loss of property due to a flood happens or when a personal tragedy occurs that makes the access to financial means an immediate necessity. Cows and floods are closely linked in people's social life because the animals have to be given specific care and attention in times of floods (see also chapter 6). The animals in the photograph have a lot of grazing ground and space to live. This abundance of land however is reduced strongly in times of floods because cows need to be transported to lands which are dry, either communal lands or rented territory. Therefore, adequate raising and care for these animals requires access to land which is a critical issue in the case study villages (see also chapter 6).

In the photograph, the producer of the picture and the animals are separated by a fence. This fence serves various functions. It keeps the cows inside the land, it keeps strangers out of the land and away from the animals and it demarcates spatial boundaries. In the discourse analysis presented above the importance to define borders and ownership of land has been identified a crucial pattern in history and present dynamics in the case study region. In the photograph it is recognised that the demarcation of landownership is broken down into a material practice of fence-making on the local level. It can be assumed that the experience with the fight for land and the need to protect ejido land against external actors has over the course of time created the need to also clearly demarcate the land of one village member from the land of another village member. In the photograph it is questionable, how much the fence serves the purpose of physical protection of the land from access of outsiders. Nonetheless, it´s symbolic force has to be emphasised. The photograph presented here points towards a more general assumption: The creation and recon-figuration of space through symbolic and material boundaries is part of the social practices of livelihood creation and of flood management in the case study region. While a fence excludes certain human and non-human beings from intrusion, at the same time it includes other humans and non-humans and aligns with the creation of a collective.

7.1.4 Finding one´s way in the social network of the village

The main topic of the picture (figure 26, colour plate p. 146) is the youth group walking together. The young generation of the case study village is represented with dynamic characteristics of movement, joy and communication. It is relevant to note that all people on the picture walk into one common direction and that they are all on the same muddy pathway. The ways in which they perform the walk are however quite different. They walk in a common direction and all of them wear the same school T-Shirt. They are classmates of the final grade of secondary school, so they might share the goal to finish school successfully. However, the different characters and backgrounds of the different people result in the fact that the ways in which they perform to approach the common goal are slightly different. Some of the youths are very active and find joy in the common activity which they express bod-ily in the act of jumping. This bodily performance can be interpreted as representing a positive orientation towards the future. Others walk their way more carefully and look to the ground, concentrating on the way or thinking about other things. Still others walk into the common direction and turn around their heads towards the girl jumping, which can be interpreted as curiosity, taking notice and caring for the other girl. As the youths have chosen to walk on the muddy pathway the impression is created that they are used to walking on muddy grounds. It is part of the social life in the village to be walking these roads, in whatever condition they are found. More-over, the practice of walking together is perceived as a joyful moment. Interestingly, all the male youths walk relatively close to each other in one line on the right part of the picture. The female youths are much more spatially separated along the way. While the first two girls have already passed by the main water hole in the middle of the pathway and seem to have walked around it on the left side of the pathway, the two girls in the centre of the picture are still in the process of taking the left way or choosing their way. The separation of boys and girls in this picture is only of temporal nature, as after the separating water hole in the pathway, girls and boys walk together again. The photograph is created in a way that emphasises the deci-sion of making one´s own way, of belonging to a group and at the same time being individual. Processes of decision making and of finding one´s own way and identity are crucial topics for any young man or woman at this age of around 15 to 18 years. In the case study villages finding one´s own way also expresses in the question

whether to leave the village after finishing secondary school and continuing education or finding a job in another place or to stay in the village and work in agriculture
and start a family. The spatial separation and different performances along the way
which are represented by the different youths on the photograph are assumed here
to indicate a decision about a change in the social and spatial setting of the village
and each one's own life in the near future. As the person who took this picture is
located on the pathway as well but at some distance to the other youths, it is not
clear which way he or she chooses. The person taking the picture might be in a
situation where he or she is aware that he has to take a decision on which way to
walk soon, but observes and in this photograph shows the decisions taken by the
others.

7.1.5 Attachment to land and adaptation to floods

Doing agriculture is the main topic of the photograph (figure 27, colour plate p.
147). In the centre of the picture are plants of *millo*, a plant that looks similar to
corn but is not produced as food but as a fibre plant used for the production of
broom sticks. In the right part of the picture, some banana plants can be recognised.
Millo plants are an important resource for the case study village in which this picture was taken, because it is a relatively water resistant plant when compared to
corn or other crops and can be harvested even several days after being immersed in
river water. The sales of broom sticks support families financially in times of the
flood season. The planting and harvest of millo and the artisanal production of
broom sticks is an important flood related practice. The type of large white clouds
in the sky and the water hole on the road in front of the field indicate that the picture
is taken in the rainy season. This is the time, when the harvest of crops becomes a
dominant activity in the villages in order to save them from getting lost in the river
water. The plant of *millo* represents a high level of adaptation to the approaching
flood. The plants are protected by a fence made of wooden sticks and a metal wire.
As in photograph 3, the fence is an important feature of the picture, delineating land
and private property. The repeated motive of the fence on several photographs taken
in the village underlines the importance of this material asset for social life in the
village. Without the creation of fences and boundaries, valuable resources are at
danger of being lost to external persons. The creation of a fence represents a certain
level of mistrust against outsiders. While the reception of external people in the
village is emphasised in parts of the discourse-oriented analysis in chapter 6, in
other cases outsiders and powerful external groups are described as invaders who
have to be kept out of the village. A high importance is attributed by village inhabitants to self-determination in the creation and reconfiguration of space within the
villages.

7.1.6 Co-presence of different temporalities

The main topic of this photograph (figure 28, colour plate p. 147) can be described as the remembrance of the past. In this case the act of remembering the past is represented in a material object, a building. This community building served as a school in former times. However, today it is not used any longer. The character of abandonment in this picture is underlined by the fact that no people can be seen in the picture. However, the building is located at a central place in the village, directly at the main road and just besides the main road there is the river shore. This indicates that the public building is likely to get flooded in times of the rainy season. Dark stains at the lower part of the building are assumed to be the effect of a severe and long-term flood in recent years. Although the building is in a state of decay, with parts of the roof missing, windows broken and vegetation growing from the inside of the building, still the doors of the building are closed, which hinders from accessing it. Among the questions this picture raises is e.g. the question what the building might be used for in current days and the question why it has not been torn down and replaced by another building. The topic of the photograph can be linked to a practice of remembering the past and holding on to achievements made in the past (see also chapter 6). As expressed in an interview by a village elder, local people remember that the provision of basic services such as water, electricity and school education had to be fought for by individuals of the village in the past. In light of the remembrance of past achievements it is not surprising that the house has not been removed. At the same time, there are high amounts of financial resources necessary in order to demolish a house and construct a new one. As long as the building is in its place, no conflict about the use of the territory has to be feared, especially no change from public to private property. It can be assumed that the land and the building are owned by the village as communal property. The fact that the building remains in its place while being in a process of decay also hints toward the fact that the property and the land are highly disputed issues by people in the village. People prefer to keep a building in its current state from changing it and risking to lose it. The building displayed on the photography can be interpreted as a metaphor that indicates towards a general pattern in social practices which is holding on to the old in order to prevent loss or dispossession. The building can be seen to represent e.g. the situation of the village in its entirety. Different ideas to carry out changes in land titles or even to resettle the village to another location have been neglected by large parts of the village population. Mechanisms of neglecting change of land and property have so far served the village to keep up the status quo.

Besides the building, the other two central objects in the photograph are the road and the two electricity lines. Both are infrastructures of high importance to the village. The road displayed in the picture is the main road leading to and from the village. The electricity line is as well the main line of electricity connection from outside. Both infrastructures are critical in the sense that they provide the basic services for the functioning of social life in the village. Especially young people rely heavily on electricity as they make strong use of electronic devices such as television, stereo systems and mobile phones. Both infrastructures are at the same time most vulnerable of being disrupted during times of flood. In times of the flood, the common infrastructures are replaced by other types of infrastructures. In the flood season, life temporarily changes from a state of connectedness with the outside world through the infrastructures displayed in the photograph to another state in which water is the main transport infrastructure and family links, personal visits and face-to-face communication replace communication with mobile devices. While the photograph presents everyday life as dependent on the infrastructures displayed, empirical research in the villages reveals that social and other material infrastructures can be just as valuable and critical as a road and electricity network. Through contrasting the material objects presented on the photograph with other material objects identified in empirical research – e.g. the *tapesco* (wooden platform for animals) or the *cayuco* (small fishing boat) – different social practices can be identified. The relevance of performing one or another type of social practice changes over the course of a year. While the road is used for transportation with different vehicles (car, motorbike, and bicycle) during the largest part of the year, transportation on the river with boats is normally a niche type of transportation which gains major importance during the times of flood. The point in time throughout the year when the road that leads out of the village is not accessible any more is a key moment within the season. It marks the change of a range of activities, e.g. the exclusive transport by boat to the town Emiliano Zapata, the increased care for children and animals including the building of *tapanco* and the initiation of the harvest of crops as well as many other *doings and sayings*.

7.1.7 Living in the water

The main topic of the photograph (figure 29, colour plate p. 148) is being in the water. In the case study villages, water space is a spatial entity as important as land space. The producer of the picture has approached water space directly, by a position that evokes the impression that one is directly inside the water or at least very close to water. Such as the plants in the middle of the picture are submerged in water, the producer of the picture is as well submerged in it or at least almost at eye level with the plants. Everyday life in the village entails the experience of being inside the water. This requires knowing how to move on and inside the water. Although many people do not know how to swim, many know how to get from one place to another in a small fishing boat. The picture was taken in a way that there is a dominance of water and clouds in the sky, while the strip of land between water and sky is reduced to a narrow line of green vegetation. The water body evokes a comparably peaceful atmosphere due to the smooth surface and strong effect of reflection; however, the sky evokes an impression of disturbance, with large white and grey clouds in several linear formations. It allows for the assumption that the clouds are rain clouds which are about to release high amounts of water through precipitation. A circular formation of small waves on the water surface indicates that raindrops fall into the water. The different levels of cloud lines produce an effect of horizontal depth in the picture. There can be seen clouds which are relatively close to the producer of the picture, around 20–50 metres away, while the smallest clouds which are closest to the horizon line can be estimated to be various kilometres away. The vast horizontal extension of space and small vertical extension of space in this photograph leads the eye of the spectator to the horizon line and the small strip between land and sky in the far distance, where few large trees can be recognised. While the body of the producer of the picture is immersed in the water and limited in movement, vision can travel through the open space. This situation and perspective created in the photograph can be assumed to point towards a desire to travel to another place. While the current situation of the flood is known for the limitation of movement, the aspiration to travel to another place can be considered as high.

The analysis of selected photographs using the documentary method allows the identification of key topics and concepts that are accessible through the material levels of social practices (table 13). The analysis especially provides information concerning relevant spatial patterns and temporal dynamics within social practices of flood management. Local understandings of proximity and distance and changes in patterns of spatial and social proximity are highly relevant for understanding flood processes in the case study region. The creation and reconfiguration of space can be identified as actions that foster self-determination. Exclusion from and inclusion into spatial entities is linked to processes of building collectives and strengthening a collective identity in which humans and non-humans are involved. Moreover, analysis of material objects through a visual approach provides access to the water spaces and the land spaces that form part of everyday life in the villages and that underlie strong changes throughout the year. The specific perception of water spaces is a key determinant for understanding flood perceptions. Here, the visual approach proves to be an indispensable part of empirical research because it provides access to an understanding of social practices, here especially the material-mental nexus. In table 13 linkages are built between topics and concepts identified through the visual approach presented in this chapter and discursive patterns identified through text-based analysis of participant observation and interviews in chapter 6. A strong link is recognised e.g. between the social practices of anticipation of the flood involving specific material aspects of practice (identified through visual approach) and the practices of living with the flood that include specific temporal dynamics and perceptions of floods (identified in discourse-based approach). Another strong link can be seen between the social practices of protection of space, represented in this chapter e.g. in the construction of material delineations in physical space and specific discursive patterns that underline the importance of land access which is described e.g. in the presence of village members in their houses even during severe flooding. The linkages identified between the material and discursive expressions of social practices gives insight into the ways in which the material-mental nexus of social practices is enacted. While in the analytical steps applied in this thesis so far, mental and material notions of social practices have been partly reviewed separately, the conceptual understanding of social practices does not sustain a dichotomy. Rather, it is argued here for an interweavement of bodily-material and mental patterns which are recognisable when dealing with the diverse empirical material gained in this thesis. Before presenting some of the interrelations identified between complex sets of social practices performed by different groups of people in the case study region, underlining the material-mental nexus of practices, the remainder of this chapter tries to enhance the picture of the context in which the social practices presented above are embedded in the case study region.

Table 13: Overview on key topics and concepts identified in the interpretation of the photographs and links to discursive patterns identified in chapter 6

Main topics and concepts identified in the photographs	Links to discursive patterns (chapter 6)
Distance towards a destination that is sought after (Photographs 1 & 7)	General link: importance of "La tierra" (chapter 6.7)
Social proximity and spatial proximity linked in times of flood (Photograph 1)	General link: importance of "La tierra" (chapter 6.7)
Travel to the village by boat on the river Usumacinta (Photograph 2)	Discursive link to practices of living with the flood; including transport and livelihood activities (chapters 6.3, 6.4)
Anticipation of the flood as a social practice (Photograph 2)	Discursive link to practices of living with the flood specific temporal dynamics and perceptions of floods (chapter 6.3)
Livelihood activities (e.g. cow rearing, production of broom sticks from Millo) as social and financial insurance components as part of flood management practices (Photographs 3 & 5)	General link: importance of livelihood activities as part of flood management (chapter 6.4)
Protection of spaces (land and property) as part of flood management practices and larger social practices (Photographs 3 & 5)	Various discursive practices concerning the importance of land access ("la tierra") and territorial dynamics (chapter 6.7)
Self-determination in the creation and reconfiguration of space (Photograph 5)	
Finding one's way – Young people's decisions for a future in or outside of the village (Photograph 4)	General link to chapter 6.4
Material realities as social realities – Holding on to the old (Photograph 6)	General link: importance of history (chapter 6.6)
The criticality of infrastructures in times of flood and in times of dry season (Photograph 6)	Discursive link to the provision with basic services (chapter 6.4)
Water spaces and land spaces – Social practices of living in and with the water (Photograph 7)	Discursive patterns described in the bodily perception of water and the flood (chapter 6.3) and importance of access to land (chapter 6.7)

7.2 SOCIO-CULTURAL CONTEXTS OF FLOOD MANAGEMENT

The following section pursues the goal of presenting some of the relevant social practices and patterns of practices that build up the social context in which flood management practices takes place in the case study villages. Selected empirical examples give an in-depth insight into aspects of social organisation and the relevance of cultural performances for social life. Understanding social practices as part of larger social dynamics, the information on the broader contexts gained in this section sheds additional light onto the core social practices of relevance in this study.

In order to analyse the socio-cultural context in a reflexive manner, representations are used that have been elaborated in a participatory manner and reflected upon by various research partners. A short number of video sequences has been chosen for interpretation as the film material is the result of long-term interaction in a team of village members, the co-researcher and the researcher. The sequences were produced as part of the collaborative video workshop carried out with a group of women and other members in the main case study village in 2015. While the process of collaborative video production and reflection was carried out throughout the entire phase of empirical research in the field and many minutes of audio-visual material were generated, only a very small part is selected for in-depth analysis in this section. Given the different layers of information provided by audio-visual material, including verbal expressions, sounds and visual impressions including movements, the analysis has to be limited to a few video frames. This enables to make adequate use of documentary method focusing first of all on the visual information and to produce a manageable amount of analytical text for further interpretation. Three short video sequences are analysed that have been recorded by a female individual, one of few members of the village who actively took part in the production of video material. The sequences analysed make part of the first edited version of the video which was produced in 2015 and presented in the village at the end of the field research phase. An analysis of the whole video is beyond the scope of this thesis as a range of different topics besides flood management are major narratives in the video. The production and the reflection on the documentary video is however a key component of the empirical process.

The focus of the analysis presented here is the visual information that can be identified in the video sequence. In order to present the steps of analysis, several still pictures are produced by taking snapshots of selected video frames. Analysis is limited to four snapshots per video sequence analysed in their chronological order (figures 30–32). Each sequence can be read starting from the picture in the upper left to the picture in the lower right. While the complete process of analysis following the documentary method is added in annex 5, the full analysis is displayed here for video sequence 1. For the other two sequences only the in-depth interpretation is presented here, including an identification of or a link to social practices. As it is decided to not analyse every frame of the video sequence, the information in audio format from the video sequences is represented separately in textual form, in a format that is similar to the representation of interview material. The information tran-scribed from audio format is used to underline the links to relevant social practices already identified.

7.2.1 Taking responsibility in the socio-cultural network

The activity represented in the snapshots (Figure 30, colour plate p. 148) is a central and important activity in the preparation of the festival of *Señor de Tila* in the case study village. The man who is the central figure of this sequence carries out the activity to light fire under a water bud in a calm and routine-like manner. He is an elder person in the village and carries out the activity alone, which emphasises his responsibility and high level of experience. Making a fire outside is an activity rel-evant in the context of livelihood practices in the villages. While in the everyday context, women use open fires in the back of their houses or in the kitchen houses to prepare meals, e.g. *tortilla* or chicken, this activity is carried out by a male village member in an open place. This indicates that it is part of a special event in the village. In the broader context of the festivities of *Señor de Tila* it was observed by the researcher that all the activities carried out during the event, including the prep-aration phase, are highly routinised as the event takes place every year. The routine includes a high level of organisation concerning the questions which group of peo-ple carries out which kind of activity in which place and at which time. All these aspects are predefined. It includes a strict preparation of activities carried out by women and those carried out by men. While their respective activities are carried out in close spatial proximity, almost all in the open space between the river and the main street or in the community hall, the tasks are clearly separated. While the male village members slaughter the pigs, the women prepare vegetables for a meal of the following day. The activity of the man shown in the snapshots is part of the sequence of activities carried out in the preparation phase of the event. More spe-cifically, the heated water is in the sequence of activities used in order to clean the skin of two pigs which are slaughtered in the village. The skin of the pigs is fried by the men subsequently and eaten by the whole group of village members who take part in the preparation of the food on the same day. The meat of the pigs is then marinated and cooked by the women in preparation for the following day, the main festive day.

Table 14: Transcript of verbal expressions from video sequence 1

Video 1 (Minutes 09:40–09:51), transcribed in Interview 2015_C_22: Lines: 134–135
[VM1]: Already you will start to have a tub of water yet on the fire.
[P1]: Yet the least.

The verbal expressions made by the video maker and the person on the picture are comments on the activity the person in front of the camera performs. While the female video maker describes the activity which can be seen and addresses the man in this description, he answers that doing this is "the least" he can do. His comment expresses an attitude of humbleness towards his own contribution to the communal work done in the village.

7.2.2 Waiting for the work to start

In the sequence of snapshots (figure 31, colour plate p. 149), a specific interaction between the women and the camera takes place. The video maker closely observes the women. There are clearly changing facial expressions of the women, from smiling to serious looks and from looking directly into the camera to looking away and hiding one´s mouth behind the hand. This can be interpreted as insecurity with the situation of being filmed and at the same time curiousness towards the camera. While women notice themselves being "observed" by the camera and recorded by their fellow female village member, they look back and also observe the camera. This provokes reactions like smiling or expressions of shyness. At the same time, women express by their bodily positions and slow movements that they are waiting for something or someone. The plastic cup in the middle of the pictures indicates that there will be something to drink. The towels which the women have placed over their shoulders allows for the assumption that it is hot and that women use to bring towels when they work because of the expected heat and sweat. The women stand together very closely, which provokes an atmosphere of closeness between them. While there are no large bodily movements between the different snapshots it can be recognised that the main activities of the pictures are waiting and observing. Women are more likely to be in observing positions in the village then men, who take initiative and move fast. However, in later sequences of the video, when women are in larger groups their movements and interactions are much more active.

Table 15: Transcript of verbal expressions from video sequence 2

Video 1 (Minutes 10:03–10:07) transcribed in Interview 2015_C_22: Lines 136–137
[VM1]: Chatting, waiting for the piggies.
[P1 & P2]: [laughter]

In the verbal expressions, the video maker comments on the activities performed by the two women. She mentions that the women are "chatting" and that they are "waiting for the piggies" to arrive and for other women to arrive in order to start their work. The women do not answer with words but start laughing. This underlines shyness towards the camera and the unusual situation of one´s routine activities being commented and questioned about. Women are not used to speaking in public gatherings and have not yet been recorded on video.

7.2.3 Performing hierarchy in the social network of men

The snapshots of the video sequence (figure 32, colour plate p. 149) represent a key moment in the activities in the preparation of the festive day of *Señor de Tila*. The transporter attracts the attention by many people who gather around its´ back door. In the transporter, the pigs are transported to the village. Men arrange around the transporter and open the door in order to get out the pigs. The high amount of people and the movement from one snapshot to the other indicates that the activity represented here is one that involves some excitement among the villagers. Every village member wants to see the pigs as they are the basic meat prepared for the food of the festivity and the physical appearance of the pigs allows for assumptions on the taste of their meat. The good taste of the food is a major criterion for the success of the festive day. There is a clear hierarchy in the different steps of taking out the pigs. One man, who wears relatively elegant clothes takes the role of the leader in this activity while delegating activities to the others. A young tall man is the most active in the opening of the door, while other elder men stand around and observe the situation. Women are seen at a larger distance of the transporter but not approaching it. The woman seen in snapshot 2 in the left part of the picture is the researcher and author of this thesis. She observes the transporter and the camera from a distant position.

Table 16: Transcript of verbal expressions from video sequence 3

Video 1 (Minutes 10:34–10:56) transcribed in Interview 2015_C_22: Lines 143–147
[P1]: So you take advantage of the camera.
[VM1]: I am here, recording.
[P2]: If you would lend me this [not understandable]
[VM1]: Stay away, stay away from there, the ant, stay away, stay away.
[P1]: Take it from there.

In the verbal expressions of the video sequence, several layers of communication are represented. The first sentence is expressed by the man on the left part of the pictures to the video maker. He addresses her in by saying "so you take advantage of the camera", which is an accusation or negative assumption. The camera filming activities in the village and especially a female village member doing the filming is a highly unusual situation in the village context. The verbal expression by the man linked with the bodily expression of superiority in snapshots 1 and 2 underline the disapproval by the man towards the filming carried out by a woman from the village. The video-maker gives response to the man, explaining her activity. The second expression by the film-maker is directed towards the researcher who stands at her back during the filming of the rest of the video sequence (from snapshot 3 onwards). She warns the researcher of the ants on the ground whose bites are painful. The man who had addressed the video-maker in the first part of the sequence does not take further notice of her and talks to the fellow male members of the group.

All three sequences represent activities which belong to the practice of preparation for the main festive day of the celebrations for *Señor de Tila* in the case study village. While in the first sequence, an activity is carried out by an individual elder of the village, in the second and third sequence various people are part of the scene. The activities carried out all link together and are performed in a highly coordinated way. People represented in the sequences perform their activities of making a fire, waiting for others and in opening a transporter in a routine-like manner. While in the first sequence no eye contact is established between the person and the camera, in the other sequences people directly look at the camera and make bodily movements directed towards the video maker. While all three sequences represent activities that belong to the social practice of the preparation of a religious festivity in the village, general patterns of interaction and organisation of practices in the case study region can be abstracted. This allows assumptions about the organisation and importance of social practices in times of the flood and in the periods of preparation for and recovery from floods. Firstly, most activities are carried out in a group. While less critical activities, like making fire to heat water, can be performed by responsible and experienced elders in the village, all the other activities involve a large group of people. However, in some groups all members carry out the same activity in a coordinated manner. Secondly, there is a strong separation between activities carried out by women and those carried out by men. This separation is not linked to a strong physical separation as men and women dwell in the same open spaces of the village. However, when it comes to performing the activities of one´s group, all members of the same gender focus on their activity not interfering with the other gender. Thirdly, a clear hierarchy among the men can be recognised. While younger men carry out the most physically demanding activities, older members delegate work or carry out activities that allow taking more time. The most challenging tasks are carried out by the most experienced villagers. In the case of the slaughtering of pigs, the activity to kill the pigs by a stab with a knife into a pig´s heart is carried out by a respectable elder man of the village while all other village members observe and wait.

The analysis of visual information in this study reveals that important information about social practices and the larger social context in which practices are performed can be identified. The knowledge gained from analysis of verbal expressions given by interview partners and of notes of participant observation written down in textual form can be fundamentally amplified. At the same time assumptions concerning social practices can be challenged or affirmed through the analysis of visual information. The narrative power that is characteristic to visual media, in this case photographs and snapshots of video sequences, helps identifying key topics and paradigms of relevance in the larger context of the research questions. Moreover, the interpretation of visual information with the documentary method allows the reconstruction of social practices that reflect in the respective narratives. The analysis of visual information in this thesis has two major outcomes. One is the identification of topics and concepts (table 13) which are linked to or can be contrasted towards the discursive practices and patterns identified in chapter 6. Moreover, the analysis of photographs and video snapshots builds the basis for in-depth analysis of the social practices of everyday life and of special events in the case study region. This allows the identification of key social practices of flood management and the material, mental, performative and spatial contexts, in which social practices take place and which they are part of.

This chapter in connection with the foregoing chapter has presented an in-depth analysis of the social practices of flood management in the case study villages focussing on discursive and material elements. Expanding the perspective even further, the following chapter introduces social practices of interaction in flood management in the case study region which involve different groups of subjects. This enables to identify important types of social practices that are related to the management and governance of floods and risks and to analyse spaces of conflicting social practices. Using the *riskscapes* approach in order to review social practices pushes analysis forward into the directions where it is possible to understand the spatial dimensions of interacting social practices of flood and risk management.

8 SOCIAL PRACTICES OF FLOOD MANAGEMENT
PERFORMED BY EXTERNAL ACTORS

As discussed in detail in the two preceding chapters, flood management comprises a range of different social practices. While it is important to underline that different groups of inhabitants in the case study villages perform flood management in different ways, it is just as relevant to point towards the fact that there are other key actors who perform flood management practices in the case study region or carry out practices that set the larger scene for these practices. There are multiple types of hierarchies and forms of interactions between local people and other actors. The term "actors" is used in a narrow sense, defining actors as the human carriers of social practices (Niccolini 2017: 105). This delineation is specifically relevant as other definitions of actors present in social theory may give an importance to aspects such as social order, structure or institutions which would lead to an overemphasis of these positions in the analysis of social practices. Following a flat ontology, social practices are used as the supreme assets of analysis. The different actors and actor groups identified in the context of this study are presented in an overview in figure 33. While in parts these actors have been addressed or mentioned in the analysis of social practices in the case study villages already, other actors are newly introduced here. It is reemphasised that the visual representation of different groups of actors on different administrative levels and arenas like "flood management" or "economy" accomplished in this figure is an abstraction, which does not compete with the supreme role given to social practices. It is the aim of this section to identify not the key actors but key social practices which intersect in flood management in the case study region. However, in order to reduce complexity and introduce the different social practices to the reader in an overview, the different carriers of the social practices are presented. The following analysis lays the basis for the identification of specific points and specific ways in which social practices interact. The identification of relevant social practices in this chapter is therefore relevant for the following step of interpretation, namely the linkage of relevant social practices in selected *riskscapes*, which is accomplished in chapter 9. The identification of social practices in this chapter is mainly carried out by an analysis of verbal expressions transcribed from interviews and published texts are the main sources of empirical information in this analysis of social practices. Furthermore, a range of material objects can be analysed in their linkage to social practices performed by external actors. The analysis of discourses dominant in these verbal expressions is a key access point to the analysis of social practices and their spatial relevance.

Figure 33 presents a simplified overview on the main actors involved in or exerting influence on practices of flood management in the case study region, here exemplified for one specific case study village. Four different main arenas of actor groups can be differentiated: Firstly, there is the flood management arena, which involves civil protection agencies at three different administrative levels, as actors from all three levels engage directly in the case study village. Similarly, the government arena, involves three administrative levels of the government development agency DIF, with two key individual actors highlighted as they perform direct interactions with the case study village. Thirdly, there is the NGO and science arena, which is grouped together only for the purpose of giving a broad overview, while there is no clear administrative hierarchy and the key topics of intervention in the village are different. The fourth arena represents economy, which includes development agencies and regional, national as well as international enterprises that relate with and exert influence on the case study region. Different practices that are performed by different actors are analysed in the following paragraphs, identifying patterns which are relevant for present and future socio-spatial flood management practices and larger social dynamics in the case study region.

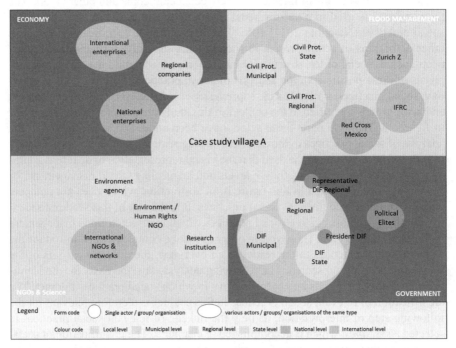

Figure 33: Actors performing social practices of relevance for flood management in one case study village
Source: author's elaboration

8.1 SOCIAL PRACTICES PERFORMED BY CIVIL PROTECTION ACTORS

Civil protection in the case study region is mainly influenced by actors on three different administrative levels: the state, the regional and the municipal level. While direct responsibility for intervention lies on the municipal level, in case of contingencies, the regional centre of civil protection for the region *Selva* becomes active as well. At state level in Chiapas, which is located in Tuxtla Gutiérrez, coordination and distribution of relevant information as well as the strategic design of campaigns and capacitation are managed. The national system of civil protection exerts influence through funds like the Fund for Natural Disasters (FONDEN) which can be accessed by government agencies at the different administrative levels in order to finance reconstruction measures, especially targeted to benefit poor people (Saldaña-Zorrilla 2007: 39). Different actors from the civil protection system were interviewed during the field research phases, including staff members and people in leading positions at the municipal, regional and state level of the Mexican civil protection system. The analysis of discourses presented below allows gaining insights into flood management practices which may be performed in the case study region. Starting with an analysis of the general paradigms and perspectives represented in the civil protection system in Chiapas, the second part of this section describes specific flood related strategies, perceptions and measures which allow the identification of some key practices of flood management.

Figure 34: Presentation of the Prevention Programme PP5
Source: adapted and translated by the author from Instituto de Protección Civil 2014

In general, considerable difference can be identified in the approaches, perspectives and resources of civil protection at state level and those at municipal and regional level. On the state level, a close orientation towards the national system of civil protection and international developments can be recognised. The design of programmes as well as the practical orientation towards awareness campaigns and capacity building allows the point that besides immediate response and recovery actions, strategic importance is given to prevention and preparedness in an integrated risk management approach (Memo_I_141014, Annex 7). The emphasis on prevention also reflects in the new civil protection law which was passed in Chiapas in September 2014 (Secretaría General del Gobierno 2014: 8ff). In a new prevention programme elaborated in Chiapas, the importance of collaboration with local communities is highlighted, making direct reference to the strategic goals of the *Hyogo Framework for Action* (HFA) (Instituto de Protección Civil 2014: 6). A slide from a presentation of the prevention programme is presented in figure 34 to underline the reference Civil Protection Chiapas makes to the international strategies at UNISDR level.

As can be recognised in the translated text of the presentation in figure 34, it is referred to a "culture of civil protection". This culture should be based in the community in order to reduce disaster risk successfully. In the presentation slide, two photographs are added, that represent collaboration between civil protection and people from rural villages. It is possible to identify two major material assets from these pictures which may be of relevance in the social practices of civil protection staff in Chiapas. The first one is a colourful chart. While the details of the chart cannot be recognised in the photograph, the classification of different colours used in the chart allows the assumption that it is a chart used to classify different alert levels of the civil protection system. This classification is a major characteristic of the way in which civil protection operates. In the case of hydro-meteorological events there exists a colour code which is used for the communication with the municipal and the local level to indicate which types of precautionary measures need to be taken (figure 35 and figure 36, colour plate p. 150).

The activity displayed on the first photograph represented in figure 34, is assumed to be the explanation of colour codes and necessary measures by civil protection staff. The second photograph refers to the communication between civil protection at municipal level and remote villages. The distribution of 2,800 radio transmitting devices is part of the prevention strategy of the civil protection institute in Chiapas (Instituto de Protección Civil 2014: 11). Communication with villages through radio technology is a key part of the social practices of flood management. This practice is promoted and technically supported by the state level and performed by civil protection staff at municipal level. As a main problem for civil protection in Chiapas, interview partners mention the high level of poverty in the state. Numbers of poverty and the indicators used to define poverty are elaborated on, underlining exceptionally high levels when compared to other Mexican states (Memo_I_141014, Annex 7).

On the municipal and regional level, contingencies are addressed with a highly limited amount of resources, including limited technical equipment, a small number of staff and low level of capacitation of staff. While concepts of disaster prevention and preparedness are known to staff members, the practical work in the villages is mainly limited to addressing flood damage and supporting recovery (I_2014_T_1;

Field notebook 1/2015 10.06.2015). In order to give insight into the different per-spectives and practices of civil protection staff at municipal level in the area of flood management, in the following paragraphs the perceptions of flood and the perfor-mance of a range of flood related measures is presented.

With regard to perspectives of flood risk and the necessary management, a range of different expressions can be identified. One staff member presents own experience with working in the case study region, at the same time giving hints towards his understanding of risk and the ways to reduce it.

Interview 2014_T_1: Lines 18–29
[P1]: If I don't attack the root of the flood, in 10 or more years the effect will maybe even be stronger. This is the part I try to tackle in my work. This is what I try to make the people under-stand. The problem is: Because here there is a lot of water, more water than before on the rivers. We have already explained it to the people and they understand that it is not our guilt or our responsibility. It has happened with earthquakes. People say, you have to help me, it is your obligation. It is my work to help, that is why I come and help. But it is not my fault that your house is flooded. Your house has been in bad conditions, there was poor maintenance, it was constructed poorly. A lot was missing. I will support you but you have to understand that the problem came from your side.

Different layers of content and meaning can be identified in the above interview sequence. Without using the term risk, the interview partner uses the terminology and logical structure of a conceptual framework for risk and vulnerability, the Pres-sure and Release (PAR) model (Wisner et al. 2004). He explains that in order to mitigate future flood effects, it is necessary to address "the root of the flood" (I_2014_T_1: 18). On the one hand he mentions flood exposure in the case study region, which according to him is more severe due to higher amounts of water in the rivers. The interview partner points towards a priority he puts in his work through the repetition of the word "understand". In order to change the situation on the ground he argues that it is necessary to make people understand, e.g the chang-ing exposure and the problems local people cause through poor construction and maintenance of houses. On the other hand, he emphasises that people need to un-derstand how civil protection operates, hinting towards difficulties civil protection staff face in interacting with local people. In the short sequence, the interview part-ner also mentions two other words important for further consideration in the context of flood management: the words "culpa", which can be translated as "guilt" or "fault", and the word "responsibility". Many of the activities carried out by civil protection staff in the case study villages aim towards "explain[ing] to the people" (I_2014_T_1: 22) and "try[ing] to make people understand" (I_2014_T_120). It is part of the social practices of flood management civil protection staff performs. Moreover, the interview partner mentions that a general approach of education is followed by civil protection at state level, including information campaigns about the disaster cycle in villages which were planned for the following months (I_2014_T_1: 37–46). A clear direction of communication can be identified in the activities of civil protection. Information is passed from civil protection staff to people in villages and it is expected that these people understand and adapt behaviour. Challenges that come up with this approach are presented in the first paragraph of the following interview sequence. Here the interview partner expresses his opinion that especially when it comes to questions of settlement "it is difficult for them to understand" (I_2014_T_1: 66).

Interview 2014_T_1: Lines 64–83

[P1]: Then we arrive there and see the house was built poorly, in a place where it should not be, there is no allowance to build in that place. So the people have to understand this part and it is difficult for them to understand.

[D1]: And how was it in Catazajá?

[P1]: Catazajá is a region that floods every year, so up to 6 months of the year the region can be flooded. They live practically at the side of the river and due to the saturation of water that is present in the ground, already with small rainfalls, the water rises and stays flooded. The problem is that earlier administrations did not give a lot of support, they didn't come there. So people lived fine, they were used to this situation. They put up their things, they went in their boat and everything was fine. This is no solution either but people don't want to live in any other place. If you tell them to be removed, they say no. So what can I do? The zone is a zone of lagoon, a giant lagoon, and this cannot be changed. So what do we do? The earlier administration, the state government gave too much assistance to these people. Even if the level of water was not high, they already gave blankets, food, all that they wanted. They installed programmes so that people would be fine and lots, lots of money for them. So what the people do is, they get used to it. And now anything that happens, they say <Listen, I have water. Listen, it already rained. I want to have food support>.

The interview partner presents the way in which civil protection staff approaches a village by saying that they "arrive there and see". A focus is given to acts of seeing, which implies collecting and analysing visual impressions. In the second paragraph of the interview sequence, the interview partner presents an analysis of problems that the municipality of Catazajá faces with regard to flood and flood management. A key actor presented by him to cause challenges to flood management in the region is the administration. It is mentioned that in the past, little assistance had been given to the villages but that "people lived fine, they were used to this situation". The expression of being used to a condition has been highlighted repeatedly in the analysis of expressions by village members and is again repeated by this external actor. Being used to little assistance in this case is evaluated as rather positive. In the more recent past administration of that municipality has changed practices of assistance, in giving, as the interview comments "too much assistance". The high level of assistance is evaluated as negative by the civil protection member as it makes local people rely heavily on this assistance and demanding for it even in situations without a major flood.

Civil protection staff in the regional centre in Palenque is in charge of taking care of contingencies and reaching out to villages on a regular basis. Besides activities reacting to emergency situations and recovery activities following a flood, a range of other activities are part of the daily work of the officers. The change of daily work during rainy season is addressed in the following interview sequence.

Interview 2015_P_2: Lines 20–36
[D1]: And now that there is or will begin the rainy season...
[P1]: Yes.
[D1]: ...how, what, how does your work change? Does your daily work change?
[P1]: Yes, yes, there already exists a contingency plan, isn´t it?
[P2]: Yes.
[P1]: There is already a contingency plan concerning the rains, the tropical storms. Don´t you have one over there? So that she could see it. I saw it yesterday. It is over there.
[P2]: Ah, yes. I don´t have it over here.
[P1]: It is a manual, right? To be prepared.
[D1]: Yes.
[P1]: Yes.
[D1]: Thank you. And this is elaborated in Tuxtla? Or here?
[P1]: No, here. Or let´s say more or less concerning the region, right?
[D1]: Aha.
[P1]: With the problems of here the municipality.
[D1]: Yes.
[P1]: Every year it is made. It is updated.

The main interview partner of this sequence describes as a major change during the rainy season, that a contingency plan is prepared, which guides intervention of civil protection in times of rains and tropical storms. This contingency plan is described as "more or less concerning the region". Besides the contingency plan, civil protection staff in the office is in charge of carrying out a risk analysis for the municipality. State government has recently requested an analysis of risks to be presented, however there is a range of challenges for the staff members to carry out the analysis. Based on the interviews it can be assumed that capacity building on the background of risk concepts as well as on the data to collect for and indicators to be used in a risk analysis is a major lack at the regional level of civil protection in Chiapas. One interview partner describes the current activities carried out in this area in the interview sequence added below.

Interview 2015_P_2: Lines 110–126
[P1]: I have here a guide of how a risk analysis is made, but it is a simple guide. This. Atlas of dangers and [...] of the municipality of Palenque. This. Right?
[D1]: Yes.
[P1]: I go collecting information and go making my survey. Here for example the level of seismicity, yes in this zone it is low, isn´t it? Of around 4, I put 4 in my list. This is an example.
[D1]: Yes.
[P1]: In hydro-meteorological questions in the low zones, flood-prone, or high zones with landslides. And this goes here. And on this I keep orienting.

> [D1]: Mhm.
> [P1]: But in itself, there is a lack of a risk atlas, personalised of the municipality of Palenque.
> [D1]: Aha, aha.
> [P1]: Or let´s say, we have this. This is from the department of risk identification in Tuxtla. And I base myself on this.
> [D1]: Mhm.
> [P1]: But there is the need for more information.

The written manual which is available in the office to elaborate a risk analysis is not sufficient in order to teach the staff how to do the analysis. As the interview partner highlights in the above sequence "there is a need for more information". Referring to a recent event in one of the case study villages, in the interview sequence presented below the interview partners give insight into the classification of different types of floods or water-related extreme events.

> Interview 2015_P_2: Lines 48, 51–60
> [P2]: No, there was no shelter made because it only was an encharcamiento.
> [P1]: So an encharcamiento is maybe less than a metre. […] Or let it be less than a metre. When it passes over one metre, flood.
> [D1]: Aha, aha.
> [P1]: Yes, yes.
> [P2]: And these are indirect effects. Yet, direct effects is when…
> [D1]: …water enters into the houses.
> [P2]: …into the houses.
> [D1]: Indirect is more like, like for example, it destroys the milpa or something like that?
> [P2]: That´s how it is.
> [D1]: It is like an indirect effect that people loose part of their harvest.
> [P2]: Or the street.

As can be seen in the interview sequence, the measures civil protection staff performs depend highly on the different classifications. Moreover, the effects of the flood are grouped into direct and indirect effects. While direct effects concern the settlement areas of the rural villages, e.g. when houses or public buildings are flooded, indirect effects concern the strips of land for agricultural production, grazing lands and the road networks.

A material remainder of classification as part of flood management practices by civil protection actors can be found in one case study village. Figure 37 (see colour plate p. 151) represents women in one of the case study villages who are leaning on a building on which a scale with lines and numbers is painted. This scale is used as a measuring device by civil protection actors from the municipal level in order to document water levels and to take measures. Measuring water levels is performed by civil protection staff with a metric scale, while people in the villages themselves measure the water level mainly using their own bodies (chapter 6).

Besides flood prevention and preparation as well as immediate help in the case of floods, civil protection staff is also involved in the organisation of recovery in affected villages. After the last major flood that occurred in the case study region in 2010 government allocated financial resources from the fund FONDEN to build new houses for those households most affected. In the case study region, the municipality of Catazajá benefited from this allocation of resources. During field research in the case study villages that belong to the Catazajá municipality, houses could be identified that had been constructed for some village members after the 2010 flood (figure 38, colour plate p. 151). The type of houses built differs considerably from houses built by villagers themselves. While local people build houses with concrete walls and aluminium roofs or built in a traditional way with wood, the new houses are made from synthetic material. For an individual or family to benefit from the house recovery scheme, civil protection staff first had to assess the severity of damage in each village and had to identify the people most affected.

Villagers comment that the houses were constructed with poor construction material which results in the fact that during heavy rainfalls which are usual in the rainy season, water passes through the walls and there is humidity in the houses (I_2014_PR_4: 2–3). According to one interview partner, the type of construction material for houses envisaged by the recovery plans was replaced by poorer material following a misuse of allocated financial resources (I_2014_PR_1: 10–11). In the interview sequence presented below, a civil protection officer presents his perspective on the problems linked with the use of new houses by local people:

Interview 2014_T_1: Lines 100–105 & 110–112
[P1]: Many people was given the opportunity to build a house, some people actually received a house like this…   [D1]: Like [name of a person]´s house.   [P1]: Yes, exactly. But the government said <I will give you a house, for sure. But now you cannot live in your first house anymore.> People many times want to live in their old house during the year and when the flood comes they want to move to their small house. […] They want a new house but they want to continue living like they did before. And they will continue to have the same problem, year after year after year.

As the interview sequence shows, with the donation of a house to people in flood-prone villages a certain change in behaviour was intended by the donors. This however did not occur in the way planned by government and people have adapted the new houses into pre-existing flood management practices.

A key solution to the flooding problems in villages of the case study region is regarded by several interview partners from civil protection to be the resettlement of population. At state level, a cost-benefit-analysis has shown that resettlement is the most economical option for small settlements in rural areas while for urban areas with larger population sizes, technical measures that protect from flooding are the most economic option (Memo_I_141014, Annex 7). Another civil protection actor who works at local and municipal level emphasises that people are reluctant to resettlement plans underlining that "[i]f you tell them to be removed, they say no. So what can I do?" (I_2014_T_1: 75). The interview partner presents a possible adaptation to floods in villages where people object to be resettled in the following interview sequence.

Interview 2014_T_1: Lines 90–97
[P1]: So we have to find a way to support them but make them understand that this is a support, the solution is that you move away from your house. You are bad in the place where you live. You have to find a dry place. But there, they have no dry places, no elevated place, everything is at the water level. So there is a lot of work to do there, a lot of work. For me one ideal form to live there would be to build houses that are resistant to the water that comes every year. Elevated houses and such. But the people don't want to. They give all kinds of reasons, for their animals, etc.

The interview partner shares his perception that those people who live in a flood prone are "bad in the place where [they] live" and that "they have to find a dry place". However, he mentions that dry land is not accessible in the case study region. Therefore, flood resistant houses, e.g. elevated houses are described by him as an option of adaptation. Another interview partner from civil protection refers to possibilities to resettle villages in schemes called "rural cities" ("ciudades rurales" in Spanish), already performed in other parts of Chiapas.

Interview 2015_P_2: Lines 170–200
[P1]: There is a zone where the government should make. As here it is flood-prone, it would only be a production project. Settlement use we will transfer it to the safe zone. But that yes they lose their lands.
[...]
[D1]: Yes, I think that this is one of the major concerns of the persons. That once they leave to live in another place that also they will use their lands.
[P1]: Yes, this is the detail. Or let´s say it is a social phenomenon.
[D1]: Mhm.
[P1]: It would be necessary to study it in detail in order to see a solution. But now it is studied, before not. Because in the years of Juan Sabines, there already were rural cities.
[D1]: Mhm. In the 80s?
[P1]: No, recently. About 10 years ago.
[D1]: Ah.
[P1]: Or let´s say, there already comes a part of investment from government. That for example they leave their properties, this has been reached in Chiapas already and where they construct a city, they have enough. Or, there is everything to make it.
[D1]: Mhm.
[P1]: Or I have already gone to Tuxtla. I have already noticed.
[D1]: And are they self-sufficient? So that they also have…
[P1]: Sustainable or rural cities.
[D1]: …fields, or? Do they have land to do agriculture?
[P1]: Yes. And the government what they try to do is invest into safe zones.

[D1]: Yes, yes. [P1]: Because every misfortune there is, it has to be put more money. And it is more expensive.

Resettlement of the case study villages, as the interview partner agrees with the researcher in the above interview sequence, could result in a loss of land property. However, the interview partner regards the concern for land property a "social phenomenon" which needs further research. He points towards the concept of *ciudades rurales*, which according to him is a realistic solution for the case study villages. He emphasises that government invests into what he calls "safe zones" while additional financial resources put in flood zones would be lost. The interview partner calls the resettlement schemes sustainable.

Under the lead of Zurich Z, a foundation directly linked to Zurich Insurance, the *Flood Resilience Program* has been installed in a flood-prone area in direct vicinity of the case study region. The paradigms, strategies and practical action promoted and performed through the programme may be of high relevance for the case study region in the present and near future. The *Flood Resilience Program* in Mexico initiated in March 2013 by Zurich, comprises a group of actors that includes the Mexican Red Cross, the International Federation of Red Cross and Red Crescent Societies (IFRC), the UK-based INGO Practical Action, as well as the US-based research institutions International Institute for Applied Systems Analysis (IIASA) and the Wharton Risk Management and Decision Processes Center (Zurich 2014). The activities in Mexico are part of a global *Flood Resilience Program* as part of Zurich's corporate responsibility strategy and involves a budget of CHF (Swiss Franc) 21 Million until 2017 (Zurich 2013). A specific flood resilience approach developed by IFRC is used in the programme: The vulnerability and capacity analysis (VCA) (I_2014_X_1: 17). According to an interview partner working in the programme, VCA is "the combination of a humanitarian and a development project" (I_2014_X_1: 31). In a publication by Zurich it is argued that it combines "two seemingly conflicting goals […] encouraging strategies that both manage risk and promote development" (Zurich 2016: 3). In an interview carried out with various programme managers from the different involved institutions, the selection criteria for the villages to take part in the programme were discussed. It was mentioned that the major selection criteria had been a marginality index, elaborated on the basis of data from the population census and other statistics, the vulnerability of the villages and the incidence of floods in recent years (I_2014_V_1: 13–15). The criteria were selected on a scientific basis and data was gathered and analysed by the involved research partners (I_2014_V_1: 3–4). The decision to select case study villages moreover was taken concerning physical–geographic characteristics (I_2014_V_1: 7–10). The main objective of the programme is to build resilience in a way that can be measured and validated. In a publication by Zurich Insurance from 2016, the conceptual framework developed for the programme is presented which highlights the specific aspects of resilience measured. The conceptual framework combines two frameworks: The 5C model developed by the UK Department for International Development (DFID) in which five major capitals (human, social, physical, natural, financial) are described and the 4R model developed by the US-based Multidisciplinary Center for Earthquake Engineering Research (MCEER) that describes a resilient system through the four criteria of robustness, redundancy, resourcefulness and rapidity (Zurich 2016: 4f). It can be identified that the criteria for village selection and the conceptual-methodological frameworks used have been chosen based on criteria developed in research institutions in the Global North.

While according to the interview partners the Mexican Red Cross staff were trained how to carry out a VCA in a participatory manner, the international staff are present in the villages for cases of "quality control" (I_2014_X_1: 37). The presence of international staff in the village and the supervision of the VCA process is part of a social practice that shapes the process of resilience building. The international experts make use of a development discourse and the necessary tools and "proxies to measure how a community fares, or how it performs after a flood" (Zurich 2016: 5), which according to them is "the outcome of resilience" (Ibid).

The analysis presented above allows the assumption that the actors involved in the *Flood Resilience Program* by Zurich perform a range of practices which represent types of practices which are relevant for the case study villages. Although no direct intervention in the case study villages is carried out, the villages of intervention lie in direct vicinity of the case study region and they are connected through the river system of the Usumacinta. The practices identified may be of relevance because they have material, spatial and social repercussions. The performance of specific interventions which include activities of humanitarian aid as well as activities of development cooperation, e.g. capacity building and formation of new local committees, reshape the socio-spatial patterns in the villages and the larger region. Moreover, the supervision of activities performed in rural areas by international experts may influence practices of national staff as well as of village members.

8.2 SOCIAL PRACTICES PERFORMED BY GOVERNMENT ACTORS

In this section, a selection of those government individuals and groups is referred to whose interventions in the case study villages have revealed high relevance for flood management. As seen in figure 33 above, a dominant group of subjects who interact with the inhabitants of one of the case study villages belong to the government agency DIF. In general, intervention of DIF in villages is performed through the municipal level of the agency. However, in the case study village activities are performed by subjects from the regional level, under the auspices of the regional manager. Activities of DIF in the village were observed by the researcher as part of participant observation. Moreover, an interview with the regional manager provided additional information concerning the social practices performed by the agency which is relevant for flood management.

Interview 2014_P_2: Lines 4–16
[P1]: I approached there the 31st of July this was my first time that I went to […] (anonymised) I didn't know either, for instructions of, of my boss the, the mum of the governor. The mum of the governor visited this community and returning from there she was to help them there and she

> brought tinaco because they had put an application that they had no drinking water. So, she
> brought them to donate their tinacos, yes in order to store their waters. When the lady presented
> herself, the lady Leti Coello, that's her name. Well, she presented to all of them and said <Look,
> my delegate>, that is me, she told them, <my delegate will be here to take care of you>. And to
> me, if you give me a task, until I fulfil it, no?
> [D1]: Sure.
> [P1]: So I said to myself, good, if she cares about the community I have to do this with a well-
> planted work here so that the people feel well and hopefully and all the communities I could visit
> them and could do the work.

As a starting point for intervention, the interview partner mentions the visit of the mother of the governor, who is president of DIF Chiapas. She gave the direct order to the regional delegate to take care of the village and to provide the villagers with their demands. This included a water tank which was distributed to each household directly by the DIF president, as well as a range of goods and services especially from the health sector (wheelchairs, appointment with medical doctors, etc.) provided by the regional delegate (I_2014_P_2: 18–19). In the last sentence of the above interview sequence the delegate points towards the fact, that the way in which this village is attended is not the usual way when compared to other villages. Special focus is put on making "the people feel well". During the interview, the reason for the special intervention carried out in this particular village remains unclear. Besides the influence political processes have on her work the interview partner highlights that municipal politicians very often put high pressures on local villages in order to gain their votes but that often they don't attend their needs (I_2014_P_2: 205). The neglect by political actors who are normally in charge of attending villages is contrasted by the interview partner with the help brought to the village under the mandate of the DIF president. The main activities performed under this mandate in the village are the organisation of a women's gardening group as well as the provision with clothes and other goods, especially the food supply (I_2014_P_2: 40–43). During empirical research in 2014, the researcher took part in several activities of the women group active in the gardening project. It was observable that a considerable amount of formal paper work was involved, which took a considerable amount of time. Every participant had to provide a copy of her identity card as well as other personal data (certificates of school education) to the students and participation lists had to be filled in almost every time a group activity was performed. The primordial importance given to document every step of work and personal information of every participant reflects the official procedures demanded from the DIF office in the state capital in Tuxtla Gutiérrez. Participation in the gardening group involves a financial compensation by DIF for each woman at the end of the project which requires the formal procedure of registration. In the research stay in 2015 the project had ended already and there was no continuation of group work by women in the gardening group. Moreover, the land that had been provided for the activity by a village farmer in 2014, was under regular cultivation again (I_2015_C_31: 3).

Besides the group activity planned as capacity building, DIF provided a series of material assets to the village, including the distribution of *despensas*. The researcher directly asked about the opinion the interview partner had about the distribution of this food support to communities. The delegate describes her view regarding the mentality in the flood–prone villages and possible changes to be stimulated in the specific case study village from outside:

Interview 2014_P_2: Lines 82–87
[P1]: And they like depending on the government and this is sad. I like to do a work to change a little bit the mentality. That is why I said if they work in the project and that they would see, this is not <Today if I feel like making a broom stick and tomorrow no.> It is a daily work. And they will have an alternative, <I like this land to harvest but not to live>. It can be: <I have my house in Zapata and come and harvest here, when the flood comes, I leave>.

The interview partner makes the assumption that people in the case study village are satisfied with a situation of dependency. She however emphasises her intension to change the mentality of people. This change of mentality would include the intensification of economic production, exemplified here in the case of the broom stick production, which so far is an additional economic activity to complement daily labour and to support the financial basis of people in times of flood or other events that influence the economic basis in the village. Moreover, the interview partner points towards a possible the transformation of the land where the case study village is located into a purely agricultural zone, which would involve the moving of the settlement to other places.

The activities carried out by the regional DIF under the mandate of the DIF president for Chiapas are part of social practices that include a discourse of development. Besides the discursive practices, the practical interventions have similarities with activities carried out in projects and programmes of international development cooperation. Development practices in this specific case involve the mentality and behaviour of people, which is intended to be altered into economic thinking and increase in productivity. Development practices also comprise transformations in the internal organisation and processes in the village. This can be identified in the intents to install new types of social groups (e.g. women gardening group or economic cooperatives) which report to external actors following bureaucratic procedures designed for government purposes. It is argued here that the social practices of development are not directed towards an increase in sustainability, in which the village would increase self-sufficiency and flood resistance. It is in contrast the realisation of a strategy composed of many different small interventions that transform life in the village in its cultural, material, political and socio-spatial spheres. Following the verbal expression of the interview partner from DIF, interventions intend to finally transform the socio-mental-spatial setup of the village in fundamental ways which would result in the abandonment of the settlement and a distribution of the population in other villages and towns.

A group of individuals, which due to their official positions belong to government but whose social practices of intervening in the case study region are not part

of official government activity, are political elites. Little information can be retrieved about the activities performed in the case study villages and the sources of
information are not official sources like government programmes, reports or public
announcements. However, spreading rumours and informal information that have
become facts in the case study villages, change local contexts and everyday activities substantially and therefore have to be taken into account in this study. The main
intervention by a group of political elites in the case study region is the acquisition
of lands in direct vicinity of one of the case study villages. While this point has
already been mentioned in the analysis in chapter 6, the specific spatial practices
known to be performed or prepared by the elites are of interest here. The first activity has been the acquisition of a *finca* and adjacent lands by the governor of Chiapas and other government representatives. Expensive cattle are reared in the *finca*,
which promises major economic benefits to the owners. Tree cutting on the land
has been mentioned already as an activity fundamentally intruding in the functioning of the local ecosystem, as it causes the destruction of an important bird habitat.
Moreover, villagers report about an increase in motor-boat activity on the lagoon,
with boats starting from a pier on the *finca* land. The rumour has spread inside the
village and beyond that an eco-tourism project is meant to be built up by the owners
of the cattle ranch. The activities described appear to follow economic interest and
stand in direct contrast to practices and interests of the inhabitants of the case study
village.

Even though national government does not directly interact with or intervene
in the case study region, recent political activities by government have indirect influence on the territorial organisation and other spatially relevant activities in the
near future. As mentioned in chapter 3, current developments in the *Proyecto Mesoamérica* (PM), a large infrastructure and development strategy for various Central
American countries, involve the construction of one or several hydro-energy projects on the Usumacinta. Plans to construct dams on the river had been developed
in the country several times before. The subject has newly gained media attention
since 2015, as the Mexican government has taken steps to establish the necessary
legal environment with Guatemala to intervene in the Usumacinta River. As a transnational river originating in Guatemala and constituting the border between Guatemala and Mexico for a considerable length, legal conditions to alter the hydrological
dynamics of the water body were not given until the recent time. While official
information on the plans is rare, information on current activities on the political
level were obtained during the field research phase in 2015. Little media information is provided in the state of Chiapas, but the newspaper *Tabasco Hoy* based
in Villahermosa, Tabasco, has accessed relevant information through an informant
(I_2015_V_1: 7). In an article of the newspaper it is mentioned, that current plans
to build dams on the Usumacinta River had been developed in 2014 and would not
be realised until 2018 (Robles García, 16.06.2015). According to the interview partner at *Tabasco Hoy* the Mexican National Institute for Transparency, Information
Access and Data Protection (INAI) demanded information from CFE concerning
the current plans to construct dams on the Usumacinta River. Even though information on dam construction plans by CFE should generally be shared publicly, so

far it is only present in INAI (I_2015_V_1: 6). The interview partner mentions that a Memorandum was signed between Mexico and Guatemala in 2015 and that a resolution is in place since July 8th 2015 (I_2015_V_1: 1–2). Following a request posed to INAI by the newspaper to obtain further information, CFE mentions that the information on recent technical studies on the Usumacinta River has been classified as confidential for a period of three years (INAI 2015: 1). The reasons for a confidentiality of information according to CFE are based on economic principles as e.g. the avoidance of comparative disadvantage (Ibid: 6). The information collected by INAI contains an overview of twelve studies that address the geo-hydrological conditions in the location of Boca del Cerro which have been investigated in relation to a feasibility of hydro-energy projects in recent decades (Ibid: 8ff). Additional technical studies were under implementation in 2015 (Ibid: 15). Moreover, communication with the Government of Guatemala has started in 2013 through the Secretary of Energy at national level, which has resulted in a Memorandum of Understanding signed in March 2015 (Ibid). This memorandum allows the completion of necessary studies for "sustainable projects for the generation of electric energy along the international course of the River Usumacinta" (Ibid: 16; translated by the author).

As the information on hydro-energy projects planned on the Usumacinta River exemplifies, current social practices performed by representatives of national government involve the creation of the necessary legal basis for the realisation of large infrastructural projects in the South of Mexico. The reorganisation of spatial entities on a legal and political basis is closely linked to the transnational infrastructure projects of the PM. In a presentation of the PM by the executive director, it is mentioned that those projects were in focus which are "of regional interest and of strategic importance [...] for an economic and socially inclusive development in the region" (Fromm Cea 2015: 2). In the discourse used in this presentation, a dominant perspective on development is put forward that promotes the simultaneous realisation of economic and social objectives (Ibid: 3). In a review of the funding sources of the programme, a strong dominance of private enterprises can be identified. Moreover, it can be identified that those sub-projects already in place are mainly those projects with a focus on economic development (Ibid: 8). Most of the economically oriented sub-projects involve fundamental transformations in the physical-material setup of the region. This can be exemplified by the System of Electric Interconnection of Central American Countries (SIEPAC) or the setup of the International Network of Highways in Central America (RICAM). It is underlined here that interlinked with the transformation of the physical space fundamental socio-spatial transformations are expected to take place. Breaking down the international programme to Mexico, the National Infrastructure Programme 2014–2018 shows the major foci of planned development in the region. One aspect highlighted in the programme is the increase of international competitiveness in different infrastructure sectors (Presidencia de la República 2014a: 33). Concerning the economic development in the South of Mexico it can be recognised that national government intends to attract national and international investment and to reshape the region, including Chiapas, the poorest of all Mexican states (Ibid: 161). A range of European companies show interest in the cooperation with Mexico, e.g. the Spanish enterprise Iberdrola, which is currently involved in the planning of the hydro-energy projects on the Usumacinta River. Linking the different international and national organisations and enterprises to the plans, discourses and activities promoted by national government reveals the complexity of interactions that are performed within specific social practices. Some of these practices are part of flood management in the case study region, others are indirectly linked to it but shape the larger conditions in which they take place.

Besides the preparation of legal frameworks and economic development plans, ongoing preparation of interventions in the Usumacinta water system can be identified from activities in the NGO and science arena. Researchers from ECOSUR (El Colegio de la Frontera Sur), a university in the South of Mexico specialised in the investigation of ecological systems, have been assigned by the national energy commission CFE (Comisión Federal de Electricidad) to carry out environmental analysis concerning the biodiversity of the Usumacinta basin. CFE is in charge of the planning for electricity production and supply in the country including the planning and supervision of hydro-energy projects (INAI 2015: 25). A researcher of ECOSUR interviewed during field research in 2014, mentions that little scientific analysis on the biodiversity of the Usumacinta ecosystem exists so far. Moreover, she explains that the analysis for CFE started in July 2014 and will continue throughout 2015, also involving case studies in the Lower Usumacinta, e.g. Pantanos de Centla in Tabasco as well as the municipalities Catazajá and La Libertad in Chiapas (I_2014_S_3: 8–12). In order to create awareness on the importance of the complex dynamics in the Usumacinta ecosystem for society at large, the interview partner underlines that "you have to give a discourse using numbers" (I_2014_S_3: 15). The national agencies and political as well as economic actors in the country require information that fits into this discourse of numbers. An example of this type of ecologic-economic discourse is the assessment of ecosystem services, carried out for the Usumacinta and other river systems in the South of Mexico (i.a. Anderson 2013; Benke 2010; Tapia-Silva et al. 2015). Two types of discourses exist in the context of dam construction on the Usumacinta: One that focuses on the production of energy and a second one that highlights that dams control floods. The interview partner mentions that in case of the Grijalva River, where four large hydro-energy projects were implemented between the 1950s and 1980s, a major discourse was made for flood control (I_2014_S_3: 18–21). Concerning the current processes, the interview partner expresses in the following way: "We don't want to intervene in this discussion. We want to show the importance of the basins and the connectivity of a river" (I_2014_S_3: 22–23). The word "connectivity" can be identified as a key word in the discursive practice of the interview partner. While not expressing her evaluation directly, she assumes that a hydro-energy project on the river Usumacinta could lead to the inhibition of connectivity. The fact that this evaluation based on scientific experience is not expressed openly but in a contained manner, points towards the sensitivity of the topic of dam construction. The interview sequence points towards a social practice of scientific actors in the country involved in the larger context of flood management: it is the application of a discourse of numbers, which avoids direct evaluation of interventions but points towards the ecological importance of a river system in its current state. Using the discourse of numbers ensures scientific recognition without losing the integrity in the scientific and political system in the country.

8.3 SOCIAL PRACTICES OF OTHER ACTORS RELEVANT FOR FLOOD MANAGEMENT

An agency involved in large-scale transformation of the Southwest Region with relevance to the social practices of flood management is, as has been discussed in the current case of planning hydro-energy projects on the Usumacinta River, the Federal Commission of Electricity (CFE). While CFE had been a public body since its foundation in 1937, the energy reform initiated in Mexico in 2013 modified article 25 of the constitution, thereby transforming CFE and Pemex (Petróleos Mexicanos) into productive enterprises of the state (Vargas Suárez 2015: 131). The CFE Law (Ley de la Comisión Federal de Electricidad) passed in 2014, expresses one of its paradigms in article 4 as following:

> "The Federal Commission for Electricity has the aim to develop entrepreneurial, economic, industrial and commercial activities in terms of its objective to generate economic value and cost effectiveness for the Mexican state as its owner" (Presidencia de la República 2014b, translated by the author).

The new law underlines the fact that CFE can make use of its subsidiary enterprises through contracts and other types of alliances with national and international actors of the public or private sector (INAI 2015: 40). In the case of hydro-energy projects, the Spanish enterprise Iberdrola is involved in the planning process. In April 2016 the enterprise was contracted by CFE in eight projects in Mexico, five of which are wind parks (Marcial Pérez, 26.11.2015; Noceda, 4.04.2016). Iberdrola being the largest private enterprise operating in the Mexican energy sector pursues a strategy of further expansion of the international market for renewable energies that involve e.g. eolic and hydro-electric technologies for the following years (Anderson, 18.07.2016). This strategy is enabled in the discourse of climate change and the strategy of reducing carbon emissions through incentivising renewable energies. The Climate Change Law passed in Mexico in 2012 and amended in 2015 underlines the gradual substitution of fossil energy through renewables as a central part of climate change mitigation (DOF 2015). Moreover, the opening of the energy sector through the energy reform, has resulted in a large range of international and national enterprises competing for the Mexican energy market. This can be exemplified in a tender for a large scale energy scheme published in 2016 that has promoted international competition, including enterprises from the sector of renewables such as Enel Power from Italy, Iberdrola and Fisterra Energy from Spain, Mota Engil from Portugal as well as US-based enterprise SunPower (Jiménez, 28.03.2016). The legal transformation of the energy market created in Mexico with the energy reform results in a gradual change of practices relating to economic and territorial features. While plans are in general made on national or state level, the new social practices which are part of these changes interact with social practices of flood management in the case study region. While examples from the hydro-energy sector in the South of Mexico are not yet explored, examples from the construction of eolic parks helps to identify key interventions by international enterprises. A prominent example is the construction of eolic parks in the Isthmus de Tehuantepec in the Mexican state of Oaxaca. Researchers from the Mexican university UNAM have underlined that the international enterprises "indirectly impacted the agendas of the Oaxacan state and municipal governments, lobbying for attracting foreign investment to install wind power plants as a priority" (Juárez-Hernández & León 2014: paragraph 12). The energy generated in wind parks is distributed mainly to enterprises associated to the producers, e.g. major industrial and commercial consumers (Ibid: paragraph 8).

More direct intervention in the case study region can be observed to be performed by some Mexican enterprises based in the region. Major relevance in regard to territorial dynamics can be ascribed to palm oil producing enterprises as they contribute to considerable transformation of land distribution and use in the municipality of Palenque and neighbouring municipalities of Chiapas and Tabasco. The palm oil sector is an example for an actor from the arena of economy that stimulates regional transformation through the integration into value chains. As palm oil enterprises (e.g. OLEOPALMA with Agroindustrias Palenque S. A.) are important employers in the region they directly interrelate with inhabitants of the case study villages. Temporal employment schemes, land acquisitions and monoculture production systems are likely to alter the socio-spatial practices in the case study villages and larger region.

Other actors that interact with current transformations of the social, political and spatial dynamics of the case study region can be subsumed in the category of Non-Governmental Organisations (NGOs). While different NGOs are present in the region whose possibilities to directly influence conditions of flood management are limited, the specific aims they pursue are of interest for this thesis. When compared to the abovementioned groups, NGOs are set apart by the level of independence in the opinions they can express and strategies they can promote. While it is not argued here, that NGOs are independent from financial constraints or larger political issues, it is however noteworthy that members of NGOs in many cases express opinions beyond the limitations that e.g. members of public agencies have. One NGO that has been active in the context of flood management and environmental issues in the case study region is ProNatura. The Mexican organisation was supported by the international NGO Conservation International in the elaboration of an impact study for the hydro-electric projects planned on the Usumacinta River. In the analysis of the report presented by ProNatura in 2007 (Amezcua et al. 2007) it can be identified that the negative consequences of a hydro-energy project on the river are focused on. In a direct quote from the report, a clear line is drawn between the possible beneficiaries and losers of a hydro-energy project.

"This analysis makes clear the potential inequality of the Tenosique project. If it was implemented in the way we understand it, the project would cause tangible costs to government and
significant harm to nature and the affected communities, at the same time as it would generate
benefits (in the optimist scenario) for the investing enterprise" (Amezcua et al. 2007: 44, translated by the author).

It is argued that in contrast to the studies carried out by ECOSUR, this international
alliance is more free to express in opposition to the construction of a dam on the
Usumacinta. The practices carried out to elaborate the report followed similar scientific procedures as ECOSUR applies, but the international support given by Conservation International provided a different level of autonomy. The major discourse
referred to in the report is ecologic sustainability (Ibid: 47). A river dam is presented
in the report as an "ecological barrier in a high-biodiversity region, interrupting a
variety of biological and social interactions" (Ibid: 14, translated by the author).
Moreover, the project is described as having "apparent shortcomings in terms of
efficiency, equity and sustainability" (Ibid: 15, translated by the author). In an interview carried out with a representative from ProNatura it is highlighted that no
comprehensive study of the ecological characteristic of the Usumacinta basin was
available but rather, a "sea of studies" had to be merged like pieces of a "puzzle"
(I_2014_S_2: 20). However, the severity of effects of a dam construction could be
recognised clearly (I_2014_S_2: 17). The interview partner mentions that since the
time when the ecological studies were carried out the number of population in the
region has increased significantly, making the effect of a possible dam to be felt
more severely (I_2014_S_2: 23). The interview partner also highlights that the regulation of the Usumacinta River was envisioned partly for reasons of flood control.
The interview partner however emphasises the experiences of failed flood control
from the Grijalva River (I_2014_S_2: 77).

Compared to ProNatura which puts forward arguments on a mainly ecosystem-
oriented scientific basis, different discursive practices can be identified to be performed by the NGO Otros Mundos Chiapas (OMC). This organisation which works
on a purely volunteer basis operates in a range of topics that concern socio-environmental effects of large-scale projects like dams or mining in Chiapas. One of the
aims of their work is described as follows: "We work to reveal practices and policies that have resulted in the destruction of the environment and to spread real alternatives in support of the earth and its people" (OMC 2008: 3). In this sentence it
can be identified that a contrast is presented by using the words "destruction" and
"alternatives". The organisation intends to develop and communicate alternative
options for processes that take place in the territories of Chiapas and other parts of
the country. The word "alternatives" is opposed to specific "practices and policies",
referring mainly to government (passing policies) as well as private enterprises
(performing certain practices). In an interview carried out with an NGO member,
the interview partner presents a range of topics in which OMC has revealed practices and policies, as well as strategies on the international level that can cause harm
on the local level. One example concerning the construction of dams is presented
in the interview sequence below:

Interview 2014_S_1: Lines 212–226 & 237–244
[P1]: But. The important part is that, in Mexico it is not needed more electricity, it has enough capacity for at least the next thirty years. Ehm, it would have to be improved the efficiency of companies of the houses of the public sector etcetera. This is not being done. You don´t look for alternatives to improve the technology which exists. Yes? [D1]: Yes. [P1]: And you can reduce the demand for energy in Mexico. And so this 40% of excess is saved and it would not be financially necessary, isn´t it? Ehm, with all the agreements that Mexico has signed of the climate change also, the dams become a very good excuse of investment. Because dams are considered as clean energy. [D1]: Yes. [P1]: And this means well that they receive carbon credits. [D1]: Mhm, for the construction of dams. [P1]: For the construction of dams. And they receive financing from the World Bank, of companies as well that these want to buy carbon credits and say this is a period of transition. Right? […] The enormous capacity of water retention, the production of electricity. These are the most harmful for the environment. There exist other sizes and types but more, the important is the production of methane, in the organic material in decomposition. [D1]: That is below the water. [P1]: Under the water. Yes. This means, you don´t produce carbon, you produce methane and the larger the projects, the more hectares of organic material is below the water that produces methane for decades.

One major point made by the interview partner is that Mexico receives carbon credits for the construction of dams because hydro-energy is regarded as clean energy (The term "clean energy" used in different official documents and reports [i.a. C2ES 2011; US Senate 2012] refers to renewable energies as described under the Clean Development Mechanism [UNFCCC 1997]). The interview partner contrasts the construction of dams triggered by government with the argument that no additional energy is needed in Mexico. Moreover, the interview partner underlines that while the system of carbon credits considers carbon dioxide saved in a hydro-power system it neglects other greenhouse gases such as methane which are produced by the new organic material which would be put under water in the area flooded for the retention basins of the project. Moreover, the interview partner describes a major practice of the NGO, namely that it identifies the logics of governments and private enterprises. This is exemplified in the following interview sequence:

Interview 2014_S_1: Lines 571–574
[P1]: Yes, exactly. But well, all the governments, including United States, finance the construction of infrastructure. New highways, new harbours, new airports. So there is everything to tell the truth. But yes, yes I think that they continue with this same logic of constructing energy so that they can install more companies in the zone.

In both sequences presented above strong separation is made between governments and people. In the first sequence two expressions can be highlighted: "they receive carbon credits" and "they receive financing", both referring to the government of Mexico. In the second sequence it is presented that "they continue with this same logic", while again it is pointed towards Mexican government and it is referred to a larger collective of "all the governments". The interview partner separates a "they" and a "we". This discursive practice is in line with discursive patterns found in text documents prepared by OMC. In several publications, the notions of defence and resistance are highlighted, e.g. in general proclamations of a "defence of biodiversity, against transgenics, transnational predation of the environment and food sovereignty" (OMC 2008: 3) or in reports on the practical activities as part of the "Programme for the defence of the earth and the territory" (OMC 2015: 4, translated by the author).

In addition to written text as the major format of presenting results of analysis OMC has elaborated a series of maps, which put interventions into a spatial context. One of these maps is presented in figure 39 (see colour plate p. 152). It is entitled "Territorial interests in Chiapas – corporate extractivism of the common goods" and contains information on the hydro-energy projects planned on the Usumacinta River, which are presented as having been declared feasible. Besides the investigation of plans of government or economic actors that do or will have detrimental effects, the organisation focuses on capacitation of local and indigenous people to defend their territories and human rights. In a manual called "No seas presa de las represas" (=Don't be a prisoner of the dams; translated by the author) (Castro Soto 2010) a range of expressions are identified which represent the discursive practices of the organisation. The manual is presented to be "the answer to those who feel insecure of what to say, how to do and when to act against any threat, [...] the sum of reasons which justify in parts the position to continue defending the freedom of our rivers" (Bregagnolo cf Castro Soto 2010: 3, translated by the author).

The education material on hydro-energy projects includes political statements as it emphasises the need for political change, e.g. represented in the slogan "Let's change the system, not the climate" (Castro Soto 2010: 1, translated by the author) and highlighting the opposition to a capitalist paradigm. Besides outreach to local communities, the organisation highlights the membership in other Mexican and Latin American organisations as for example the Latin American Network against Dams and in Defence of the Rivers, their Communities and Water (REDLAR), the Latin American Network against Monoculture of Trees (RECOMA) or the Campaign for the Demilitarisation of the Americas (CADA). Moreover, the NGO is active on the international level through collaboration with a range of INGOs such as International Rivers (IR) or the Movement of Victims and Affected people by Climate Change (MOVIAC) (OMC 2008: 3f). The social practices of the NGO include active engagement with solidarity-based grassroots movements and international networks. These practices are directly linked with the critical attitude the NGO holds towards government and the inspiration it takes from local paradigms of "the good life" and international anti-capitalist ideologies. These social practices are of relevance to flood management practices in the case study region as they establish an important basis for interaction between villages along river systems and government as well as the private sector. Capacitation of local people for resistance against the capitalist system and defence of common goods may also involve specific practices linked to practices of flood management. These practices are directed e.g. towards interventions of actors such as civil protection in the definition of risks and vulnerabilities, the overall management of floods and links between DRR measures and development interventions.

8.4 SYNTHESIS OF EMPIRICAL RESULTS CONCERNING SOCIAL PRACTICES OF FLOOD MANAGEMENT

The analysis of the selected actor groups presented above leads to the identification of a range of different social practices with direct or indirect effects on flood management in the case study region. Some of the actors interact directly with people and within the spatial realms of the case study villages while other actors perform practices that influence the broader socio-spatial context. The social practices involve different levels of analytical and practical relevance. Table 17 presents a selective overview on the social practices identified for the different actors and actor groups presented in this chapter and the two earlier chapters (6 and 7). For analytical reasons, the selected practices of the different actors are broken down in the table in the categories practices, discourses, interests, spatial dimensions and risk perspectives. The selection of these practices is made in order to present some of the key social practices in an overview and to allow the reader to identify possible contrasts in the practical activities carried out, in the discourses and in the interests identified. The table triggers the identification of rather known contrasts between different actors, e.g. contrasts between the interests of local population and the interests of private enterprises. While the identification of actors and their practices, discourses and interests is a relevant step in the analysis it has to be highlighted that it presents only a preliminary step towards the understanding of the complexity of social practices involved in flood management. The use of categories may hinder the identification of specific interrelations of practices which occur independently from the specific actors who perform them or it may hinder the identification of contradicting practices among one and the same group of actors. A reconnection of the empirical results with the conceptual basis of this thesis is necessary to gain an understanding of the interrelated and complex patterns of social practices and their spatial and temporal relations, constraints and repercussions. In the following chapter, the theoretical basis of this study is reconsidered in connection with the empirical material and local conceptualisations and an attempt is made to partially represent the complexity of flood management practices in the case study region.

Table 17: Overview of selected actors and their relevant social practices and other dimensions o flood management in the case study region

Actors	Practices	Discourses	Main interests	Spatial dimensions	Risk perspectives
Local population	Observe water level and weather in regular and detailed manner Build tapanco for small animals (poultry) to live on during flood Organise transport on cayucos (small fishing boats) and lanchas (boats with external motor) Wait for despensas (external food supply) Fight for land and resist resettlement	"We are used to it"; "The Good Life"; "Moderni-sation"	Ensure livelihood basis for this and next generation Self-determination and self-sufficiency Provide a better life for children	Local and regional (Chiapas, Tabasco)	Flood is not a risk but a yearly characteristic
Civil protection Regional level	Measure and evaluate floods with local and state information Explain to local people the risks of flood and ways of living Perform measures according to fixed flood classification Elaborate risk analysis with statistical data	"Protection of population"; "Efficient contingency management"; "Resettlement"	Protection of population with limited resources Educate and change mind-set of local people	Regional (Municipal-ities Palenque, Catazajá and La Liertad)	Flood as risk (Risk = exposure x vulnerability) Reduce exposure (resettlement)
Civil protection State level	Monitor hydro-meteorological events and give early warning Capacitate staff concerning prevention and risk analysis Give technical support and organise distribution of food aid during mejor contingencies Give support to regional level through regional supervisor	"Prevention"; "Modern DRR"; "Climate Change"	Protection of population Educate most vulnerably groups Professionalise civil protection in Chiapas	State (and regional)	Flood as risk (Risk = exposure x vulnerability) Main cause of vulnerability: poverty
Zurich/IFRC/Red Cross Mexico	Carry out VCA in rural villages Control quality of VCA through international staff Build resilience that can be measured scientifically Invest money in Mexico as part of corporate social responsibility (Zurich Z)	"Resilience"; "Develop-ment"; "DRR"	Build resilience that can be measured Promote VCA approach internationally Promote member institutions	Local, National, Inter-national	Flood as risk (Risk = exposure x vulnerability) Reduce vulnerability and build resilience

Table 17 (continued)

Actors	Practices	Discourses	Main interests	Spatial dimensions	Risk perspectives
DIF Regional level	Carry out intensified development activities in case study village Keep direct contact to DIF president in Chiapas Closely monitor activities in case study village and provide files for documentation in state capital Avoid contact to DIF municipal level due to political rivalry	"Development"; "Eradication of poverty"	Perform activities in obedience to DIF president and in order to win favour of local people Change mind-set of people in villages	Regional, Local, State	Flood as risk (Risk = exposure x vulnerability) Reduce exposure (resettlement)
National government	Carry out large scale infrastructure development in case study region as part of Plan Mesoamérica Create legal conditions for transnational hydro-energy project with Guatemala on the Usumacinta Promote Climate Change Law and renewable energy sector	"Economic and social development"; "Climate change"	Attract international enterprises to invest in the South of Mexico	National, Inter-national	Flood as risk (Risk = exposure x vulnerability) Dams as flood control measure
Political elites	Aquire lands near case study village to perform economic activities Transform land to gain access to lagoon and make eco-tourism project Keeot indirect contact to village inhabitants (DIF)	"Privileges of political elites"; "Eradication of poverty"	Gain access to strategically and economically important territories as privates property	National	No information
Mexican enterprises (energy sector)	Interaction with international enterprises of the renewable energy sector Finance scientific and technical studies on Usumacinta Prevent information on dam projects to be known to the public Make bids for energy projects on international level	"Climate change"; "Clean energy"; "Comparative advantages"	Raise the production of electricity in the South of Mexico Attract international enterprises from the renewables sector	National, International	No information
International enterprises (energy sector)	Carry out renewable energy projects in Mexico (e.g. windparks in Tehuantepec)	"Climate change"; Clean energy"	Gain access to Mexican and Central American markets	International	No information

Table 17 (continued)

Actors	Practices	Discourses	Main interests	Spatial dimensions	Risk perspectives
Researchers	Elaborate environmental studies concerning biodiversity on the Usumacinta for energy enterprises and government Provide reports with quantitative data on biodiversity	"Connectivity of river system"; "Ecosystem services"	Develop reports based on state-of the-art scientific methods Secure access to research funding	Regional, National	Flood as risk (Risk = exposure x vulnerability)
NGO (environment)	Consult different actors on conservation of biodiversity on the Usumacinta through reports and participation in commissions Carry out projects on climate change mitigation in collaboration with actors from international development cooperation	"Biodiversity"; "Climate change"; "Development"	Promote ecosystem-based approaches to mitigate effects of climate change	State, National, Inter-national	Flood as risk (Risk = exposure x vulnerability) Importance of EcoDRR
NGO (human rights)	Analyse projects of economic intervention and socio-environmental effects in Chiapas Develop and communicate alternatives to economic development projects Capacitation of rural people in support of autonomy and resistance against government projects	"Mother Earth"; "Anti-Capitalism"; "Indigenous people and knowledge"	Present failures of government and the capitalist system Support local villages in defence of human rights Link with (inter-) national networks for support	Local, State, Inter-national	No information

9 *RISKSCAPES* – SOCIAL PRACTICE PATTERNS OF FLOOD MANAGEMENT IN THE SOUTH OF MEXICO

Social practices, the flood and flood risk as well as space and spatiality are core terms and concepts that are used in this thesis to conceptualise socially relevant phenomena in the South of Mexico. The specific nexus between the three theoretical concepts – social practices, risk and space – gives access to a deeper understanding of processes that shape everyday life in the case study region and are at the same time of general social and spatial relevance. In order to present a differentiated and comprehensive answer to the main research question of this study (chapter 2) it is the goal of this chapter to link the richness of empirical results with the theoretical basis and to identify key patterns that can help to further develop conceptual thought in social practice theory. Linking the main theoretical concepts with the empirical results leads to identifying key social practices of flood management and their local conceptualisations. Moreover, it is possible to connect these conceptualisations and social practices with larger social dynamics. Following Niccolini (2017: 105), the study of large-scale phenomena can be accomplished through the analysis of social practices and their multiple interrelations. Adopting a flat ontology, it is considered that the analysis of social practices and their mutual relationships is the adequate way to account for social phenomena such as flood management (Ibid: 100). In this study, the *riskscapes* approach is used in order to analytically review the complex interrelations between social practices of flood management in the South of Mexico. These interrelations are presented in different social practice accounts as "textures" or "rhizomes" (Ibid: 103ff) in order to underline the interwoven character of processes in *the social*.

   The following section discusses the main insights gained through the previously accomplished empirical work on the three theoretical strands on social practices, risk and space. Building up on this, the specific conceptualisation of *riskscapes* of flood management are presented exemplified by selected dynamics between patterns of social practices identified in the field research. This is followed by a discussion of the relevance of the *riskscapes* approach for the larger domains of social practice theory and social geography.

9.1 INSIGHTS ON THE MAIN THEORETICAL CONCEPTS

Chapters 6, 7 and 8 present a large variety of social practices carried out by different individuals and groups in the case study region. An in-depth analysis has proved useful, as the focus on social practices allows the identification of patterns of a material and immaterial kind which contribute to and are part of the dynamic interaction within and beyond social groups. Those social practices have been selected which build up the larger process relevant in this research: flood management. Practices have been highlighted as *doings and sayings* (Schatzki 2001b: 51) and have been analysed as a nexus of mental processes and bodily performances, including discursive patterns and material repercussions of *doings and sayings*. The bodily experience of the flood by village members is expressed e.g. in the measuring of the water level with the own body (the hands, the waist, etc.). Interaction with the flood is revealed to intrinsically involve bodily and material notions, e.g. in the use of small boats for transport in flood times starting immediately at the door of people's houses, or the fear from falling from the boat into the water for those who cannot swim. Flood is an experience that involves all senses, including the aural (e.g. the sound of river water filling small streams). The social practices of flood management performed e.g. by civil protection staff involve distinct materialities, e.g. small dams to retain water in a river bed and metric scales to measure the flood. This example underlines the co-presence of different social practices that relate to one and the same social phenomenon: the flood.

Another social practice of flood management in one of the case study villages is an increased care taking of animals. This practice involves bodily performances such as the construction of *tapancos*, floating wooden platforms for poultry, which is carried out when the water level rises. Inseparably linked to bodily performances, mental processes are an important feature of a social practice. Using the analytical frame taken from the work of Schatzki (2002: 77) (figure 40a), four categories can be distinguished that describe how a social practice is organised. The analysis of the abovementioned social practice of caring for animals is presented in an abstract visual format in figure 40b. Among (1) the *practical understandings* in this case is the ability of a person to build a *tapanco*, which involves knowledge, experience and skill as well as relevant context knowledge to decide what makes sense to do. (2) The *rules* involved in this practice include e.g. the socio-cultural conventions in the village on when and how to build a *tapanco* and who in a family may be allowed or responsible to build it. (3) *Teleoaffective structures* organise the practice insofar as those people who decide to build a *tapanco* pursue specific "acceptable or correct ends" (Schatzki 2001b: 60) with their performance, namely to protect their property and to save animals from drowning. Directly linked to these ends are specific affectual aspects, e.g. the close emotional link people have with these animals and the embedding of these non-human species within a collective identity. Finally, all these mentally mediated processes are linked to (4) *general understandings*, which approve the practice to make sense to be performed. In this case, the building of a *tapanco* makes sense, e.g. due to the general understanding of village members that they are part of nature rather than separate from it. Giving up on the animals and leaving them to drown would contradict general understandings of life and death in the case study region. This example shows the complex internal patterns one social practice involves. Larger complexity is revealed when considering that one social practice is always closely linked to other practices. The practice of *care for animals* is closely linked to the practice *observation of the water level and the weather*.

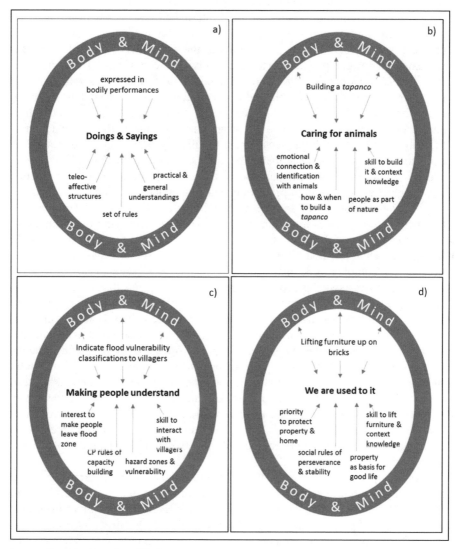

Figure 40: a) Analytical model of social practices on the micro level elaborated on the basis of Schatzki (2002); b)-d) selected practices and their relevant bodily performances and mental processes identified in this study
Source: author's elaboration partly based on accounts by Schatzki (2002)

People build a *tapanco* only if and when it makes sense to them to do so. In order to make sense and become a relevant performance, the knowledge on the water level is necessary, which can only be provided by that other social practice. More-over, a specific temporal relation of these interlinked practices is in place, as the specific point in time, when the *tapanco* is built depends strongly on the temporal characteristics involved in the observation of the water level. Another example is presented for a social practice performed by civil protection staff. The social prac-tice of *making people understand* involves bodily performances as well as mental processes. Among the observable bodily performances can be named the travelling of staff to rural villages in a civil protection car, the indication of flood and vulner-ability classifications through large paper charts or the verbal explanation of measures designed for flood management to village groups. Linked to this are so-cially and materially mediated mental processes of flood and vulnerability classifi-cations (figure 40c). These mental aspects relevant in the organisation of a practice, e.g. the practical skill of civil protection staff to interact with village members and the formal and informal rules for capacity building that are part of the civil protec-tion working procedures highlight the importance of identifying the cultural and social context in which social practices are carried out and the ways in which the linkage of mental and bodily processes contribute to a social practice.Social prac-tices are too complex to be analysed and explained in all details. However, it can be noted that some social practices seem to be more complex than others. The high complexity social practices can have is best exemplified for this study by the social practice pattern *we are already used to it*. This practice pattern is at the same time a key expression and local concept identified in empirical research. It is an extended social practice that involves a range of *doings and sayings* in the case study villages. It involves a multitude of bodily performances which in parts are performed in the same manner since decades but at the same time can underlie dynamic adaptation to new circumstances. Moreover, a myriad of mental processes is linked to the bod-ily performances and forms part of the social practice. One visible performance involved in this social practice is that local people lift the furniture in their house on bricks once the water level rises to a level that threatens to enter the house (figure 40d).

Another performance closely linked to this is the act of male family members to remain in the house even if a house gets flooded for a long period of time. One element of both performances is a *teleoaffective structure* which gives priority to the protection of property and home. This is linked to the *general understanding* of material property and access to land as the basis for a good life. It is a socio-cultural understanding which is closely linked to traditions, the history and experiences of the people as well as their religious beliefs and moral values. This *general under-standing* links to the social *rules* of perseverance and stability which have devel-oped in the case study region over time and reflect in the material world, e.g. in architectural characteristics of houses and villages. A *practical understanding* rel-evant in this example is the skill to lift furniture on bricks without causing damage and the relevant context knowledge on e.g. what makes sense to do when electricity fails and electronic devices such as fridges and television do not work anymore.

This is a strongly limited selection of practical knowledge relevant for the social practice *we are used to it*, while in chapter 7 a range of other aspects are discussed. While presenting these examples is elucidating in the identification of relevant patterns in the social organisation and values of the case study villages, it is only in its´ complexity that the social practice and the relevant context of which this practice forms part can be fully understood. The social practice theory developed by Schatzki stands out through the way in which it embraces complexity and in which it enables micro-level (Schatzki 2002) as well as macro-level analysis (Schatzki 2011; Niccolini 2017) of social practices. Understanding "the social as a mesh of human practices and material arrangements" *(Schatzki 2003: 195)* allows for analytical insight on the phenomena addressed in the case study. Moreover, it is the ontological claim made in *The site of the social* that opens up the analytical view for a myriad of social practices that are part of and form the larger conditions of social life. One of the key insights gained through the confrontation of Schatzki´s accounts with the social phenomena considered in this empirical research lies in identifying the larger context, e.g. economic, historical and political conditions of flood management in Chiapas within one specific social practice while vice versa identifying the dynamic composition of *doings and sayings* and social practices within a larger pattern of social practices. Linking this in-depth insight into *the social* with the concepts of the construction of space is one of the main conceptual developments presented in this thesis.

The term *risk* is a key term used repeatedly in the thesis. It has been presented in chapter 4 as a concept that organises society and is used e.g. as a political instrument but which involves perceptions and decisions by a multitude of actor groups in society. Furthermore, risks have been presented to be formed through a range of social practices. However, the crucial question that resulted from empirical research was the question, which relevance risk as a concept had in the lives of the people involved in the empirical study and consequently which relevance it should have in the concepts of this thesis. It is argued here that it is necessary to regard risk not as a given characteristic but as a term that reflects a process of social construction and socio-political negotiation. In order to describe the social practices related to the dealing with floods the term flood management is used. On the basis of the empirical results it is argued that social practices related to flooding include different conceptualisations of a flood, different knowledges and different values and beliefs connected to the larger realm of flood processes. Only some of these practices directly involve the concept of risk. However, the concept of risk is not dismissed from this thesis as key social practices identified in empirical research reveal that a more recent introduction of specific risk concepts, discourses and risk reduction strategies has fundamental influence on the social practices already in place in the case study region. By pointing towards risk concepts we can identify new patterns of social interaction and processes of transformation that span across different geographic scales. It is e.g. the definition of risk from disaster risk science which can be identified to trickle down from national civil protection agencies to the regional staff of civil protection based in Palenque, Chiapas. The concept of flood risk, which is made measurable through the elaboration of certain indicators fed by data on the physical geographic environment and the population, stands in contrast to other concepts of flood and other concepts of risk that exist in the case study villages and larger region. What a risk is and what it is not for a human being or social group depends fundamentally on the context in which this human being or group is embedded. Moreover, it is crucial to differentiate between risks described by a person for him- or herself and risks attributed to others. The terms used locally to describe the flood situation are rich in information about the perceptions and conceptualisations of a flood. The ambivalent evaluation of flood events as something positive and beautiful on the one hand and something negative on the other hand is, as mentioned before, not a contradiction but a complementation in a large picture of flood phenomena. Moreover, the flood is something people are afraid of and suffer from, but it is at the same time something that benefits. If this notion was to be translated into a concept of risk, it is obvious that here risk is both a threat and an opportunity. Similar results concerning the positive side of the flood is presented by Collins (2009: 599) for environmental hazards in an urban context in the El Paso (USA) – Ciudad Juárez (Mexico) border metropolis. Besides the multitude of conceptualisations of the flood in the local context of the case study villages in this thesis, the concept of risk as a possible negative outcome is attributed mainly to other processes than floods. Loss of harvest, loss of access to land or degradation of the lagoon ecosystem are pressing issues that are mentioned by local people to be among the major risks. These processes are not linked mainly to the flood but to other dynamics. The discursive power of risk concepts that originate in reports and policy papers of the global DRR community is present in the case study region and exerts influence on different administrative levels and sectors. Moreover, the influence of risk concepts and risk related practices reflect in the transformation of material arrangements (e.g. the Grijalva–Usumacinta river system, DRR measures as part of the *Proyecto Mesoamérica*). In summary, it is argued that risk concepts play a subordinate role besides the other theoretical features of this study; however, the practices related to specific risk discourses are of major relevance in order to understand the social phenomena analysed. Therefore, risk concepts cannot be discarded but they are put into a context in which room is given to other conceptualisations related to floods.

As it has been possible to show in the empirical data presented in this thesis, in one geographical location *space* is produced through various processes and by different actors. In line with Lefebvre (1991: 26), it is underlined that it is most relevant to take an in-depth look into the processes of production of space rather than focusing only on the resulting spatial configurations. In these processes, organised by social practices, it is possible to recognise key patterns of *the social*. In taking the triadic concept of the production of space presented by Lefebvre as analytical tool, it is possible to retrace different lines and traditions of production processes. In the empirical material, a large range of social practices are identifiable that can be connected with the three different concepts of *espace*. On the level of *espace perçu* we are able to identify a range of social practices that involve bodily-material productions of space. In the case study villages, the body is a major agent in the perception of space. Social practices of flood management entail the production of a space which encompasses flora, fauna, built and natural environment (houses, water bodies, soil, air, etc.) and amidst them human beings. This part of *perceived space* is directly linked to and overlaps with *lived space*, as it is described further below in this section. The material objects that are part of this spatial dimension point towards a specific characteristic of *espace perçu*. A range of material objects such as the wooden platforms on which furniture is placed or the small boats that lie in front of the villagers' houses have been described as tangible remnants of flood related practices. Besides reminding people of the past they point towards the future. These objects materialise and produce space through processes of anticipation. More than being just material objects, in them a junction of mental and bodily-material processes becomes visible. This provides an empirical example of how perception of space, anticipation of dynamics and production of space in the mental and material world are connected.

The level of *espace conçu* consists of different representations of space. In the empirical case these representations are a result of a series of knowledge producing processes, which entail e.g. various social practices performed by staff members of civil protection in the case study region and in the state of Chiapas. As part of their responsibility, civil protection staff travels to flood-prone villages to gather information and to share information on floods with inhabitants. Following the expressions in the interviews, staff members primarily engage in an activity of "to come and see", a rapid evaluation of flood risk which according to interview partners is possible through first visual impressions gained. Following Lefebvre (1991: 93), these acts of seeing space are acts of creating space in which the visual has a prominent role compared to the other senses. Another activity that contributes to *espace conçu* is carried out by civil protection staff in the tool of risk and vulnerability analysis. This activity is based on statistical and geographic data retrieved from official sources such as INEGI and other providers of quantitative data. It is elaborated by staff members on their computers in the regional office in Palenque. Besides a written report as part of the risk and vulnerability analysis, risk maps are created. On state level, hazard and risk maps already exist and can be retrieved openly from the website of the civil protection agency in Chiapas (Secretaría de Protección Civil 2015b). These maps represent risk in physical space in an abstract manner. Other actors that are mainly active on the level of *espace conçu* are international actors involved in the *Flood Resilience Program* managed by Zurich Insurance. They contribute to the creation of space through social practices that involve the attribution of risk to a physical environment and the people that inhabit it. Risk categories and indicators are built following scientific procedures developed at the involved research institutions in the UK and the US. This example shows how the introduction of conceptualisations made in the academic sector in one world region can prepare the conceptual ground and provide arguments for socio-spatial transformation in another world region.

The third level presented by Lefebvre is *espace vécu* or *lived space*. On this level we understand that everyday social practices produce space as a *lived* and *performed space*, as an interaction between the abstract and the material. It is only in the interaction between both that space can be produced. In the case study villages, local conceptualisations of space and the flood are closely linked to the material entities people have produced in the past and create permanently or on a regular basis. Harvesting *millo* during times of floods and producing broom sticks with this fibre plant is more than a practical measure of adaptation to floods. It is a socio-spatial practice that involves abstract conceptualisations about livelihood and larger economic processes in the region; it involves the reshaping of physical space through the use of material objects and the own body as well as the reshaping of the social space through bodily and verbal expressions that produce meaning in different social groups at different scales. This example underlines that one social practice like harvesting *millo* and producing broom sticks, contributes to the production of a complex *lived space*. This practice is part of a larger social practice pattern called *we are already used to it*. The social practice related to *millo* builds up the larger pattern and is at the same time influenced by it. Changes in other social practices that are part of this pattern can result in the transformation of this social practice. If, as assumed here, the social practice of working outside of the village as wage labourers is performed more in the future so that more people leave the village on a temporal basis, the production of broom sticks will most likely decrease and the level of organisation of the broom stick sales will change.

Interdependency however does not only exist between the social practices performed within one social group but also between social practices of different actors at different scales. This interdependency can be identified in the social practice of creating abstract space by actors from civil protection at state level in Chiapas and in the ways in which this social practice breaks down in decisions on different administrative levels that influence the material and social conditions in the case study region and villages. The atlas of hazards and risks for Chiapas (Secretaría de Protección Civil 2015b) informs regional and local decisions on practical measures to be performed by civil protection staff from Palenque. Decision making interacts in a specific way with the social practices of flood management in the case study villages. Thereby these social practices create a specific *lived space*, in which different processes are performed. The idea of municipal government to resettle local villages from the physical locations along the river to other strips of land is informed and influenced by the different social practices described above. In *lived space,* actors construct hazard zones and reproduce them by locating villages in these zones. This involves material processes such as the donation of food stuff or the withdrawal of basic services that would consolidate permanent settlement in the villages. In the case study villages, the permanence of settlement during flood times, e.g. by not leaving one´s house, can be described as an act of resistance that is part of *lived space*. At the same time however, actors at the village level adopt *doings and sayings* performed by civil protection staff and integrate some of these into their own social practices. As an example the discourse of poverty and the need for external assistance is identified as a relatively new notion in social practices of flood management in the villages. Again parallels can be drawn to the empirical case presented by Collins (2009: 600) for the Paso del Norte. The author describes that local people respond to the specific social practices of marginalisation on the one hand through compliance and on the other hand through resistance (Ibid).

Another important aspect of *lived space* is the notion of time. As Stuart Elden reads the work of Lefebvre, the French author made the attempt to write a "history of space" (Elden 2001: 817), which requires "a radicalising of the notion of history so that it becomes spatialised" (Ibid). Different narratives of history in the case study region guide conceptualisation of space and everyday life. Flood events take place in a large continuum of flood phenomena and other socio-environmental processes and they form part of the larger historical and political processes that have formed and reshaped space since early times of human presence in the case study region until today. We find remnants of various historical events and times in the social practices performed today. It is the importance of access to land which has developed throughout the history of the *lived space*, which involves space in a physical and social sense. As notions of time and temporality are important features of the *riskscapes* identified in this study, this point is discussed in more detail at the end of this chapter.

The triadic concept that describes the production of space has enabled the analysis of a range of social practices of flood management in their temporal and spatial relevance. It is argued that flood management practices are more than fixed measures and strategies but they entail complex social conceptualisations, material objects, bodily performances and mental processes. The work of Lefebvre which is carefully linked with the social practice theory of Schatzki allows the identification of key contradictory practices which result in conflicting processes in the production of space. The identification of "new contradictions" ascribed to Lefebvre and applied to empirical cases by different authors (i.a. Carlos 1999; Gómez Soto 2008; Lima 2012) furthermore provides important conceptual input for further developing the concept of *riskscapes*. The *riskscapes* concept allows getting hold of some of the relevant empirical phenomena observed and analysed in the field. In the following sections, flood management practices in the South of Mexico are reviewed and linked through the analytical lens of *riskscapes*. This allows the identification of synergies and contradictions between social practices which prove relevant not only for the case study region. The empirical findings are believed to be of value for the identification of similar patterns in other parts of Mexico.

9.2 A *RISKSCAPE* OF FLOOD MANAGEMENT IN THE SOUTH OF MEXICO – CONFLICTING SOCIAL PRACTICES AND CONTESTED SPACES

When looking at the social practices of flood management in the case study region, it is recognisable that the concept of risk holds an ambivalent position. On the one hand the concept of risk is not the main feature of the conceptualisations of flood by local people. Important other everyday conceptualisations of the flood are in place and are performed in the *doings and sayings* of village inhabitants. Flooding is part of everyday life as people are "already used to it" and see it as "natural" (see also chapter 6). It is moreover part of the *rhythm* that their lives have, a pattern in time and space that is related with positive and negative characteristics at the same time. On the other hand, the concept of risk is a crucial one when considering the social dynamics and the planning of interventions in the larger case study region. The concepts of risk that different actors such as development agencies, international research institutions, enterprises or civil protection organisations do not only exist in the minds of people but they are performed on discursive and bodily-material levels through specific social practices. These social practices produce and reproduce spaces in the triadic sense of space presented above. Even though risk as a concept does not play a role in the everyday conceptualisations of all actors looked at in this research, it influences the larger conditions in which relevant social practices of flood management are performed. Linking this observation with the theoretical accounts of Schatzki (i.a. 2001, 2002) that have been discussed in this thesis, we argue that risk discourses and more generally risk practices are part of the larger set of social phenomena of which the specific social practices inhabitants of the case study villages perform are part of. Different risk practices are highly relevant because they shape the social and physical spaces in the case study region and because they interact with various practices of flood management.

This point of interaction between different social practices is where the *riskscapes* concept develops its specific value. The concept developed by Müller-Mahn (2013) offers an analytical basis to identify the form and quality of interaction between different relevant social practices. Following the assumptions made by Schatzki it is believed that there are manifold ways of interaction, where one practice e.g. makes

> "courses of action easier, harder, simpler, more complicated, shorter, longer, ill-advized, promising of gain, disruptive, facilitating, obligatory or proscribed, acceptable or unacceptable, more or less relevant, riskier or safer, more or less feasible, more or less likely to induce ridicule or approbation" (Schatzki 2002: 226).

In this case, different social practices related to flood management reveal patterns of synchrony and asynchrony, of strengthening or weakening each other and they contradict or complement each other. Social practices in their specific interrelations create patterns that influence the dynamics of social phenomena, which can reach from the local up to the global scale.

In the following paragraphs some examples of patterns of social practices of flood management from the case study are presented, named hereafter *riskscapes*. Focus is put away from the different actors and instead a look is taken at the myriad of different social practices performed. By doing this the approach refrains from drawing a picture of the structures of the social but focuses on the performances and dynamic contexts of performances. The aim is to overcome too narrowly drawn lines that separate different actors with their particular activities and strategies. The narrow understanding of actors as human carriers of social practices enables to look at the practices rather than putting practices into ordered schemes or structures. It is intended to identify the specific points on the level of social practices where singular practices, their mental and material elements, relate to each other in supportive, opposing, competing, parenthetic or other ways. The word *risk* used in the *riskscapes* presented here can be understood not as a superposition of a risk concept over local concepts, but emphasis is put on possible new risks produced by the interaction of social practices which are of high relevance for the involved case study villages and which at the same time are characterised by a considerable level of uncertainty.

The elaboration of a multidimensional *riskscape* is based on the empirical results generated in this study and especially on the overview of results presented in table 17. In the table, different social practices identified in the case study region are presented and linked to crucial discourses, interests, spatial dimensions and risk perspectives. While the vast amount of empirical material would allow the identification of a large number of patterns within the *riskscape*, in this study three patterns are selected which show specific types of interaction between social practices and are therefore of major importance in relation to the research question.

9.2.1 *Riskscape* of flood management a) with focus on prevention and preparedness

The *riskscape* of flood management is built up by a range of social practices that contribute in different ways to prevention or preparedness towards floods in the case study region. These practices are performed by different groups of subjects, some of them who are physically located in the case study villages, others who perform practices that indirectly influence or interfere with those social practices performed locally. A selection of social practices that form part of the *riskscape* are represented in figure 41 and discussed in the following section. The various blue layers represent different social practices and the layers without description indicate that there are a large range of other social practices that constitute the *riskscape* of flood management in the case of this study. Figures 42 and 43 further below follow the same scheme of representing social practices, but lay a focus on other topics within the *riskscape* of flood management.

A key practice of flood management identified in the case study villages is the regular and detailed observation of the water level in the river and the weather, carried out by almost all village inhabitants. This observation is closely related to the bodily experience of past flood events and the bodily perception of interaction of the human subjects with space. Measurement of water levels is performed by referring to the own body, e.g. in the estimation with hand-widths of how much water is lacking until the river will run over the river bed and enter one's house. Observation and measurements are performed moreover in a collective manner within the case study villages. In addition to the discussions taking place among the villagers of how much is lacking until a house is flooded, the subjects ask for additional observations of rainfall and river level by relatives in villages and towns upstream the river. People indicate that this information from places upstream allows them to estimate how and when a flood will occur in the village.

Besides the multi-faceted practice of individual and collective observation and exchange of information, another practice performed by local people is an important part of flood management. It is the building of floating wooden platforms called *tapancos* on which small animals like poultry can survive during flood times. This practice has been discussed in detail in chapter 7, as it reveals the inclusion of animals into the collective identity of local people and as it is as a practice with visible impacts in the material world. Different to the wooden platforms on which furniture are put and sometimes remain even in times where there is no flood, the *tapanco* does not endure after a flood. It is built directly at the point in time when the water level rises and when animals are threatened to drown in the water. After a flood, a *tapanco* is discarded. Different social practices are carried out in preparation for or early phase of a flood by staff members of the regional civil protection organisation.

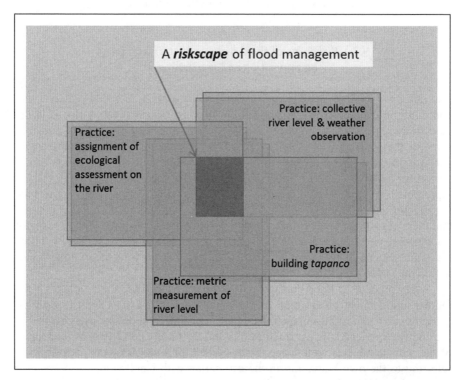

Figure 41: Riskscape of flood management with focus on prevention and preparedness
Source: author's elaboration

The measurement of the water level takes place by using metric scales which have been painted on buildings near the river or the lagoon in the villages by the staff members. Support measures are provided by civil protection staff according to the water level measured by the metric scale. The measurements in the villages are complemented by hydro-meteorological data the staff members obtain from mete-orological agencies through civil protection colleagues from the state level, passed in Tuxtla Gutiérrez. At state level, staff members monitor hydro-meteorological events and run a system for early warning. The social practices of the village inhab-itants and of the civil protection staff members exist almost entirely separate from another. While civil protection staff members communicate with village members through radio in order to obtain the water levels in the river and adjacent water bodies, the measures to be taken locally are not discussed and local practices like using *tapanco* for animals and wooden structures for the furniture are not integrated into civil protection practices. Moreover, the social practices of flood management performed by civil protection staff include the elaboration of a risk analysis. This procedure of risk analysis prescribes the use of statistical data generated by popu-lation census to describe the vulnerability of local people and links them with phys-ical-geographical data which represent the exposure towards hazards like floods or landslides. While the investigation of local specifics of vulnerability is recom-mended in the risk analysis procedures, it is not part of the practice of civil protec-tion staff so far.

Table 18: Selected social practices that form part of the riskscape

- Observe water level and weather in regular and detailed manner [LP]
- Build tapanci for small animals (poultry) to live on during flood [LP]
- Measure and evaluate floods with local and state information [CPR]
- Monitor hydro-meteorological events and give early warning [CPS]
- Capacitate staff concerning prevention and risk analysis [CPS]
- Promote Climate Change Law and renewable energy sector [NG]
- Finance scientific and technical studies on Usumacinta River [E]
- Carry out projects on climate change mitigation in collaboration with actors from interna-tional development cooperation [ENGO]

Abbreviations	CPS: civil protection state level	ENGO: environmental NGO
CPR: civil protection re-gional level	E: energy sector	LP: local population

A social practice that connects indirectly to the before mentioned social practices of flood management is the practice of the Federal Energy Commission (CFE) to assign different research institutions with the elaboration of environmental assess-ments along the Usumacinta River. The knowledge generated in these investiga-tions is intended to support the assessment of environmental impacts the construc-tion of hydro-energy projects on the river would have. The construction of dams for hydro-energy production on other rivers in Chiapas has been framed as a promotor of regional economy and at the same time as a solution to flooding. In current sci-entific debate, socially and environmentally detrimental effects are discussed to re-sult from hydro-energy projects. The practice of assigning research institutions to provide quantitative data of specific environmental services of the Usumacinta

River is in contrast to holistic conceptualisations of a complex, integrated and transnational river system.

9.2.2 *Riskscape* of flood management b) with focus on development and the "good life"

The *riskscape* of flood management involves a range of practices that deal with the paradigms of development and alternative concepts like "the good life" or "la vida buena" and which perform these concepts in processes of the production of space. Various types of relationships can be identified between different social practices, including supportive relationships as well as strong contradictions.

On the level of the case study villages, two major social practices have been identified which stand in contrast to each other. In the presence of these two contradicting social practices a first trace of an ongoing process of social transformation can be identified. The first social practice involves the act of local people from the case study villages to expect and wait for food supply that is given to them by aid organisations and civil protection during floods. People hardly store food for the times of flooding but expect government agencies to take care of their lack of food.

On the other hand, a different social practice is performed in the case study villages which is denominated "self-sufficiency". Many households emphasise this independence from others which is based on the fact that they produce agricultural goods which they sell and use for their own consumption and from which they also make provisions in years of a large harvest. These two practices contradict each other on the level of performance, as they involve action or non-action in preparation of the floods. Moreover, the practices also reflect different self-understandings and identities. The notion of self-sufficiency and autonomy is contrasted by the self-understanding of deficit and dependence. On a spatial level, we can identify an ongoing process of transformation which results in the fact that the use of the land to produce food is slowly replaced by a non-use of land or sales of land to others.

Another social practice interacts with the social practice of self-sufficiency. Staff members from civil protection at the regional level describe their interventions in the case study villages as explaining the risks of flood and local ways of living. They emphasise the act of "making people understand" (see also chapter 8) which represents the core intension of intervention. The self-understanding of the civil protection agency as being able to and responsible for inducing a change of mind of local people stands in direct contrast to the self-understanding of those locals who emphasise their autonomy in thought and action. On the spatial level a contrast exists between those who have a high level of mobility and can go to the villages at any time on one side and the village members whose mobility is limited due to a lack of resources on the other side. Moreover, the acts of "making people understand" includes the intension of civil protection staff to convince villagers of a resettlement of the village which would involve the abandonment of the land currently inhabited. As described above, local people resist interventions on space by remaining in their houses and protecting their land and property even under conditions of high levels and long durations of floods.

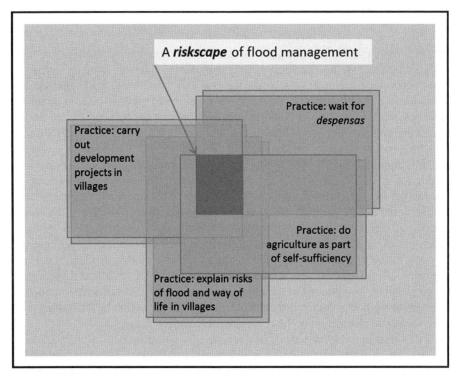

Figure 42: Riskscape of flood management with focus on development and the "good life"
Source: author´s elaboration

Table 19: Selected social practices that form part of the riskscape

- Wait for *despensas* (external food supply) [LP]
- Do agriculture in a self-sufficient manner [LP]
- Explain to local people the risks of flood and ways of living ("make people understand") [CPR]
- Carry out VCA in rural vilages [Z]
- Build resilience and development that can be measured scientifically [Z]
- Carry out intensified development activities in case study villages [DIF]
- Closely monitor activities in case study village and provide files for documentation in state capital [DIF]
- Carry out large scale infrastructure development in case study region as part of Plan Mesoamérica [NG]
- Keep indirect contact to village population (through DIF) [PE]
- Develop and communicate alternatives to economic development projects [HNGO]

Abbreviations	DIF: development agency	NG: national government
CPR: civil protection re-gional level	regional level	PE: political elites
	HNGO: human rights NGO	Z: Zurich Insurance Group
	LP: local population	

Interventions carried out by the governmental development agency DIF in one case study village include the donation of food stuff during times of flood as well as in times when there are no floods. Since the year 2014, interventions in that village have shown a strong increase, especially the support of women-centred development projects and donations given in times of absence of flooding. It can be assumed that this social practice of development intervention stimulates social transformation in the village by making people get used to food support by government and not relying on small-scale agriculture for subsistence any longer. Agricultural production is stimulated by DIF to take place in form of cash crop production of few marketable agricultural products. The question why development intervention has increased in one specific village can be answered when considering the social practice of a group of political elites and especially the spatial transformations in the case study region. The family link of the main political subject, the governor of Chiapas, to the president of DIF, namely the governor's mother, is identified as an important factor that facilitates a type of intervention, which in formal ways would not be possible. The interest of the political elites in land at strategically important locations along the Usumacinta River stimulates the intervention by DIF which provoke social transformation in the respective case study village.

The social practices described in this *riskscape* so far mainly emphasise social transformation in the direction of a development paradigm and the increase in dependence of local people from government actors. One social practice can be identified however, which supports local practices of self-sufficiency. The NGO Otros Mundos Chiapas (OMC) carries out activities in rural villages of Chiapas to strengthen the self-understanding of autonomy. This intervention is part of a social practice performed by various NGOs and INGOs whose goal is to develop and communicate alternatives to economic development projects and which emphasise the traditions and values of indigenous people, underlining that those groups already inhabited the region of Chiapas before arrival of colonial actors.

9.2.3 *Riskscape* of flood management c) with focus on territorial organisation and transformation

As underlined in chapter 8, the challenges of flood management in rural areas in Chiapas are closely linked to questions of territorial organisation, especially to conflicts concerning access to land. The case study region holds strategic value not only for village inhabitants but for a range of individuals and groups from the regional level as well as political elites in Chiapas. The analysis of the specific patterns of territorial organisation in this *riskscape* involves a large range of different social practices that are performed on different geographical scales. It is described in this *riskscape* how interaction between diverse social practices of creating and performing space takes place and it is indicated which the possible consequences are with regard to the socio-spatial dimensions of everyday life. Practices of territorial organisation and of territorial changes can be identified which underline the argument that what can be observed are processes that may contribute to socio-spatial transformation.

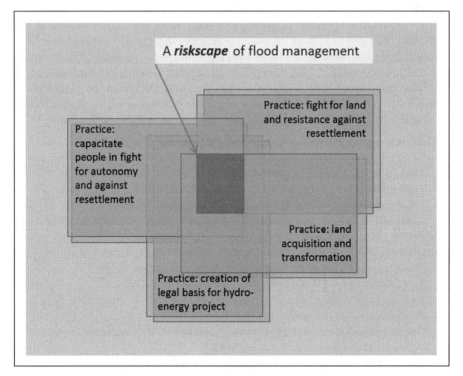

Figure 43: Riskscape of flood management with focus on territorial organisation and transformation
Source: author's elaboration

Some of the *doings and sayings* identified in the villages can be summarised in a social practice of fight for land and resistance. The struggles to obtain land experienced by elderly village members and their ancestors in the past link with current experience that access to land is disrupted by others. A direct connection to flood management can be identified in the fact that staff members from civil protection and government authorities try to intervene in the villages through plans and practical measures of resettlement. As the reason for resettlement it is mentioned that the area where people live is a flood zone and that people are at risk of losing their lives or property due to the regular flood events. Members of the case study villages however resist resettlement or have returned to their original settlement after resettlement had been initiated. Protecting the access to community lands in the case study villages, which are protected under the *ejido* land titles, is an important part of the social practices of everyday life. It is through social practices like cultivating the land with varieties of corn, *millo*, beans, cabbage, chilli and other crops, and remaining in the houses during floods, that space is *lived* in the sense of what Lefebvre calls *espace vécu*.

The two social practices described here exemplify how resistance in space is being performed and promoted locally. In order to understand resistance in this *riskscape*, we need to look at the social practices other groups of people perform. Political elites intervene on the local level e.g. through the practice of land acquisition and transformation. The destruction of habitats of endemic bird species on newly acquired lands in the vicinity of one case study village and increased use of motor boats in the lagoon are examples where the social practice in relation to land contradicts or undermines those social practices village inhabitants perform.

Table 20: Selected social practices that form part of the riskscape

- Fight for land and resist resettlement [LP]
- Perform measures according to fixed flood classifications [CPR]
- Invest money in Mexico as part of corporate social responsibility [Z]
- Create legal conditions for transnational hydro-energy project with Guatemala on the Usumacinta [NG]
- Aquire lands near case study village to perform economic activities [PE]
- Transform land to gain access to lagoon and make eco-tourism project [PE]
- Prevent information on dam projects to be known to the public [E]
- Carry out renewable energy projects in Mexico (e.g. Windparks in Tehuantepec) [IE]
- Provide reports with quantitative data on biodiversity [R]
- Consult different actors on conservation of biodiversity on the Usumacinta River through reports and participation in commissions [ENGO]
- Analyse projects of economic intervention and socio-environmental effects in Chiapas [HNGO]
- Capacitate rural people in support of autonomy and resistance against government projects [HNGO]

Abbreviations	HNGO: human rights NGO	PE: political elites
CPR: civil protection regional level	IE: international enterprises in the energy sector	R: research institutions
DIF: development agency regional level	LP: local population	Z: Zurich Insurance Group
	NG: national government	

The analytical tools provided by Schatzki allow the identification how these social practices are conflicting on the different bodily-material and mental levels. As mentioned before, the separation between the mental and bodily-material is an artificial step in analysis; however, it allows the development of a deeper understanding of the multifaceted character of social practices and the levels on which social practices can interact. In this case, conflict can be identified in almost all aspects of the practices.

Starting with the bodily-material aspects of living with land and performing changes on land it has been described that political elites acquire land and cut trees without knowing or ignoring the fact that endemic bird species find their natural habitat in the trees. Whereas it is difficult to get access to information on the *teleoaffective structures* involved in the social practices performed by political elites from the empirical results, information for village inhabitants are available. There is a direct interaction of local people with the animals that live in the village surroundings. The incorporation of animals into the collective identities of people creates a basis, where the aims people pursue (*telos*) and their affective relationship with other beings guide and are part of social practice. In this line, we can understand that the destruction of the habitat of endemic birds has direct effects on the levels of identity and the material world. How to perform changes in the land and water bodies is, if the conceptualisations made by Schatzki are followed, part of the *practical understandings* of a group of people. In addition to the question how intervention is performed it has to be asked what makes sense to people to do. Whether it makes sense to e.g. cut trees and threaten a bird habitat, is linked to a set of *rules* the specific social group has established. While the set of rules the inhabitants have established do not allow the cutting of the trees under these circumstances, it appears that the political elites have different *practical understandings* and a different set of *rules*. This example shows that the performance on the bodily-material level that can be observed by the human eye is only a small aspect of what the social practices of territorial organisation and changes in land entail. In order to understand bodily-material performances and how they stand in conflict to each other it is necessary to make visible some of the other aspects which are part of social practices.

Another social practice related to territorial organisation can be observed to take place on another geographic scale but to interact with social practices in the case study region. The claiming of interest in the land of the case study region is performed through the signing of a Memorandum of Understanding between the Government of Mexico and the Government of Guatemala. This document prepares the legal basis for transformation of land and the river system of the Usumacinta which will occur if a planned hydro-energy project is realised. Empirical work presented by Aalders (2015: 13) highlights the importance of considering scales and the power inequalities that can be produced by supremacy of global or international scales over local scales. The creation of international agreements and the global interest in the Mexican market of renewable energies has to be considered in this *riskscape*. This can be done through identifying the different social practices from the global to the local level that interact in the *riskscape* of flood management. Besides the legal basis created at an international scale, the scientific basis for intervention is laid through the elaboration of environmental assessments by research institutions. Researchers contracted by CFE mention that what is demanded from them is to indicate through quantitative data the importance of biodiversity and the economic value of the Usumacinta river system. Space is produced in these assessments through abstractions, e.g. in models of ecosystem services elaborated with quantitative methods. This process represents the production of what Lefebvre calls *espace conçu* or *represented space*. This *represented space* of models and maps interacts with the spaces of representation, the *lived space*. In this case again we identify contradictions between the abstract models that ascribe values to strips of land and the social practices performed by village inhabitants which encompass the value people give to their land through *teleoaffective structures*, *rules* and *understandings*. This specific appreciation of land results in acts of resistance on the local level. A social practice that closely interlinks with local resistance is the support given by the NGO Otros Mundos Chiapas to villages in the larger case study region. Besides capacitations given for an increase in autonomy and self-determination, the NGO also produces abstractions of space, e.g. through maps created to display territorial interests in Chiapas (see figure 39 in chapter 8, colour plate p. 152). With these maps, a basis is provided for researchers and activists to make visible possible and ongoing interventions by using the language and symbols of science. While the village inhabitants do not know and therefore cannot make use of these maps, the ideological support provided by the NGO members influences the possibilities for future resistance against resettlement. While the plans of the construction of the hydro-energy project are not made public yet, it can be assumed that by the time they get to the public, social practices of resistance on the different geographical scales will increase. Added to that it is likely that conflicts between contradictory practices will become more visible once the plans for hydro-energy projects are discussed publicly.

9.3 INTERRELATIONS OF SOCIAL PRACTICES IDENTIFIED THROUGH THE *RISKSCAPE* APPROACH

The *riskscape* concept and the analytical approach linked to it makes visible how social practices complement each other and in some cases show strong contrasts between the performances of individuals and groups of people who intervene in the case study region. The identification of different interests is not enough to retrace the dynamic and complex ways in which practices interact and change. Breaking

down interests, paradigms and what is often called "actors" into specific social practices and the nexus of bodily-material and mental processes allows getting hold of
dynamic processes in an empirical as well as conceptual manner and possibly even
anticipate patterns that prepare and predefine social transformation. The *riskscapes*
concept moreover provides insight into the ways in which floods are embedded in
larger temporal and socio-spatial patterns, which comprise flood events in the past
and in the future. Further above the aspect of anticipation has been introduced as an
important characteristic of the social practices of flood management of local people
in the villages. What Wunderlich (2008: 92) states for urban environments can be
transferred to the rural case study region: "social, spatial and natural rhythms influence, shape and characterise everyday life […] and are responsible for the perception of time in places and feelings of identity" (Ibid). Referring to the concept of
rhythm coined by Lefebvre (2004: 6) we can observe notions of repetition and difference in the social practices related to the flood. Just as social practices of flood
management involve different spatialities, they may involve different rhythms.
More light is shed by Schatzki (2009: 36) on aspects of time. Schatzki argues that
there exist "interwoven timespaces [that] are fundamental to human society" (Ibid).
He differentiates *timespaces* from mere objective spatial and temporal properties of
social phenomena (Ibid). His understanding of *timespaces* involves an existential
temporality which conceptualises past, present and future as three dimensions of
temporality which do not order events but are rather general features of activity
(Ibid: 37). The social practices of flood management involve all three temporal dimensions which cannot be represented by specific events in the past, present or
future, but by a continuity of human activity. Feeding this conceptualisation back
into the *riskscape* concept it is crucial to identify not only different spatial levels at
which actors perform social practices but also different temporalities that coexist
and interrelate in specific ways. The repercussions different social practices have in
different temporal dimensions, e.g. how practices performed in the present pave the
way for other practices to be performed in the future can only be assumed with high
levels of uncertainty. Pahl et al. (2014: 377) describe for climate change related
processes that specific social practices can induce a temporal transfer of risks to the
future. Building on their arguments, it is assumed here that certain temporal trajectories for the future of the case study region are produced through the specific patterns of social practices described in the *riskscape* above. Social practices performed and newly introduced in the case study region in current decades can directly and indirectly increase the fragility of ecosystems and vulnerability of local
people towards floods. The empirical and conceptual results underline the relevance
of temporal dimensions of social practices. The integration of temporal trajectories
and processes of social transformation into the *riskscapes* concept is therefore a key
task for future work in geography.

Besides the advantage of the *riskscape* approach to support the analysis of complex patterns of interrelation between social practices, the approach links social
practices to larger patterns of *the social*. A *riskscape* is not one spatial and temporal
complex separate from other complexes of everyday life. It is fundamentally con-

stituted by various characteristics and processes that originate in large-scale phenomena e.g. employment and livelihood, territorial property rights, elections and
the provision of goods and services, or family relations, religion and celebrations.
The dialectic process of firstly taking a detailed look into specific social practices
and secondly considering the complex context of these social practices allows the
identification of new links. Other authors describe this dialectic process as a "zooming in […] and zooming out" and underline its´ usefulness for "representing large-
scale phenomena" (Niccolini 2017: 107). This study underlines that social research
can empirically "see" and conceptually "grasp" how large-scale phenomena represent in selected social practices and how at the same time one social practice reflects
in and constitutes the larger textures of social life. Looking into the different dimensions of flood management as a social phenomenon teaches us lessons not only
of how risk is socially constructed in space but it allows a glimpse into the constitution of social phenomena at large.

10 CONCLUSION – SOCIAL GEOGRAPHIC RESEARCH IN A CONTEXT OF RISK AND DEVELOPMENT

This study presents an enquiry into social phenomena linked to floods along the lower part of the Usumacinta River in the state of Chiapas in the South of Mexico. Starting research with an interest in risk concepts linked to flood events in a specific local setting in Mexico, this study from the beginning on had to review critically those concepts of risk dominant in disaster risk science. I have argued that thinking about floods and the positive and negative repercussions they have in people´s lives requires involving local perspectives and conceptualisations of floods. There are various conceptualisations and perspectives of flooding in the case study region and most importantly, there are multiple social practices that are created and performed around and as part of flood phenomena. The research question developed at the beginning of this study, addresses the fact that there are different social practices and different actors who perform them, which results in negotiations and in some cases in conflicts. These discrepancies reflect in the larger social context of the case study region and they have major repercussions for the lives of local people, way beyond processes of river floods.

In order to get access to an understanding of floods as social phenomena with specific spatial relevance I have chosen to make use of social practice theory. The specific version of social practice theory presented by philosopher Theodore Schatzki (1996, 2001a, 2001b, 2002) provides important delineations of how social practices can be understood, how they are constituted mentally, bodily and materially and how they matter in grasping conceptually what happens in the larger realms of *the social*. The insight into the organisation of social practices through what are called *practical understandings*, *teleoaffective structures*, *rules* and *general understandings* (Schatzki 2002: 77) is of considerable value for the conceptual outline as well as the empirical design of this study. Moreover, the specific version of a social ontology argued for by Schatzki (2002, 2003), allows this study to link to broader perspectives of social theory that ask how the social itself is constituted and how aspects of social life can and do change. In sum, this study is based on the argument that social practices are the key components of socio-spatial phenomena and that it is through social practices that social research can understand processes that happen in real life. Putting specific spatial dynamics at the centre of this study, I introduce the triadic concept of space developed by sociologist Henri Lefebvre (1991). His argument that space is produced through social practices (Lefebvre 1991: 294) feeds into a specific conceptual and analytical approach applied in this study: *Riskscapes*. The *riskscapes* concept developed by Müller-Mahn (2013) in social geography is related to a social-constructivist tradition of risk research and provides a specific entry point into research on floods for this study. Building up on work by Müller-Mahn and Everts (2013), this study develops an own version of a *riskscapes* conceptualisation that accounts for the complex and interwoven character of flood related social practices. This specific *riskscapes* approach which is informed by Schatzki´s version of social practice theory, a critical deconstruction of risk concepts and a triadic concept of space (Lefebvre 1991), is then translated into the design of empirical research in a case study region in Chiapas, Mexico.

The methodology designed for empirical research is influenced by paradigmatic and ethical considerations as well as by the specific research interest linked to flood phenomena. This results in a methodology which is fundamentally based on local perspectives, conceptualisations and an active role of research partners from case study villages as well as a co-researcher from Chiapas. Moreover, questions of ownership of the empirical data that were generated throughout field research are addressed in a transparent manner, which contributes to the sharing of intellectual property rather than the limitation to one researcher or institution alone. In line with an ethnographic orientation of empirical research, a range of qualitative research methods were developed and implemented. Empirical research was carried out in nine months of field research in Mexico split up in two phases, one in 2014 and the other in 2015. The major part of field research involved empirical work in a small number of case study villages in the municipalities Catazajá and Palenque in the North of Chiapas. In addition to that, interviews were carried out with actors relevant for flood management and its larger context in the South of Mexico. This information was complemented by literature research that was carried out in archives, libraries and private collections of chroniclers in the South of Mexico as well as the capital, Mexico D.F. While a variety of research methods was applied, primary importance was given to participant observation, qualitative interviews and different audio-visual methods. The empirical data gained through the application of these methods provided the major part of relevant information on social practices.

The use of audio-visual methods, especially in participatory photography and video workshops carried out in rural settlements along the river Usumacinta, provides crucial and novel information on the characteristics and dynamics of social practices. Audio-visual methods such as participatory video production and photography allow the identification of mental as well as bodily-material processes that form a social practice. Moreover, the visual medium is a highly sensitive medium concerning patterns in the material world. As this research has shown, material objects can "tell a story" about the social practices which they are part of or which they were part of in the past and might be in the future. The visual medium therefore is apt to make visible different temporal dimensions which coexist in the physical world in the present. In addition to the novel information the visual and audio-visual methods provide, they can furthermore complement information obtained through interviews and participant observation. The characteristics of social practices identified through discourse-oriented analysis of verbal expressions can be further explored through a visual approach. It is especially valuable in a context of grounded theory, as local conceptualisations can be informed by both languages, the verbal and the visual. Visual methods are therefore regarded as an important component of qualitative social research, which in specific cases enhances the research methodology, as it is the case for the study presented here. These methods are used in geographic research to a limited extent. However, as this thesis underlines, relevant research questions in the discipline, e.g. those related to social geographic questions on space, materiality and embodiment of human activity, can profit from the use of visual and audio-visual methods in many ways.

This study has generated a series of conceptually and practically relevant results. A myriad of social practices that are part of flood management and its larger socio-spatial context have been identified. The identification of social practices has been mainly informed by the analysis of discursive practices, e.g. through verbal expressions and symbologies obtained from field research as well as through the analysis of bodily-material aspects of social practices accessed through visual media such as photographs and videos which were elaborated in a participatory manner. The social practices identified are performed by a range of different actors, such as inhabitants of the rural villages along the Usumacinta River, staff members of civil protection at different administrative levels, governmental development agencies, NGOs and INGOs as well as economic actors, political elites and many others. Not all actors who perform relevant social practices are physically present in the case study region but they interact with it on different geographic scales. Linking empirical data with social practice theory and the *riskscape* concept has allowed the extraction of relevant patterns of social practices, thereby not focussing on the actors primarily but on the practices, that are performed. The specific *riskscape* approach developed in this study provides the necessary analytical view to identify specific types of interactions between social practices. Thereby, the diverse types of interrelations between social practices described by different authors of practice theory (i.a. Schatzki et al. 2001; Schatzki 2002; Shove et al. 2012; Hui 2017; Niccolini 2017) are reiterated and empirically grounded.

In the municipalities of Catazajá and Palenque some of the social practices of flood management complement or support each other, others exist without major interference and again others stand in contrast or compete to be performed. A highly relevant example of two different sets of practices is the identification of two different systems of flood level measurement that exist in parallel. On the one hand, there are local measurements that involve social practices which give major importance to the human body, e.g. through the use of parts of the body as measurement devices. On the other hand, there are measurement systems by civil protection staff that involve social practices that employ metric scales for measurement on the local level and remote measurement techniques such as rainfall predictions by national meteorological agencies. While both sets of social practices coexist and are perpetuated by the different actors, decisions for support to local villages by civil protection staff are taken based solely on the information gained through metric measurement. Another example hints towards two sets of social practices that coexist in the case study villages and compete for performance. The provision with food and other livelihood resources is performed in the case study villages in social practices of self-sufficient agricultural production including the storage of food for times of flood. However, another type of social practices has been introduced in the villages more recently. This other type involves the buying of ready-made food from stores located outside of the villages, the absence of food storage practices and the expectations of villagers to be supplied with food from civil protection and other external actors in times of floods. Both sets of social practices are connected to a large range of other social practices, including discursive practices of self-sufficiency and resistance on the one hand or discourses of development and Western lifestyle on the other hand. Through these social practices the village inhabitants connect to different actors from outside the villages, e.g. development agencies, political elites or NGOs. I argue that in this empirical example of social practices it is possible to identify complex patterns of interaction that indicate towards processes of social transformation. While in this case, new practices slowly transpire into people´s lives on the level of concepts of life, namely discourses of Western lifestyle, other examples underline changes in the material world which can also result in social transformations.

The *riskscapes* concept and the empirical analysis of the *riskscape* of flood management approaches the complex nature of social practices of flood management. Different social practices do not only link different spatial entities but within one social practice or within one practice pattern different temporal dimensions – past, present and future – coexist and refer to each other. Moreover, specific social practices of flood management can be linked to large phenomena and their performances, e.g. processes of climate change and related policies, international disaster risk reduction strategies, or transformations in the social-ecological systems of Mexico. Within the *riskscape* concept, the analysis of social practices allows us to get one step closer to embracing complexity of social life. Besides a description of social practices, linkages and their effects, the analysis of social phenomena is relevant insofar as it points towards social inequalities. I argue that inequalities come into being and are perpetuated first and foremost on the level of social practices. The thesis provides a detailed analysis of flood management processes in Chiapas, Mexico with local, regional and national relevance. Flood management is characterised as an entangled net of different performances and material patterns, which reflect social, cultural and political aspects, which are characteristic for Mexico and Central America. Moreover, it can be seen that performances and material patterns link different scales. Global processes contribute to the context in which flood management in Chiapas is performed in various ways. In some cases, global dynamics perpetuate conditions of inequality and in other cases they reconfigure them and allow for a more equal distribution of possibilities and resources. Considering the power global dynamics can have in local and regional contexts, this thesis addresses not only challenges of a specific empirical case in Mexico but it provides an insight into the entanglement of different world regions and patterns found repeatedly in relations between the global North and South. Recently planned infrastructural megaprojects among which the construction of a hydro-energy project on the Usumacinta is one case point towards socio-spatial reconfigurations. As recent empirical research from a political ecology perspective shows, dynamics in the global infrastructure sector contribute to the formation of "hydro-social territories" (Boelens et al. 2016; Duarte-Abadía & Boelens 2016; Swyngedouw & Williams 2016). This thesis, observing similar dynamics, provides a detailed analysis of the patterns of practical performance linked to water and water infrastructures in Mexico. Social research that aims towards the identification of and a change in relations of inequality that transgresses geographic scales, needs to provide the necessary analytical tools to bring about change in the social practices humans perform.

The conceptual and empirical approach chosen in this study also involves some limitations. On the conceptual side, the visual models prepared in order to represent the key processes involved in social practices as *doings and sayings* go along with simplifications. It is acknowledged that complex patterns of activity and social practices cannot be represented sufficiently in two-dimensional visualisations. However, I have chosen to prepare visualisations in chapters 4 and 9 in order to put emphasis on specific features of social practices and their interactions. Thereby I firstly underline that it is possible to identify single entities that contribute to the formation of a social practice and secondly I show using the symbology of different layers how different social practices interconnect in a *riskscape*. Linked to the application of a *riskscape* perspective I recognise another limitation of this study which is mainly linked to the empirical design and analytical approach of this study. The comparison of social practices performed by different actors that directly or indirectly interact in the case study region, involves different methodological ways in which social practices have been identified. This results from the fact that it was not possible to carry out participant observation and interviews with all relevant carriers of social practices. While this may be a general challenge in qualitative research given the fact that time periods for field research are influenced by funding and other constraints, it does not undermine the results won in this research. Rather, I argue that even in the absence of a uniform methodological approach towards all social practices, the multitude of sources of data and information and the complementation of methods is believed to have contributed to a versatile empirical research.

Moreover, I regard that it has not been fully possible to make transparent all ways in which the *epistemology of the heart* was implemented throughout this study. As it is mentioned as an epistemological perspective, it guides general thinking and research procedures in this study and therefore would not require to be broken down onto the level of methods. However, given the fact that this "alternative epistemology" is hardly used in geographic publications, research requires more reflection throughout the whole process. A lot of information retrieved from the emotional and other non-verbal expressions of the research partners have only been represented shortly in this study. Therefore, it is one aim for further research to address more in detail the new insights that can be gained in the linkage of "feeling" and "thinking", including the development of the necessary epistemological terminology.

This study refers to various political implications of flood management. However, it does not explicitly apply concepts developed in political geography. While these are considered in the larger context described as a crucial feature in interacting with social practices, it is beyond the scope of this study to address them in detail. Conditions which are of high relevance for political dynamics in Chiapas, such as the EZLN movement, are therefore not conceptualised explicitly in this study. I regard that it is not within the main line of the conceptual and empirical approach chosen here to take up these processes. Notwithstanding, I consider them as most relevant for the South of Mexico in general and they may gain importance in the future given the fact that dominant discourses and interventions by international and multi-national enterprises increase potential conflict with various local discourses, social practices and practice patterns. I argue that the present thesis which puts a different focus can stimulate new works in political geography that could complement the disciplinary perspectives and create mutual benefit for a conceptual development and most importantly for revealing processes of dominant political intervention in different world regions and moreover in various spatial and temporal dimensions.

For the future, it is highly relevant to further develop empirical research on social practices in social geography, thereby firstly contributing to a better understanding and analysis of socio-spatial transformation and secondly influencing processes of socio-spatial transformation in ways that support sustainable practices. While this study has contributed to the conceptual and methodological development of social practice research, more work is needed in order to address recent socio-ecological and technological transformations or the anticipation of processes alike. This concerns on the one hand the research questions to be answered and on the other hand it concerns the ways we as geographers do research including epistemological, methodological and pragmatic orientations. Moreover, improvement should be made in the ways in which geography communicates research results on social phenomena linked to risk concepts, like floods and flood management, with different stakeholders in society.

To begin with, a range of critical questions have been identified in this research. One of them concerns the conceptual nature of social practices and how they can be addressed more precisely. It is the question when a nexus of *doings and sayings* starts to be a social practice. How many single and connected activities are needed to make for a social practice? The challenge of defining what "count[s] as 'a practice'" (Shove et al. 2012: 121) is mentioned by different researchers that deal with social practice theory. This challenge which at first sight is of conceptual nature becomes relevant in empirical research, e.g. when a researcher tries to identify a social practice in the field. As one activity – e.g. pouring water into a bucket – can be part of different social practices at the same time, it can be relevant in the empirical cases studied to identify analytical borders between these different practices. Another relevant question concerns the different temporal dimensions of social practices. While temporality can be understood in a chronological sense, giving importance to specific events in history or in a rhythmic sense of regularly occurring events, authors like Schatzki (2009: 37) highlight the advantages of conceptualising temporality in a co-existence of past, present and future within social practices. While a comparison of the different perspectives on temporality can be conceptually intriguing, it is necessary to translate them into empirical research. While the present thesis has provided empirical insight into different perspectives of temporality, it brings up new conceptual questions. As an example, the difficulty to make distinctions between past and future in a conceptualisation that argues for a co-existence of different temporalities is one conceptual challenge that has to be solved in research in the future. Social science research on risk is a specific area that can come up with conceptually and empirically relevant insights into temporality, as the different concepts of risk and closely related concepts like uncertainty involve diverse perspectives on temporality (i.a. Alaszewski & Burgess 2007; Brace & Geoghegan 2010; Schetter 2013). Moreover, researchers in the discipline of geography, which throughout the history of the discipline have dealt with questions on space and on time, are challenged to contribute to further empirical and conceptual contributions that address temporality.

Besides relevant conceptual questions for future research, this study has brought up questions concerning the empirical research process. Involving a co-researcher in empirical field research is an approach that can challenge a doctoral dissertation. Who owns the intellectual property rights of conceptualisations generated as an outcome of a collaborative research process? How can the synergies built through collaborative work be accounted for in academic research without re-introducing hierarchies among the different research partners? If we want to take seriously the ethical principles for participatory research, members of the scientific communities need to create new options for labelling collectively generated knowledge as such and for further sharing authorship. Another set of questions concerns the introduction of audio-visual and visual media into the methodology and presentation of results in human geography. While visual material is acknowledged to support spoken and written words in empirical research and the presentation of results, the visual as an independent medium for the presentation of results is not yet part of academic tradition. How then can information that is fundamentally visual be communicated to other researchers in a way that it has scientific impact? What is necessary in our disciplines to introduce appropriate visually-based formats of exchange about research results? While it is still tried to fit visual media into conventional formats of research presentation, I am convinced that many other formats are "out there" which could become fruitful instruments of communicating about our research in the future. Performances like those found in creative arts, exhibitions that invite the visitor to physically interact with the exhibit objects or virtual reality labs that make research colleagues get actively involved in a virtual research environment are only some examples of already existing initiatives. It is an open question whether these could be future formats of doctoral dissertations or if they will continue to be only accepted to communicate results to lay people. The options and limitations of these formats must be further explored and I argue for unfearful initiatives among junior and senior academics.

While the questions in the above paragraphs concern future activities in the research community, this section addresses communication of research results and exchange of different kinds of information between researchers, practitioners and civil society. I do not intend to repeat arguments for a better science-policy dialogue in the fields of risks and disasters which are presented in detail by many researchers and practitioners (i.a. Aitsi-Selmi et al. 2015; Ford et al. 2013; Hinkel 2011) Rather, I believe that our societies do not lack arguments for but the social practices of

dialogue and mutual learning in science-policy exchange. Furthermore, local peo-
ple´s sets of knowledge for disaster risk reduction (DRR) have been presented at
sufficient extent (i.a. Reichel & Frömming 2014; UNISDR 2008). International organisations argue for the integration of different types of knowledges into DRR
strategies and measures (IFRC 2015; UNISDR 2008, 2015a). I argue that it is at the
level of practices, namely the mental and bodily-material performances including
risk perceptions and discourses, allocation of financial resources and spatial planning – to present only a short selection – that changes in DRR can take place. As
an outcome of this thesis I present the overall recommendation that the perspective
of social practices needs to be passed on to those subjects who are responsible for
taking decisions for a large group of people, whether grouped in singular settlements, federal states or transnational formations. I argue in line with Shove et al.
(2012: 19) who emphasise that "policy makers need to intervene in the dynamics
of practice if they are to have any chance of promoting healthier, more sustainable
ways of life". It is policy makers and it is those individuals and institutions who
implement policies who so far predominantly lack an understanding of the role social practices play in our social dynamics and processes of transformation. As this
thesis is focused on the topic of flood management and the social construction of
flood risks in disaster risk management, the following points give recommendations
for practitioners who work in these areas. While some of the points present rather
practical ideas that could be implemented without long preparation, others are intended to provide food for thought and thereby stimulate changes, that may take
more time.

a)   Flood perception and discourses on responsibility

International strategies for prevention and preparedness highlight the need to involve local perceptions of risk and of the environment. This study has revealed a
multitude of bodily perceptions that are involved in the perception of the flood. The
human-nature divide is not a major feature in the perceptions of local people as their
bodily perceptions do not only involve the human but extend to non-human beings,
such as the animals living in people´s backyard and the water flowing in the river.
It is believed relevant to provide and protect the necessary ground for these perceptions to take place and develop. As bodily perceptions are key parts of the social
practices of flood management in and beyond the case study villages, I argue that
allowing social practices of flood perception to develop further is a key process of
prevention. If slogans of capacity building and strengthening local knowledge are
to be taken seriously, they need to enable these social practices to reproduce over
space and time. Accomplishing this is a key challenge, because it requires a change
of thinking about local knowledges in the DRR community. Rather than presenting
local perceptions and knowledges as "exotic", their relevance in explaining social
dynamics needs to be taken into account. The example of an ambivalent evaluation
of the flood by local people in this study underlines the argument that phenomena
such as flooding should always be regarded within their context. It is by reflecting
on the context of flood phenomena that practitioners can develop viable long-term

strategies to reduce vulnerability. Another aspect linked to flood perception is the discourse on responsibility for negative effects of flooding. Collins (2009: 600) argues that "naturalizing risk [...] is a narrative strategy for denying responsibility". I argue that it is relevant to regard different discourses related to flood and risk which involve questions of responsibility and which answer them in different ways. Rather than dismissing one and promoting the other, it is useful to see which social practices are linked to the different discourses on responsibility and which social and spatial consequences result from their performance.

b)   Flood management and other practices

The empirical case shows that social dynamics that take place in a contested area where specific types of discourse have gained dominance can hardly be influenced by approaches of only arguing for participatory approaches. Dominant discourses demand for new and relevant information to be provided in specific formats. Provided in other formats, they do not find consideration and cannot reshape the discourse. Transferring this argument to the level of social practices, this study reveals that dominant social practices involve specific bodily-material processes as well as verbally and visually performed processes which cannot be easily reshaped unless new elements are introduced in the same formats already used. It is my argument at this point that it is not realistic to narrow down the myriad of social practices to only few types in order to make the social practices of flood management more participatory. In contrast, I underline the need of expanding the spectrum of accepted formats in these larger sets of social practices. For the case of flood management, it is crucial to approach transformation of related practices from two angles: (1) already existing practices can be influenced and transformed through introducing new knowledges, new understandings and different materialities and (2) practitioners need to reformulate interventions in the disaster cycle and to develop strategies of flood management as part of other practices, as e.g. those related to self-sufficiency. To give one example for the ways in which already existing practices could be reshaped we look at the social practice called "making people understand" which is performed by civil protection staff in Chiapas. While it is neither possible nor reasonable to dismiss the practice completely in the short run, what is possible is to reshape the knowledges used for and the understandings promoted by "making people understand". Instead of presenting e.g. charts with colour codes for alert levels developed in a rationale of emergency management, new material formats and ways of knowledge gaining and sharing could be introduced which would not oppose the general aspiration of the practice but transform it from the inside out. The different ways of knowledge production and sharing presented in this study could be introduced into this practice, among them scientific formats, as well as bodily, material and immaterial formats encountered in the case study villages. It is at this point where the expertise of researchers from the social sciences is asked, not only the conceptual but rather the practical expertise in dealing with various social practices and the actors involved in flood management.

c)   History and current social practices

This study has revealed the importance of considering the history of space. Space is performed through social practices and is a result of earlier performances as well as current ones. Especially on the material level remnants of former social practices can be identified which still influence ongoing dynamics today. Disaster risk management (DRM) needs to take into consideration that change of behaviour which is often a major aim of intervention on local scales, is a process that takes time. Studying changes in social practices from the past and the presence of past practices in the current time is indispensable for collaboration with local people. If the history of a location is ignored, it is almost impossible to support the creation of a future that does not cause major ruptures in the social organisation. On a more general level, the concept of time needs to be reconsidered in DRM. While conceptualisations of risk often imply immediate action and monitoring of specific dynamics in fixed time intervals, local concepts can differ considerably from this and include aspects of circularity and rhythm in the perception of time. A flood understood by some individuals and groups as a singular event that can happen again over time, can alternatively be understood as an expression of the rhythmic character of a river or of life in general. Experience, a bodily and mental pattern that refers to the past can at the same time be understood as anticipation, the bodily and mental perception of things on the way. It is not by integrating local knowledge into DRM measures that these different perceptions of time and the importance of history can be accounted for adequately. It is believed that organisations that are in charge of DRM at different administrative levels need to dismiss unidirectional logics of time and should redesign procedures and time plans consequently.

d)   Spatial planning and mega-projects

As has been argued in this study, flood management is closely linked to questions of access to and management of land. For example, a range of linkages exists between the options and constraints of flood management and those of governmental efforts in spatial planning. Introducing social practice thinking and analysis into the approaches of spatial planning is not only helpful but necessary, because spatial planning guides the development of important socio-spatial dynamics for the future. This is also relevant at different scales as the empirical example of the planned hydro-energy projects on the river Usumacinta indicates. Practices of spatial planning at national and international level that are performed as part of the *Proyecto Mesoamérica* (PM) and the *National Infrastructure Programme* (PNI) interact with practices of land management at sub-national levels. Discourses of economic development and those of flood protection are used in order to promote the construction of dams as part of the hydro-energy projects. Current examples such as the case of Oroville dam underline however that large hydro-energy projects are often not in the necessary conditions to cope with extreme precipitation dynamics (The Guardian, 14.02.2017), phenomena related to climate change that may increase in different parts of the world including the South of Mexico in the future. While social practice thinking may not directly contribute to better construction or planning techniques, it provides options for reshaping those social dynamics that guide discourses of mega-projects and decision-making processes.

e)   Social inequalities and transformations

Building up on to the foregoing points I want to underline that by using a social practice perspective it is possible to make visible inequalities that are produced in the performance of specific social practices patterns. The *riskscape* approach can support the analysis and visualisation of interactions between social practices (and the actors who perform them), making visible otherwise invisible relations, dependencies and even material constraints. Policy makers seek for different types of knowledge to base their decisions on. Why not provide those who have started acknowledging the limitations of behavioural approaches with a practice theory approach that supports reflection of current practices to which policy makers contribute (Shove et al. 2012: 141; Niccolini 2017: 113)? I argue that it is relevant for social researchers to approach practitioners more actively in order to promote social practice thinking. The breaking down of practice theory and the *riskscape* approach into language of practitioners requires simplifications that may be regarded critically in the academic world. However, I argue that by not introducing our conceptual developments into policy making, an important chance to influence processes of social transformation which are ongoing with or without practice theorists would be lost. Some policy areas, in climate change fora like the IPCC or in DRR fora like the Global Platform for Disaster Risk Reduction (GPDRR) search for adequate contributions from the social sciences to enhance our understandings on the options of shaping transformation processes in societies.

While the recommendations presented here have been developed keeping in mind the empirical case of this study, I believe that they can be useful for any person or group that is involved in decision-making, implementation of policies and strategies as well as participation in processes of social transformation. The understanding of the formation and changes of social practices and larger patterns of practices is indispensable for adequate decision-making as well as for practical action in the future. This can touch on issues of climate change mitigation and adaptation, refugee movements and subsequent socio-spatial dynamics, organisation of transnational energy infrastructures as well as a whole range of other processes which the reader of this thesis may consider the pressing issues of present and coming generations. Scholars and practitioners, who consider the potential of a social practice perspective can, as I believe, contribute to shaping recent and future processes of social transformation.

BIBLIOGRAPHY

Aalders, J. T (2015).: Fluid Risks. The Politics of Risk-Scaling at Urban Rivers in Nairobi. Master thesis, University Lund.

Adam, B., Beck, U. & Loon, J. van (Eds.) (2000): The risk society and beyond. Critical issues for social theory. London, Thousand Oaks, New Delhi: Sage Publications.

Aitken, S. C. (2007): Poetic child realism: Scottish film and the construction of childhood. In: Scottish Geographical Journal 123 (1): 68–86.

Aitsi-Selmi A., Blanchard K., Al-Khudhairy, D., Ammann, W., Basabe, P., Johnston, D., Ogallo, L., Onishi, T., Renn, O., Revi, A., Roth, C., Peijun, S., Schneider, J., Wenger, D. & Murray, V. (2015): UNISDR STAG 2015 Report: Science is used for disaster risk reduction. n. P. Retrieved from http://preventionweb.net/go/42848, Accessed 23.02.2017.

Alaszewski, A. & Burgess, A. (2007): Risk, time & reason. In: Health, Risk & Society 9 (4): 349–358.

Alston, L., Mattiace, S. & Nonnenmacher, T. (2008): Coercion, culture and debt contracts: the henequen industry in Yucatan, Mexico, 1870–1915. NBER Working paper series. National Bureau of Economic Research. Retrieved from http://www.nber.org/papers/w13852.pdf, Accessed 03.02.2017.

Amezcua, I., Carreón, G., Marquez, J., Vidal, R. M., Burgués, I., Cordero, S. & Reid, J. (2007): Tenosique: análisis económicoambiental de un proyecto hidroeléctrico en el Río Usumacinta. Conservation Strategy Fund, Conservación Estratégica, Serie Técnica No. 10.

Anderson, B. (2016): "Producimos 15% de la energía de México": Iberdrola. In: Milenio, 7/18/2016. Retrieved from
http://www.milenio.com/negocios/ftmercados-Iberdrola_Mexico-negocios_0_775722616.html, Accessed 02.02.2017.

Anderson, E. P. (2013): Hydropower development and ecosystem services in Central America. Inter-American Development Bank. Retrieved from http://www19.iadb.org/intal/intalcdi-/PE/2013/11508.pdf, Accessed 23.02.2017.

Anderson, R. (1998): Intuitive inquiry: a transpersonal approach. In: Braud, W. & Anderson, R. (Eds.): Transpersonal research methods for the social sciences: Honoring human experience. Thousand Oaks: Sage Publications: 69–94.

Anderson, R. (2004): Intuitive inquiry: an epistemology of the heart for scientific inquiry. In: The Humanistic Psychologist 32 (4): 307–341.

Angulo, F. (2006): Construcción social del riesgo y estrategias adaptativas frente a El Niño. El caso de Tlacotalpan y Cosamaloapan en la cuenca baja del Papaloapan, Veracruz. Documento en proceso. Tesis de Maestría. Universidad Autónoma de la Ciudad de México, México D.F.

Antwi-Agyei, P., Fraser, E. D. G., Dougill, A. J., Stringer, L. C. & Simelton, E. (2012): Mapping the vulnerability of crop production to drought in Ghana using rainfall, yield and socio-economic data. In: Applied Geography 32 (2): 324–334.

Appadurai, A. (1990): Disjuncture and difference in the global cultural economy. In: Theory, Culture & Society (Vol. 7): 295–310.

Arreguín Cortés, F. I., López Pérez, M. & Marengo Mogollón, H. (2011): Los retos del agua en México en el siglo XXI. In: Oswald Spring, Ú. (Ed.): Retos de la investigación del agua en México. Cuernavaca: UNAM: 19–33.

Arrivillaga Cortés, A. (1996): Chicle, chicleros y chiclería. Sobre su historia en El Petén. In: Centro de Estudios Superiores de México y Centroamérica (CESMECA) (Ed.): Anuario 1996 del

Centro de Estudios Superiores de México y Centroamérica. Tuxtla Gutiérrez: CESMECA: 362–398.

Ashley, S. T. & Ashley, W. S. (2005): Flood fatalities in the United States. In: Journal of Applied Meteorology and Climatology (47): 805–818.

Banco Interamericano de Desarrollo (BID) (2014): Plan de Adaptación, Ordenamiento y Manejo integral de las cuencas de los ríos Grijalva y Usumacinta (PAOM). Diagnóstico integrado con identificación de áreas prioritarias. Programa de adaptación a las consecuencias de cambio climático en la provisión de servicios de las cuencas. Resumen Ejecutivo. Bethesda. Retrieved from https://publications.iadb.org/handle/11319/6459?locale-attribute=en, Accessed 23.02.2017.

Bankoff, G. (2001): Rendering the world unsafe: 'vulnerability' as Western discourse. In: Disasters 25 (1): 19–35.

Bankoff, G., Cannon, T., Krüger, F. & Schipper, L. A. (2015): Introduction: exploring the links between cultures and disasters. In: Krüger, F., Bankoff, G., Cannon, T., Orlowski, B. & Schipper, L. A. (Eds.): Cultures and disasters. Understanding cultural framings in disaster risk reduction. London, New York: Routledge: 1–16.

Bankoff, G., Frerks, G. & Hilhorst, D. (Eds.) (2004): Mapping vulnerability. Disasters, development and people. Milton Park, New York: Earthscan.

Banks, M. (1995): Visual research methods. In: Social Research Update 11.

Bastian Duarte, A. I. & Berrío Palomo, L. R. (2015): Saberes en diálogo: mujeres indígenas y académicas en la construcción del conocimiento. In: Leyva Solano, X., Alonso, J., Hernández, R. A., Escobar, A., Köhler, A. & et al. (Eds.): Prácticas otras de conocimiento(s). Entre crisis, entre guerras. San Cristobal de Las Casas: Cooperativa Editorial Retos: 107–132.

Beck, U. (1986): Risikogesellschaft. Auf dem Weg in eine andere Moderne. Frankfurt a. M.: Suhrkamp.

Beck, U. (2006): Living in the world risk society. A Hobhouse Memorial public lecture given on Wednesday 15 February 2006 at the London School of Economics. In: Economy and Society 35 (3): 329–345.

Bedford, T. & Cooke, R. (2001): Probabilistic risk analysis: foundations and methods. Cambridge: Cambridge University Press.

Benjamin, T. (1981): El trabajo en las monterías de Chiapas y Tabasco 1870–1946. In: Historia Mexicana 30 (4): 506–529.

Benke, A. C. (2010): Streams and rivers of North America: Western, Northern and Mexican basins. In: Likens, G. E. (Ed.): Rivers ecosystems ecology: a global perspective. San Diego, Burlington, London, Amsterdam: Elsevier; Academic Press: 373–384.

Benke, A. C. & Cushing, C. E. (Eds.) (2005): Rivers of North America. Burlington: Academic Press; Elsevier.

Berg, B. L. (2009): Qualitative research methods for the social sciences. 7. Ed. Boston: Allyn & Bacon.

Berger, P. L. & Luckmann, T. (1967): The social construction of reality. A treatise in the sociology of knowledge. Harmondsworth, Middx., Engl.: Penguin Books.

Berz, G., Kron, T., Loster, T., Rauch, E., Schimetschek, J., Schmieder, J., Siebert, A., Smolka, A. & Wirtz, A. (2001): World map of natural hazards – a global view of the distribution and intensity of significant exposures. In: Natural Hazards (23): 443–465.

Birkmann, J. (2013): Measuring vulnerability to promote disaster-resilient societies and to enhance adaptation: conceptual frameworks and definitions. In: Birkmann, J. (Ed.): Measuring vulnerability to natural hazards. Towards disaster resilient societies. 2. Ed. Tokyo, New York: United Nations University: 9–79.

Blaikie, N. (2010): Designing social research. The logic of anticipation. 2. Ed. Cambridge: Polity Press.

Blaikie, P., Cannon, T., Davis, I. & Wisner, B. (1994): At risk: natural hazards, people's vulnerability, and disasters. London, New York: Routledge.

Blom, F. (1956): Por "asi es la cuenca del río Usumacinta o Mono Sagrado". In: Compañ Pulido, E. (Ed.): Asi es la cuenca del río Usumacinta o Mono Sagrado. Villahermosa: Compañia Editora Tabasqueña: 9–11.

Blom, F. (1954): Ossuaries, cremation and secondary burials among the Maya of Chiapas, Mexico. In: Journal de la Société des Américanistes 43: 123–136. Retrieved from http://www.persee.fr/doc/jsa_0037-9174_1954_num_43_1_2418, Accessed 23.02.2017.

Boelens, R., Hoogesteger, J., Swyngedouw, E., Vos, J. & Wester, P. (2016): Editorial: Hydrosocial territories: a political ecology perspective. In: Water International 42 (1): 1–14.

Bogner, A., Littig, B. & Menz, W. (2014): Interviews mit Experten. Eine praxisorientierte Einführung. Wiesbaden: Springer VS.

Bohnsack, R. (2009): Dokumentarische Bildinterpretation. Am exemplarischen Fall eines Werbefotos. In: Buber, R. & Holzmüller, H. H. (Eds.): Qualitative Marktforschung. Konzepte – Methoden – Analysen. 2. Ed. Wiesbaden: Gabler Verlag: 951–978.

Bohnsack, R., Nentwig-Gesemann, I. & Nohl, A.-M. (2013): Die dokumentarische Methode und ihre Forschungspraxis. Einleitung. In: Bohnsack, R., Nentwig-Gesemann, I. & Nohl, A.-M. (Eds.): Die dokumentarische Methode und ihre Forschungspraxis. Grundlagen qualitativer Sozialforschung. 3. Ed. Wiesbaden: Springer VS: 9–32.

Bongaerts, G. (2007): Soziale Praxis und Verhalten – Überlegungen zum Practice Turn in Social Theory. In: Zeitschrift für Soziologie 36 (4): 246–260.

Bourdieu, P. (1972): Esquisse d'une théorie de la pratique: Précédé de trois études d'éthnologie kabyle. Geneva: Droz.

Bourdieu, P. (1985): The social space and the genesis of groups. In: Theory and Society 14 (6): 723–744.

Bourdieu, P. (1991): Physischer, sozialer und angeeigneter physischer Raum. In: Wentz, M. (Ed.): Stadt-Räume. Frankfurt a. M.: Campus: 25–34.

Bourdieu, P. (1998): Praktische Vernunft. Zur Theorie des Handelns. Frankfurt a. M.: Suhrkamp.

Boyd, S. (2013). Vulnerability, resilience, and adaptive capacity in response to climate change in Chiapas, Mexico. Oregon State University. Retrieved from https://www.researcEd.ate.net/publication/254558576_Vulnerability_Resilience_and_Adaptive_Capacity_in_Response_to_Climate_Change_in_Chiapas_Mexico, Accessed 20.01.2017.

Brace, C. & Geoghegan, H. (2010): Human geographies of climate change: landscape, temporality, and lay knowledges. In: Progress in Human Geography 35 (3): 284–302.

Breuer, F. (2010): Reflexive Grounded Theory. Eine Einführung für die Forschungspraxis. Unter Mitarbeit von Barbara Dieris und Antje Lettau. 2. Ed. Wiesbaden: VS Verlag für Sozialwissenschaften.

Briones, F. (Ed.) (2012): Perspectivas de investigación y acción frente al cambio climático en Latinoamérica. Número especial de Desastres y Sociedad en el marco del XX Aniversario de LA RED. Mérida: LA RED.

Brouwer, R., Akter, S., Brander, L. & Haque, E. (2007): Socioeconomic vulnerability and adaptation to environmental risk: a case study of climate change and flooding in Bangladesh. In: Risk Analysis 27 (2): 313–326.

Brown, A. (2016): Hydrology: Amazon flooding. In: Nature Climate Change 6 (3): 232.

Browne, K., Bakshi, L. & Law, A. (2010): Positionalities: It´s not about them and us, it´s about us. In: Smith, S. J., Pain, R., Marston, S. A. & Jones III, J. P. (Eds.): The SAGE Handbook of social geographies. London, Thousand Oaks, New Delhi, Singapore: Sage Publications: 586–604.

Brydon-Miller, M. (2008): Ethics and action research: deepening our commitment to principles of social justice and redefining systems of democratic practice. In: Reason, P. & Bradbury, H. (Eds.): The SAGE handbook of action research. Participative inquiry and practice. 2. Ed. London, Thousand Oaks, New Delhi, Singapore: Sage Publications: 199–210.

Calvi, G. M., Pinho, R., Magenes, G., Bommer, J. J., Restrepo-Vélez, L. F. & Crowley, H. (2006): Development of seismic vulnerabililty assessment methodologies over the past 30 years. In: ISET Journal of Earthquake Technology 43 (3): 75–104.

Cambridge University Press: Cambridge dictionary 2017. Retrieved from http://dictionary.cambridge.org/, Accessed 11.02.2017.

Cameron, J. (2016): Focusing on the focus group. In: Hay, I. (Ed.): Qualitative research methods in human geography. 4. Ed. Don Mills: Oxford University Press: 203–224.

Cannon, T. & Müller-Mahn, D. (2010): Vulnerability, resilience and development discourses in context of climate change. In: Natural Hazards 55 (3): 621–635.

Canter, R. L. (2007): Rivers among the ruins: The Usumacinta. In: The PARI Journal 7 (3): 1–24. Retrieved from
http://www.mesoweb.com/pari/journal/archive/PARI0703.pdf Accessed 18.12.2016.

Capdepont Ballina, J. L. (2011): Mesoamérica o el Proyecto Mesoamérica: la historia como pretexto. In: LiminaR. Estudios sociales y humanísticos. 9 (1): 132–152.

Carlos, A. F. A. (1999): "Novas" contradições do espaço. In: Damiani, A. L., Carlos, A. F. A. & Carvalho de Lima Seabra, O. (Eds.): O espaço no fim de século, a nova raridade. São Paulo: Contexto: 62–74.

Carvajal, F. S. (1951): Exposición del representante del Gobierno de Tabasco en la controversia sobre limites con Chiapas (Abril 15, 1908). Publicaciones del Gobierno del Estado. Villahermosa: Cia. Editora Tabasqueña.

Castellanos-Navarrete, A. & Jansen, K. (2013): The drive for accumulation. Environmental contestation and agrarian support to Mexico's oil palm expansion. LDPI Working Paper 43. Retrieved from http://www.plaas.org.za/sites/default/files/publications-pdf/LDPI43Navarrete%26Jansen.pdf, Accessed 20.12.2016.

Castro Soto, G. (2010): No seas presa de las represas. Manual para mejor conocer y combatir esta plaga. San Cristobal de Las Casas. Retrieved from
http://www.otrosmundoschiapas.org/index.php/temas-analisis/39-39-represas/853-manual-no-seas-presa-de-las-represas, Accessed 23.12.2016.

Center for Climate and Energy Solutions (C2ES) (2011): Clean energy standards: state and federal polica options and implications. Arlington. Retrieved from https://www.c2es.org/docUploads/Clean-Energy-Standards-State-and-Federal-Policy-Options-and-Implications.pdf, Accessed 02.02.2017.

Centro Nacional de Prevención de Desastres (CENAPRED) (n. Y.): Website: Conócenos. Antecedentes. Retrieved from
http://www.cenapred.unam.mx/es/dirQuienesSomos/Antecedentes/, Accessed 20.12.2016.

Centro Nacional de Prevención de Desastres (CENAPRED) (2004): Inundaciones. Serie Fascículos Mexico D.F.

Centro Nacional de Prevención de Desastres (CENAPRED) (2014): Atlas nacional de riesgos. Mexico D.F.

Chakravarty, S., Ghosh, S. K., Suresh, C. P., Dey, A. N. & Shukla, G. (2012): Deforestation: causes, effects and control strategies. In: Okia, C. A. (Ed.): Global perspectives on sustainable forest management.: InTech: 3–28. Retrieved from http://www.intechopen.com/books/global-perspectives-on-sustainable-forest-management/deforestation-causes-effects-and-control-strategies, Accessed 02.02.2017.

Clandinin, D. J., Huber, J., Steeves, P. & Li, Y. (2011): Becoming a narrative inquirer. Learning to attend within the three-dimensional narrative inquiry space. In: Trahar, S. (Ed.): Learning and teaching narrative inquiry. Travelling in the borderlands. Amsterdam: Benjamins: 33–52.

Cloke, P., Cook, I., Crang, P., Goodwin, M., Painter, J. & Philo, C. (2004): Practising human geography. London, Thousand Oaks, New Delhi: Sage Publications.

Collier, G. A. & Lowerry Quaratiello, E. (1994): Basta! Land and the Zapatista Rebellion in Chiapas. Oakland: The Institute for Food and Development Policy.

Collins, A. E. (2009): Disaster and development. London: Routledge.

Collins, T. W. (2009): The production of unequal risk in hazardscapes: An explanatory frame applied to disaster at the US–Mexico border. In: Geoforum 40 (4): 589–601.

Comisión Federal de Electricidad (2014): Informe anual 2014. México D.F. Retrieved from http://www.cfe.gob.mx/ConoceCFE/1_AcercadeCFE/Paginas/Informe_Anual_2014.aspx, Accessed 23.02.2017.

Comisión Nacional del Agua (CONAGUA) (2012): Libro Blanco CONAGUA-O1. Programa Integral Hídrico de Tabasco (PIHT). Retrieved from http://www.conagua.gob.mx/conagua07/contenido/Documentos/LIBROS%20BLANCOS/CONAGUA-01%20Programa%20Integral%20de%20Tabasco%20(PIHT).pdf, Accessed 23.02.2017.

Comisión Nacional del Agua (CONAGUA) (2014): Estadísticas del agua en México. Mexico, D.F.

Comisión Nacional del Agua (CONAGUA) (2015a): Reporte del clima en México. Reporte anual 2015. Comisión Nacional del Agua. México D.F.

Comisión Nacional del Agua (CONAGUA) (2015b): Monitor de sequía en México. Agosto 31, 2015. Mexico D. F.

Comisión Nacional del Agua (CONAGUA) (2015c): Monitor de sequía en México. Septiembre 30, 2015. Mexico D. F.

Comisión Nacional del Agua (CONAGUA) (2015d): Monitor de sequía en México. Octubre 31, 2015. Mexico D. F.

Comité Estatal de Información Estadística y Geográfica de Chiapas (CEIEG) (2015): Región XIII Maya. Mapa regional. Tuxtla Gutierrez. Retrieved from http://www.ceieg.chiapas.gob.mx/home/wp-content/uploads/downloads/productosdgei/info_geografica/MapasRegionales/13-MAYA.pdf, Accessed 20.12.2016.

Compañ Pulido, E. (Ed.) (1956): Asi es la cuenca del río Usumacinta o Mono Sagrado. Villahermosa: Compañia Editora Tabasqueña.

Consejo Nacional de Evaluación de la Política de Desarrollo Social (CONEVAL) (2010): Chiapas. Pobreza municipal. Retrieved from http://www.coneval.org.mx/coordinacion/entidades/chiapas/Paginas/pob_municipal.aspx, Accessed 21.02.2017.

Corbin, J. & Strauss, A. (1990): Grounded theory research: procedures, canons, and evaluative criteria. In: Qualitative Sociology 13 (1): 3–21.

Cornwall, A. & Jewkes, R. (1995): What is participatory research? In: Social Science & Medicine 41 (12): 1667–1676.

Corson, M. W. (1999): Hazardscapes in reunified Germany. In: Environmental Hazards 1 (2): 57–68.

Cosgrove, D. (2008): Geography & vision. Seeing, imagining and representing the world. London, New York: I. B. Tauris & Co Ltd.

Costa, F. S., Raupp, I. P., Damázio, J. M., Oliveira, P. D. & Guilhon, G. F. (2014): The methodologies for the flood control planning using hydropower reservoirs in Brazil. In: 6th International Conference on Flood Management, September 2014, São Paulo. Retrieved from http://www.abrh.org.br/icfm6/proceedings/papers/PAP014381.pdf, Accessed 02.02.2017.

Crang, M. (2010): Visual methods and methodologies. In: DeLyser, D. (Ed.): The SAGE handbook of qualitative geography. London, Thousand Oaks, New Delhi, Singapore: Sage Publications: 208–224.

Crate, S. A. (2011): A political ecology of "water in mind": attributing perceptions in the era of global climate change. In: Weather, Climate & Society 3: 148–164.

Crush, J. (1995): Introduction. Imagining development. In: Crush, J. (Ed.): Power of development. London, New York: Routledge: 1–22.

Curran, D. (2013): Risk society and the distribution of bads: theorizing class in the risk society. In: The British Journal of Sociology 64 (1): 44–62.

Curti, G. H. (2008): The ghost in the city and a landscape of life: a reading of difference in Shirow and Oshii's Ghost in the Shell. In: Environment and Planning D 26: 87–106.

Dainton, B. (2014): Time and space. 2. Ed. Montreal, Kingston: McGill- Queen´s University Press.

Darmawi, Sipahutar, R., Bernas, S. M. & Imanuddin, M. S. (2013): Renewable energy and hydro-power utilization tendency worldwide. In: Renewable and Sustainable Energy Reviews 17 (1): 213–215.

Davies, N. (1981): Human sacrifice. In history and today. London, Basingstoke: Macmillan.

Dean, M. (1998): Risk, calculable and incalculable. In: Soziale Welt 49: 25–42.

Deleris, L. A., Elkins, D. & Paté-Cornell, M. E. (2004): Analyzing losses from hazard exposure: a conservative probabilistic estimate using supply chain risk simulation. In: Ingalls, R. G., Rossetti, M. D., Peters, J. S. & Smith, B. A. (Eds.): Proceedings of the 2004 Winter Simulation Conference. Washington D. C.: 1384–1391.

DeLyser, D. (2010): Writing qualitative geography. In: DeLyser, D. (Ed.): The SAGE handbook of qualitative geography. London, Thousand Oaks, New Delhi, Singapore: Sage Publications: 341–358.

Deutsches Komitee Katastrophenvorsorge (DKKV) (2012): Detecting disaster root causes. A framework and an analytic tool for practitioners. Bonn (DKKV Publication Series 48, 48). Retrieved from
http://www.preventionweb.net/files/globalplatform/entry_bg_paper~studydetectingdisaster-rootcausesweb.pdf, Accessed 23.02.2017.

Diario Oficial de la Federación (DOF) (2014): Programa Nacional de Infraestructura 2014-2018. 29.04.2014. Mexico D. F.

Diario Oficial de la Federación (DOF) (2015): Ley General del Cambio Climático. 6.06.2012. Última reforma 02.04.2015. Mexico D. F.

Diario Oficial de la Federación (DOF) (2016): Ley de Aguas Nacionales. 24.03.2016. Mexico D. F.

Diemont, S. A. W. (2006): Ecosystem management and restoration as practices by the indigenous Lacandon Maya of Chiapas, Mexico. Dissertation, Ohio State University.

Dietz, G. (2012): Reflexividad y diálogo en etnografía colaborativa: el acompañamiento etnográfico de una institución educativa "intercultural" mexicana. Reflexivity and dialogue in collaborative ethnography: the ethnographic accompaniment of a Mexican "intercultural" educational institution. In: Revista de Antropología Social 21: 63–91.

DiSalle, R. (2006): Understanding space-time. The philosophical development of physics from Newton to Einstein. Cambridge, New York: Cambridge University Press.

Dittmer, J. (2010): Textual and discourse analysis. In: DeLyser, D. (Ed.): The SAGE handbook of qualitative geography. London, Thousand Oaks, New Delhi, Singapore: Sage Publications: 274–286.

Dodds, K. (1996): The 1982 Falklands War and a critical geopolitical eye: Steve Bell and the if... cartoons. In: Political Geography 15 (6–7): 571–592.

Domínguez R. & Sánchez, J. L. (1990): Las inundaciones en México. Proceso de formación y formas de mitigación. Mexico D.F.: CENAPRED.

Doocy, S., Daniels, A., Murray, S. & Kirsch, T. D. (2013): The human impact of floods: a historical review of events 1980-2009 and systematic literature review. In: PLOS Currents Disasters, 16.04.2013. Retrieved from http://currents.plos.org/disasters/article/the-human-impact-of-floods-a-historical-review-of-events-1980-2009-and-systematic-literature-review/, Accessed 23.02.2017.

Douglas, M. (1992): Risk and blame. Essays in cultural theory. London, New York: Routledge.

Douglas, M. & Wildavsky, A. (1982): Risk and culture: an essay on the selection of technological and environmental dangers. Berkeley, Los Angeles, London: University of California Press.

Duarte-Abadía, B. & Boelens, R. (2016): Disputes over territorial boundaries and diverging valuation languages: The Santurban hydrosocial highlands territory in Colombia. In: Water International 41 (1): 15–36.

Duncan Golicher, J., Ramirez-Marcial, N. & Levy Tacher, S. I. (2006): Correlations between precipitation patterns in Southern Mexico and the El Niño sea surface temperature index. In: Interciencia 31 (2): 80–86.

Dunn, K. (2016): Interviewing. In: Hay, I. (Ed.): Qualitative research methods in human geography. 4. Ed. Don Mills: Oxford University Press: 149–188.

Durand, F., Vernette, G. & Augris, C. (1997): Cyclonic risk in Martinique and response to Hurricane Allen. In: Journal of Coastal Research. Special Issue No. 24 Island states at risk. Global climate change, development and population: 17–27.

Eche, D. (2013): Land degradation, small-scale farms´ development, and migratory flows. The case of Tapachula, Chiapas. Kassel: Kassel University Press. Retrieved from http://www.uni-kassel.de/upress/online/frei/978-3-86219-478-0.volltext.frei.pdf, Accessed 23.02.2017.

Eckstein Raber, S. (1966): El ejido colectivo en México. Mexico D.F.: Fondo de cultura económica.

Egner, H. & Pott, A. (Eds.) (2010a): Geographische Risikoforschung. Zur Konstruktion verräumlichter Risiken und Sicherheiten. Stuttgart: Franz Steiner Verlag.

Egner, H. & Pott, A. (2010b): Risiko und Raum: Das Angebot der Beobachtungsperspektive. In: Egner, H. & Pott, A. (Eds.): Geographische Risikoforschung. Zur Konstruktion verräumlichter Risiken und Sicherheiten. Stuttgart: Franz Steiner Verlag: 9–31.

El País: Desastres Naturales. Retrieved from http://elpais.com/tag/desastres_naturales/a, Accessed 20.12.2016.

Elden, S. (2001): Politics, philosophy, geography: Henri Lefebvre in recent Anglo-American scholarship. In: Antipode 33 (5): 809–825.

Elden, S. (2007): Governmentality, calculation, territory. In: Environment and Planning D 25 (4): 562–580.

Elliott, E. D. (1983): Risk and culture: An essay on the selection of technical and environmental dangers. Faculty Scholarship Series Paper 2192. Yale Law School. Retrieved from http://digitalcommons.law.yale.edu/fss_papers/2192, Accessed 23.02.2017.

Elsner, J. B., Kossin, J. & Jaggar, T. H. (2008): The increasing intensity of the strongest tropical cyclones. (455): 92–95.

Elverfeldt, K. von, Glade, T. & Dikau, R. (2008): Naturwissenschaftliche Gefahren- und Risikoanalyse. In: Felgentreff, C. & Glade, T. (Eds.): Naturrisiken und Sozialkatastrophen. Berlin: Spektrum Akademischer Verlag: 31–46.

Endfield, G. H. (2008): Climate and society in colonial Mexico. A study in vulnerability. Malden: Blackwell Publishers.

Englehardt, J. (2010): Crossing the Usumacintca: stylistic variability and dynamic boundaries in the Preclassic and Early Classic Perios Northwest Maya lowlands. In: Archeological Review from Cambridge 25 (2): 57–76.

Entrena Duran, F. (1990): Efectos politico-sociales y economicos de la reforma agraria en Mexico. In: Estudios rurales latinoamericanos 13 (1–2): 153–169.

Escher, A. & Zimmermann, S. (2006): Visualisierungen der Landschaft im Spielfilm. In: Franzen, B. & Krebs, S. (Eds.): Landscape culture on the move. Microlandscapes. Münster: Westfälisches Landesmuseum für Kunst und Kulturgeschichte: 254–264.

Esselmann, P. C. & Oppermann, J. J. (2010): Overcoming information limitations for the prescription of an environmental flow regime for a Central American river. In: Ecology and Society 15 (1). Retrieved from http://www.ecologyandsociety.org/vol15/iss1/art6/, Accessed 02.02.2017.

Etzold, B. (2014): Raumaneignungen, Regeln und Profite in Dhakas Feld des Straßenhandels – Sozialgeographische Erklärungsversuche auf Grundlage von Bourdieus Theorie der Praxis. In: Geographica Helvetica 69 (1): 37–48.

Ewald, F. (1991): Insurance and risk. In: Burchell, G., Gordon, C. & Miller, P. (Eds.): The Foucault effect. Studies in governmentality. Hempel Hempstead: Harvester-Wheatsheaf: 197–210.

Fairclough, N. (1992): Discourse and social change. Cambridge, Malden: Polity Press.

Fals Borda, O. (2009): Una sociología sentipensante para América Latina. Bogotá: Siglo del Hombre Editores.

Fekete, A. (2010): Assessment of social vulnerability for river-floods in Germany. Dissertation, University Bonn. Retrieved from http://hss.ulb.uni-bonn.de/2010/2004/2004.pdf, Accessed 12.02.2017.

Felgentreff, C. & Dombrowsky, W. (2008): Hazard-, Risiko- und Katastrophenforschung. In: Felgentreff, C. & Glade, T. (Eds.): Naturrisiken und Sozialkatastrophen. Berlin: Spektrum Akademischer Verlag: 13–30.

Felgentreff, C. & Glade, T. (Eds.) (2008): Naturrisiken und Sozialkatastrophen. Berlin: Spektrum Akademischer Verlag.

Fernández Eguiarte, A., Romero Centeno, R. & Zavala Hidalgo, J. (2012): Atlas climático de México y áreas adyacentes. Volumen 1. Universidad Autónoma de México (UNAM) & Comisión Nacional del Agua (CONAGUA). Mexico D. F. Retrieved from http://atlasclimatico.unam.mx/ACM/, Accessed 23.02.2017.

Few, R. (2003): Flooding, vulnerability and coping strategies: local responses to a global threat. In: Progress in Development Studies 3 (1): 43–58.

Folke, C. (2006): Resilience: The emergence of a perspective for social-ecological systems analyses. In: Global Environmental Change 16 (3): 253–267.

Ford, J. D., Knight, M. & Pearce, T. (2013): Assessing the 'usability' of climate change research for decision-making: a case study of the Canadian International Polar year. In: Global Environmental Change 23 (5): 1317–1326.

Fornet-Betancourt, R. (Ed.) (2010): Gutes Leben als humanisiertes Leben. Vorstellungen vom guten Leben in den Kulturen und ihre Bedeutung für Politik und Gesellschaft heute. Dokumentation des VIII. Internationalen Kongresses für Interkulturelle Philosophie. Aachen: Wissenschaftsverlag Mainz.

Foucault, M. (1994): Dits et écrits. IV. 1980–1988. Paris: Gallimard.

Foucault, M. (2007): Security, territory, population. Lectures at the Collège de France, 1977–78. Translated by Graham Burchell. Basingstoke: Palgrave Macmillan.

Fromm Cea, L. (2015): Proyecto Integración y Desarrollo Mesoamérica. Hacia una integración Mesoamericana para un desarrollo eficaz e inclusivo., 2015. Retrieved from http://www.proyectomesoamerica.org/joomla/images/Documentos/ppt%20general%20PM%20con%20cifras%20$%2013%20agosto%2015%20VF.pdf, Accessed 23.02.2017.

Funke-Wieneke, J. (2008): Sich Bewegen in der Stadt. Eine Besichtigung mit Maurice Merleau-Ponty. In: Funke-Wieneke, J. & Klein, G. (Eds.): Bewegungsraum und Stadtkultur. Sozial- und kulturwissenschaftliche Perspektiven. Bielefeld: Transcript Verlag: 75–97.

Gaillard, J. (2015): People's response to disasters in the Philippines: vulnerability, capacities and resilience. New York: Palgrave Macmillan.

Galeano, E. (2004): Las venas abiertas de América Latina. 76. Ed. Mexico D.F.: Siglo XXI.

Gall, M., Nguyen, K. H. & Cutter, S. L. (2015): Integrated research on disaster risk: Is it really integrated? In: International Journal of Disaster Risk Reduction 12: 255–267.

Gallopín, G. C. (2006): Linkages between vulnerability, resilience, and adaptive capacity. In: Global Environmental Change (16): 293–303.

García, E. (1988): Modificaciones al sistema de clasificación climática de Köppen. Mexico D.F.

García, J. A. (2001): Ecología, migración y mestizaje en el Petén. In: Amérique Latine Histoire et Mémoire. Les Cahiers ALHIM. (2). Retrieved from http://alhim.revues.org/595, Accessed 23.02.2017.

García Acosta, V. (Ed.) (1996): Historia y desastres en América Latina. Volumen I: Bogotá: Tercer Mundo Editores.

García Acosta, V. (Ed.) (1997): Historia y desastres en América Latina. Volumen II. Lima.

García Acosta, V. (2005): El riesgo como construcción social y la onstrucción social de riesgos. In: Desacatos. Revista de Antropología Social (19): 11–24.

García Acosta, V., Audefroy, J. F. & Briones, F. (Eds.) (2012): Estrategias sociales de prevención y adaptación. Social strategies for prevention and adaptation. Mexico D.F.: CIESAS.

Garfinkel, H. (1967): Studies in ethnomethodology. Englewood Cliffs: Prentice-Hall.

Garrett, B. L. (2011): Videographic geographies: using digital video for geographic research. In: Progress in Human Geography 35 (4): 521–541.

Garschagen, M. (2014): Risky change? Vulnerability and adaptation between climate change and transformation dynamics in Can Tho City, Vietnam. Stuttgart: Franz Steiner Verlag.

Gavin, M. C., Wali, A. & Vasquesz, M. (2007): Working towards and beyond collaborative resource management: parks, people and participation in the Peruvian Amazon. In: Kindon, S., Pain, R. & Kesby, M. (Eds.): Participatory action research approaches and methods: connecting people, participation, and place. Milton Park, New York: Routledge: 60–70.

Gelling, L. & Munn-Giddings, C. (2011): Ethical review of action research: the challenges for researchers and research ethics committees. In: Research Ethics 7 (3): 100–106.

Gelman, O. & Macías, S. (1984): Toward a conceptual framework for interdisciplinary disaster research. In: Ekistics. Natural Hazards and Human Settlements Disasters — II: Research, Planning and Management. 51 (309): 507–510.

Giddens, A. (1979): Central problems in social theory: action, structure and contra diction in social analysis. Berkeley, Los Angeles: University of California Press.

Giddens, A. (1984): The constitution of society: outline of the theory of structuration. Cambridge: Polity Press.

Giddens, A. (1990): Consequences of modernity. Cambridge: Polity Press.

Giddens, A. (1991): Modernity and self-identity: self and society in the ate modern age. Cambridge: Polity Press.

Glaser, B. G. & Strauss, A. L. (1967): The discovery of grounded theory. Strategies for qualitative research. New Brunswick, London: Aldine Transaction.

Glasze, G., Husseini, S. & Mose, J. (2009): Kodierende Verfahren in der Diskursforschung. In: Glasze, G. & Mattissek, A. (Eds.): Handbuch Diskurs und Raum. Theorien und Methoden für die Humangeographie sowie die sozial- und kulturwissenschaftliche Raumforschung. Bielefeld: Transcript Verlag: 293–314.

Glasze, G. & Mattissek, A. (2009): Die Hegemonie- und Diskurstheorie von Laclau und Mouffe. In: Glasze, G. & Mattissek, A. (Eds.): Handbuch Diskurs und Raum. Theorien und Methoden für die Humangeographie sowie die sozial- und kulturwissenschaftliche Raumforschung. Bielefeld: Transcript Verlag: 153–179.

Gobierno del Estado de Chiapas (1982): Obras del gobierno de Juan Sabines Gutiérrez. Tuxtla Gutierrez: Gobierno del Estado de Chiapas.

Gobierno del Estado de Chiapas (2013): Plan estatal de desarrollo Chiapas 2013–2018. Tuxtla Gutierrez. Retrieved from http://www.chiapas.gob.mx/media/ped/ped-chiapas-2013-2018.pdf, Accessed 23.02.2017.

Gobierno del Estado de Chiapas (2014): Instrumentos Normativos para la formulación del anteproyecto de presupuesto de egresos. Capítulo XXIII. Estadística de población.

Gombrich, E. H. (1972): Symbolic images. Studies in the art of the Renaissance. London, New York: Phaidon.

Gómez-Soto, W. H. (2008): Espaço e política em Lefebvre. In: Pensamento Plural. Pelotas. 3: 179–185.

Gómora Alarcón, J. (2014): La ribera mexicana del Río Suchiate, territorio fronterizo en extinción. Conflictos generados por la abundancia del recurso hídrico. In: Pueblos y fronteras 9 (17): 59–77.

González Pacheco, C. (1995): Los bosques de México y la banca internacional. Mexico D.F.: UNAM.

González Villarreal, F. J. (2009): Evaluación de la vulnerabilidad del sistema de presas del río Grijalva ante los impactos del cambio climático. Instituto Nacional de Ecologia, Instituto de Ingeniería & UNAM. Mexico D.F. Retrieved from http://www.inecc.gob.mx/descargas/cclimatico/ine_a1-027_2009.pdf, Accessed 23.02.2017.

González-Espinosa, M. & Brunel Manse, M. C. (Eds.) (2015): Montañas, pueblos y agua. Dimensiones y realidades de la Cuenca Grijalva. Mexico D. F.: Juan Pablos Editor.

Gracida Galán, J. N. (2011): El rumor de las aguas advierte lo que vendrá. In: Romero Rodríguez, L. del C. (Ed.): Tabasco: Entre el agua y el desastre. Expresiones sociales en torno a sus inundaciones. México D.F.: Clave Editorial: 53–55.

Grant, J., Nelson, G. & Mitchell, T. (2008): Negotiating the challenges of participatory action research: relationships, power, participation, change and credibility. In: Reason, P. & Bradbury, H. (Eds.): The SAGE handbook of action research. Participative inquiry and practice. 2. Ed. London, Thousand Oaks, New Delhi, Singapore: Sage Publications: 589–601.

Guerrero Arias, P. (2011): Corazonar la dimensión política de la espiritualidad y la dimensión espiritual de la política. In: Alteridad. Revista de Ciencias Humanas, Sociales y Educación (10): 21–39.

Guha-Sapir, D. & Ph. Hoyos, R. B. – EM-DAT: Country profile Mexico. In: The CRED/OFDA International Disaster Database – www.emdat.be – Université Catholique de Louvain. Brussels., Accessed 24.11.2016.

Hahn, H. P., Hornbacher, A. & Schönhuth, M. (2008): „Frankfurter Erklärung" zur Ethik in der Ethnologie. Deutsche Gesellschaft für Völkerkunde. Frankfurt a. M. Retrieved from http://www.dgv-net.de/wp-content/uploads/2016/07/DGV-Ethikerklaerung.pdf, Accessed 23.02.2017.

Hammersley, M. (2006): Ethnography: problems and prospects. In: Ethnography and Education 1 (1): 3–14.

Hartley, J. & Benington, J. (2000): Co-research: a new methodology for new times. In: European Journal of Work and Organizational Psychology 9 (4): 463–476.

Hauptmanns, U., Herttrich, M. & Werner, W. (1987): Technische Risiken. Ermittlung und Beurteilung. Berlin, Heidelberg: Springer.

Heffernan, M. (2009): Histories of geography. In: Clifford, N. J., Holloway, S., Rice, S. P. & Valentine, G. (Eds.): Key concepts in geography. London, Thousand Oaks, New Delhi, Singapore: Sage Publications: 3–20.

Herbert, S. (2000): For ethnography. In: Progress in Human Geography 24 (4): 550–568.

Hernández, F. (2008): La investigación basada en las artes. Propuestas para repensar la investigación en educación. In: Educatio Siglo XXI 26: 85–118.

Heron, J. & Reason, P. (2008): Extending epistemology within a co-operative inquiry. In: Reason, P. & Bradbury, H. (Eds.): The SAGE handbook of action research. Participative inquiry and practice. 2. Ed. London, Thousand Oaks, New Delhi, Singapore: Sage Publications: 366–380.

Hewitt, K. (1983): The idea of calamity in a technocratic age. In: Hewitt, K. (Ed.): Interpretations of calamity from the viewpoint of human ecology. Boston: Allen & Unwin: 3–32.

Hinkel, J. (2011): Indicators of vulnerability and adaptive capacity: towards a clarification of the science-policy interface. In: Global Environmental Change 21 (1): 198–208.

Holling, C. S. (1097): Resilience and stability of ecological systems. In: Annual Review of Ecology and Systematics 4: 1–23.

Hopf, C. (2012): Qualitative Interviews. Ein Überblick. In: Flick, U., Kardorff, E. von & Steinke, I. (Eds.): Qualitative Forschung. Ein Handbuch. 9. Ed. Reinbek bei Hamburg: Rowohlt: 349–360.

Hörnqvist, M. (2010): Risk, power and the state. After Foucault. Milton Park, New York: Routledge.

Howard, P. (1998): The history of ecological marginalization in Chiapas. In: Environmental History 3 (3): 357–377.

Hudson, P. F., Hendrickson, D. E., Benke, A. C., Varela-Romero, A., Rodiles-Hernández, R. & Minckley, W. (2005): Rivers of Mexico. In: Benke, A. C. & Cushing, C. E. (Eds.): Rivers of North America. Burlington: Academic Press; Elsevier: 1030–1084.

Hui, A. (2017): Variation and the intersection of practices. In: Hui, A., Schatzki, T. & Shove, E. (Eds.): The nexus of practices: connections, constellations, practitioners. London, New York: Routledge: 52–67.

Huizer, G. (1969): Peasant organization in the process of agrarian reform in Mexico. Social Science Institute Washington University, St. Louis.

Hunt, M. A. (2014): Urban photography/cultural geography: spaces, objects, events. In: Geography Compass 8 (3): 151–168.

Huxley, M. (2007): Geographies of governmentality. In: Crampton, J. & Elden, S. (Eds.): Space, knowledge and power: Foucault and geography. Aldershot, Burlington: Ashgate Publishing: 185–204.

Ingalls, R. G., Rossetti, M. D., Peters, J. S. & Smith, B. A. (Eds.) (2004): Proceedings of the 2004 Winter Simulation Conference. Washington D. C.

Ingold, T. (1993): The temporality of the landscape. In: World Archaeology 25 (2): 152–174.

Ingold, T. (2000): The perception of the environment: essays in livelihood, dwelling and skill. London: Routledge.

Ingold, T. & Vergunst, J. (Eds.) (2008): Ways of walking: ethnography and practice on foot. Aldershot, Burlington: Ashgate Publishing.

Instituto de Protección Civil (n.Y.): PROCEDA. Procedimiento estatal de alerta por lluvias. Tuxtla Gutierrez.

Instituto de Protección Civil (2014): PP5 – Programa Preventivo de Protección Civil. Tuxtla Gutierrez.

Instituto Nacional de Estadística y Geografía (INEGI) (2010a): Prontuario de información geográfica municipal de los Estados Unidos Mexicanos. Emiliano Zapata, Tabasco. Clave geoestadística 27007. INEGI. Retrieved from http://www3.inegi.org.mx/sistemas/mexicocifras/datos-geograficos/27/27007.pdf, Accessed 23.02.2017.

Instituto Nacional de Estadística y Geografía (INEGI) (2010b): Censo de población y vivienda. INEGI. Retrieved from http://www.inegi.org.mx/est/contenidos/proyectos/ccpv/cpv2010/Default.aspx, Accessed 23.02.2017.

Instituto Nacional de Transparencia, Acceso a la Información y Protección de Datos Personales (INAI) (2015): Recurso de revisión. RDA 1726/15, Comisión Federal de Electricidad, Folio: 1816400023815. Mexico D.F.

Instituto para el Federalismo y el Desarrollo Municipal (INAFED) (2010): Enciclopedia de los municipios y delegaciones de Mexico. Estado de Tabasco. n. P.

Intergovernmental Panel on Climate Change (IPCC) (2001): Climate change 2001: impacts, adaptation, and vulnerability. Contribution of Working Group II to the Third Assessment Report of the Intergovernmental Panel on Climate Change. Cambridge: Cambridge University Press.

Intergovernmental Panel on Climate Change (IPCC) (2007): Climate Change 2007: Synthesis Report. Contribution of Working Groups I, II and III to the Fourth Assessment Report of the Intergovernmental Panel on Climate Change [Core Writing Team, Pachauri, R.K and Reisinger, A. (Eds.)]. Geneva.

International Federation of the Red Cross and Red Crescent Societies (IFRC) (2006): What is VCA? An introduction to vulnerability and capacity assessment. International Federation of the Red Cross and Red Crescent Societies. Geneva. Retrieved from http://www.ifrc.org/Global/Publications/disasters/reducing_risks/VCA/whats-vca-en.pdf, Accessed 23.02.2017.

International Federation of Red Cross and Red Crescent Societies (IFRC) (2014): World disasters report. Focus on culture and risk. Geneva.

International Federation of Red Cross and Red Crescent Societies (IFRC) (2015): World disasters report. Focus on local actors, the key to humanitarian effectiveness. Geneva.

Jammer, M. (1969): Concepts of space: the history of theories of space in physics. 2. Ed. Cambridge: Harvard University Press.

Janesick, V. J. (2001): Intuition and creativity: a Pas de Deux for qualitative research. In: Qualitative Inquiry 7 (5): 531–540.

Jarvis, D. S. L. (2007): Theorizing risk: Ulrich Beck, globalization and the rise of the risk society. Lee Kuan Yew School of Public Policy, Singapore.

Jeschke, S. & Jakobs, E. M. (2013): Einführung in den Band. In: Jeschke, S., Jakobs, E. M. & Dröge, A. (Eds.): Exploring Uncertainty: Ungewissheit und Unsicherheit im interdisziplinären Diskurs. Wiesbaden: Springer: 7–16.

Jiménez, G. (2016): Enel, Iberdrola y el Sindicato Mexicano de Electricistas compiten en la primera subasta eléctrica. In: Economía Hoy, 3/28/2016. Retrieved from www.economiahoy.mx/empresas-eAm-mexico/noticias/7449152/03/16/Enel-Iberdrola-y-el-Sindicato-Mexicano-de-Electricistas-compiten-en-la-primera-subasta-electrica.html, Accessed 23.02.2017.

Jiménez Pérez, J. A. & Köhler, A. (2012): Producción videográfica y escrita en co-labor. Un camino donde se encuentran y comparten conocimientos. In: Kummels, I. (Ed.): Espacios mediáticos: cultura y representación en México. Berlin: Tranvía: 319–346.

Jonkmann, S. N. (2005): Global perspectives on loss of human life caused by floods. In: Natural Hazards 34 (2): 151–175.

Juárez-Hernández, S. & León, G. (2014): Wind energy in the Isthmus of Tehuantepec: development, actors and social opposition. In: Problema del Desarrollo. Revista latinoamericana de economía 45 (178). Retrieved from http://www.probdes.iiec.unam.mx/en/revistas/v45n178/body/v45n178a6_1.php, Accessed 02.02.2017.

Kant, I. (1922): Critique of pure reason. Translated into English by F. Max Molar. 2. Ed. London: Macmillan.

Kapur, A. (2010): Vulnerable India: a geographical study of disasters. New Delhi: Sage Publications.

Karmalkar, A. V., Bradley, R. S. & Diaz, H. F. (2011): Climate change in Central America and Mexico: regional climate model validation and climate change projections. In: Climate Dynamics 37 (3–4): 605–629.

Kasperson, R. E. (2009): Coping with deep uncertainty: challenges for environmental assessment and decision-making. In: Bammer, G. & Smithson, M. (Eds.): Uncertainty and risk. Multidisciplinary perspectives. London, Sterling: Earthscan: 337–347.

Kauppinnen, K. (2012): Subjects of Risk. Neoliberale Gouvernementalität in einer gegenwärtigen Frauenzeitschrift. In: Dreesen, P., Kumięga, Ł. & Spieß, C. (Eds.): Mediendiskursanalyse. Diskurse – Dispositive – Medien – Macht. Wiesbaden: Springer VS: 189–206.

Kearns, R. A. (1997): Constructing (bi)cultural geographies: research on, and with, people of the Hokianga district. In: New Zealand Geographer 53 (2): 3–8.

Kearns, R. A. (2016): Placing observation in the research toolkit. In: Hay, I. (Ed.): Qualitative research methods in human geography. 4. Ed. Don Mills: Oxford University Press: 313–333.

Keck, M. & Sakdapolrak, P. (2013): What is social resilience? Lessons learned and ways forward. In: Erdkunde 67 (1): 5–19.

Kelman, I., Gaillard, J. C., Lewis, J. & Mercer, J. (2016): Learning from the history of disaster vulnerability and resilience research and practice for climate change. In: Natural Hazards 82 (1): 129–143.

Kelman, I., Gaillard, J. C. & Mercer, J. (2015): Climate change's role in disaster risk reduction's future: beyond vulnerability and resilience. In: International Journal of Disaster Risk Science 6 (1): 21–27.

Kemmis, S. (2008): Critical theory and participatory action research. In: Reason, P. & Bradbury, H. (Eds.): The SAGE handbook of action research. Participative inquiry and practice. 2. Ed. London, Thousand Oaks, New Delhi, Singapore: Sage Publications: 121–138.

Kindon, S., Pain, R. & Kesby, M. (Eds.) (2007a): Participatory action research approaches and methods: connecting people, participation, and place. Milton Park, New York: Routledge.

Kindon, S., Pain, R. & Kesby, M. (2007b): Introduction: connecting people, participation and place. In: Kindon, S., Pain, R. & Kesby, M. (Eds.): Participatory action research approaches and methods: connecting people, participation, and place. Milton Park, New York: Routledge: 1–6.

Kjerfve, B., Magill, K. E., Porter, J. W. & Woodley, J. D. (1986): Hindcasting of hurricane charac-teristics and observed storm damage on a fringing reef, Jamaica, West Indies. In: Journal of Marine Research 44 (1): 119–148.

Krause, F. (2016): Making space along the Kemi River: a fluvial geography in Finnish Lapland. In: cultural geographies: 1–16.

Kron, T. (2013): "Uncertainty" – Das ungewisse Risiko der Hybriden. In: Jeschke, S., Jakobs, E. M. & Dröge, A. (Eds.): Exploring Uncertainty: Ungewissheit und Unsicherheit im interdisziplinä-ren Diskurs. Wiesbaden: Springer: 55–82.

Kummels, I. (Ed.) (2012): Espacios mediáticos: cultura y representación en México. Berlin: Tranvía.

Kundzewicz, Z. W., Kanae, S., Seneviratne, S. L., Handmer, J., Nicholls, N., Peduzzi, P., Mechler, R., Bouwer, L. M. & et al. (2013): Flood risk and climate change: global and regional perspec-tives. In: Hydrological Sciences Journal 59 (1): 1–28.

Lammel, A., Goloubinoff, M. & Katz, E. (Eds.) (2008): Aires y lluvias. Antropología del clima en México. Mexico D.F.: CIESAS.

Lamnek, S. (2010): Qualitative Sozialforschung. 5. Ed. Weinheim, Basel: Beltz.

Landa, R., Magaña, V. & Neri, C. Agua y clima: elementos para la adaptación al cambio climático. Mexico D.F.: SEMARNAT; UNAM.

Latournerie Lastra, J. R. (1985): Playas de Catazaja. Monografía municipal. Playas de Catazajá.

Laurier, E. (2008): How breakfast happens in the café. In: Time & Society 17 (1): 119–134.

Lee, J. & Ingold, T. (2006): Fieldwork on foot: perceiving, routing, socializing. In: Coleman, S. & Collins, P. (Eds.): Locating the field. Space, place and context in anthropology. Oxford, New York: Berg: 67–85.

Lees, L. (2004): Urban geography: discourse analysis and urban research. In: Progress in Human Geography 28 (1): 101–107.

Lefebvre, H. (1974): La production de l'espace. In: L'Homme et la société 31 (1): 15–32.

Lefebvre, H. (1991): The production of space. Translated by Donald Nicholson-Smith. Oxford, Cambridge: Blackwell Publishers.

Lefebvre, H. (Ed.) (2004): Rhythmanalysis. Space, time and everyday life. Translated by Steward Elden and Gerald Moore. London, New York: Continuum.

Lévi-Strauss, C. (1966a): The scope of anthropology. In: Current Anthropology 7 (2): 112–123.

Lévi-Strauss, C. (1966b): Anthropology: Its´ achievements and future. In: Current Anthropology 7 (2): 124–127.

Lewis, J. (2015): Cultures and contra-cultures. Social divisions and behavioural origins of vulnera-bilities to disaster risk. In: Krüger, F., Bankoff, G., Cannon, T., Orlowski, B. & Schipper, L. A. (Eds.): Cultures and disasters. Understanding cultural framings in disaster risk reduction. Lon-don, New York: Routledge: 109–122.

Leyva Solano, X. (2010): Caminando y haciendo o acerca de prácticas decoloniales. In: Köhler, A., Leyva Solano, X., López Intzín, X., Martínez Martínez, D. G., Watanabe, R., Chawuk, J., Jiménez Pérez, J. A., Hernández Cruz, F. E., Estrada Aguilar, M. & Icó Bautista, P. A. (Eds.): Sjalel kibeltik. Sts'isjel ja kechtiki'. Tejiendo nuestras raíces. Mexico D.F.: Orê y Xenix Filmdistribution: 366–402.

Leyva Solano, X. (2015): Breve introducción a los tres tomos. In: Leyva Solano, X., Alonso, J., Hernández, R. A., Escobar, A., Köhler, A. & et al. (Eds.): Prácticas otras de conocimiento(s). Entre crisis, entre guerras. San Cristobal de Las Casas: Cooperativa Editorial Retos: 23–34.

Leyva Solano, X., Alonso, J., Hernández, R. A., Escobar, A., Köhler, A., Cumes, A.et al. (Eds.) (2015): Prácticas otras de conocimiento(s). Entre crisis, entre guerras. San Cristobal de Las Casas: Cooperativa Editorial Retos.

Likens, G. E. (Ed.) (2010): Rivers ecosystems ecology: a global perspective. San Diego, Burlington, London, Amsterdam: Elsevier; Academic Press.

Lima, J. R. de (2012): Contradições na produção do espaço rural brasileiro: modernização do campo, espacialização da pobreza e resistência. In: Geonordeste 23 (1): 136–156.

Lincoln, Y. S., Lynham, S. A. & Guba, E. G. (2011): Paradigmatic controversies, contradictions, and emerging confluences, revisited. In: Denzin, N. K. & Lincoln, Y. S. (Eds.): The SAGE handbook of qualitative research. 4. Ed. Thousand Oaks, London, New Delhi, Singapore: Sage Publications: 97–128.

Lippuner, R. (2007): Sozialer Raum und Praktiken: Elemente sozialwissenschaftlicher Topologie dei Pierre Bourdieu und Michel de Certeau. In: Günzel, S. (Ed.): Topologie. Zur Raumbeschreibung in den Kultur- und Medienwissenschaften. Bielefeld: Transcript Verlag: 265–277.

Littlechild, R., Tanner, D. & Hall, K. (2015): Co-research with older people: perspectives on impact. In: Qualitative Social Work 14 (1): 18–35.

López Intzín, J. (2015): Ich´el-ta-muk´: la trama en la construcción del Lekil-kuxlejal. Hacia una hermeneusis intercultural o visibilización de saberes desde la matricialidad del sentipensar-sentisaber tseltal. In: Leyva Solano, X., Alonso, J., Hernández, R. A., Escobar, A., Köhler, A. & et al. (Eds.): Prácticas otras de conocimiento(s). Entre crisis, entre guerras. San Cristobal de Las Casas: Cooperativa Editorial Retos: 181–198.

Lorda, M. A. (2015): La comprensión del territorio a partir del modelo de formación socio- espacial desde la práctica de la horticultura en el periurbano de bahía Blanca, Argentina. In: Interespaço 1 (3): 32–55.

Lossau, J. & Lippuner, R. (2004): Geographie und spatial turn. In: Erdkunde 58 (3): 201–211.

Luhmann, N. (1991): Soziologie des Risikos. Berlin, New York: de Gruyter.

Lukinbeal, C. & Zimmermann, S. (Eds.) (2008): The geography of cinema – A cinematic world. Stuttgart: Franz Steiner Verlag.

Lupton, D. (1991): Introduction: risk and sociocultural theory. In: Lupton, D. (Ed.): Risk and sociocultural theory: new directions and perspectives. Cambridge, New York: Cambridge University Press: 1–11.

Lupton, D. (2006): Sociology and risk. In: Mythen, G. & Walklate, S. (Eds.): Beyond the risk society. Critical reflections on risk and human security. Berkshire: Open University Press: 11–24.

Lupton, D. (2013): Risk. 2. Ed. London: Routledge.

MacGregor, D. G. & Godfrey, J. R. (2011): Observations on the concept of risk and Arab culture. Defense Technical Information Center & U.S. Department of Defense. Retrieved from http://www.dtic.mil/dtic/tr/fulltext/u2/a553206.pdf, Accessed 23.02.2017.

Macleod, M. (2008): Pueblos indígenas y la "buena vida": descentrando los discursos del desarrollo. In: Estudios Latinoamericanos, nueva época 22: 59–78.

Magaña, V., Pérez, J. L., Vázquez, J. L., Carrisoza, E. & Pérez, J. (2004): El Niño y el clima. In: Magaña, V. (Ed.): Los impactos de El Niño en México. Mexico D.F.: 23–68.

Marcial Pérez, D. (2015): Iberdrola abre su quinto parque eólico en México por 130 millones. In: El País, 11/26/2015. Retrieved from http://economia.elpais.com/economia/2015/11/25/actualidad/1448485719_595285.html, Accessed 23.02.2017.

Markowitz, H. (1952): Portfolio Selection. In: The Journal of Finance 7 (1): 77–91. Retrieved from http://links.jstor.org/sici?sici=0022-1082%281952903%297%3A1%3C77%3APS%3E2.0.CO%3B2-1, Accessed 23.02.2017.

Martínez Assad, C. (1996): Breve historia de Tabasco. Mexico D.F.: El Colegio de México.

Martínez Ruiz, J. L., Murillo Licea, D. & Martínez Ruiz, J. (2007): Pueblos indígenas de México y agua: los Mayas prehispánicos. In: United Nations Educational, Scientific and Cultural Organization (UNESCO) (Ed.): Atlas de culturas del agua en América Latina y el Caribe. n. P.

Mason, J. (2002): Qualitative researching. 2. Ed. London, Thousand Oaks, New Delhi: Sage Publications.

Massey, D. (2005): For space. London, Thousand Oaks, New Delhi: Sage Publications.

Matten, D. (2004): The impact of the risk society thesis on environmental politics and management in a globalizing economy – principles, proficiency, perspectives. In: Journal of Risk Research 7 (4): 377–398.

Mattissek, A., Pfaffenbach, C. & Reuber, P. (2013): Methoden der empirischen Humangeographie. Braunschweig: Westermann.

Mejido Costoya, M. (2013): Latin American post-neoliberal development thinking: the Bolivian 'turn' toward Suma Qamaña. In: European Journal of Development Research 25 (2): 213–229.

Mendozah, R. (1955): Breve historia del estado de Tabasco. Villahermosa: El Colegio de México.

Merriam, S. B., Johnson-Bailey, J., Lee, M.-Y., Kee, Y., Ntseane, G. & Muhamad, M. (2001): Power and positionality: negotiating insider/outsider status within and across cultures. In: International Journal of Lifelong Education 20 (5): 405–416.

Merriman, P., Jones, M., Olsson, G., Sheppard, E., Thrift, N. & Tuan, Y. (2012): Space and spatiality in theory. In: Dialogues in Human Geography 2 (1): 3–22.

Miggelbrink, J. & Schlottmann, A. (2009): Diskurstheoretisch orientierte Analyse von Bildern. In: Glasze, G. & Mattissek, A. (Eds.): Handbuch Diskurs und Raum. Theorien und Methoden für die Humangeographie sowie die sozial- und kulturwissenschaftliche Raumforschung. Bielefeld: Transcript Verlag: 181–198.

Miller, F., Osbahr, H., Boyd, E., Thomalla, F., Bharwani, S., Ziervogel, G., Walker, B., Birkmann, J., van der Leeuw, S., Rockström, J., Hinkel, J., Downing, T., Folke, C. & Nelson, D. (2010): Resilience and vulnerability: complementary or conflicting concepts? In: Ecology and Society 15 (3): 11. Retrieved from http://www.ecologyandsociety.org/vol15/iss3/art11/, Accessed 23.02.2017.

Mistry, J., Bignante, E. & Berardi, A. (2016): Why are we doing it? Exploring participant motivations within a participatory video project. In: Area 48 (4): 412–418.

Mitchell, R. (2010): Health, risk and resilience. In: Smith, S. J., Pain, R., Marston, S. A. & Jones III, J. P. (Eds.): The SAGE Handbook of social geographies. London, Thousand Oaks, New Delhi, Singapore: Sage Publications: 329–350.

Morchain, D. & Kelsey, F. (2016): Findings ways together to build resilience. The vulnerability and risk assessment methodology. Oxford: Oxfam.

Müller-Mahn, D. (Ed.) (2013): The spatial dimension of risk. How geography shapes the emergence of riskscapes. London: Routledge.

Müller-Mahn, D. & Everts, J. (2013): Riskscapes: the spatial dimensions of risk. In: Müller-Mahn, D. (Ed.): The spatial dimension of risk. How geography shapes the emergence of riskscapes. London: Routledge: 22–36.

MunichRe (2004): Megacities – Megarisks. Trends and challenges for insurance and risk management. Munich.

Mustafa, D. (2005): The production of an urban hazardscape in Pakistan: modernity, vulnerability, and the range of choice. In: Annals of the Association of American Geographers 95 (3): 566–586.

Mythen, G. & Walklate, S. (Eds.) (2006): Beyond the risk society. Critical reflections on risk and human security. Berkshire: Open University Press.

Nathan, F. (2008): Risk perception, risk management and vulnerability to landslides in the hill slopes in the city of La Paz, Bolivia. A preliminary statement. In: Disasters 32 (3): 337–357.

National Geographic (n. Y.): Natural disasters. Retrieved from http://environment.nationalgeographic.com/environment/natural-disasters/, Accessed 23.02.2017.

Navarrete Cáceres, C. (Ed.) (2013): En la diáspora de una devoción. Acercamientos al estudio del Cristo Negro de Esquipulas. Mexico D.F.: UNAM.

Neumayer, E., Plümper, T. & Barthel, F. (2014): The political economy of natural disaster damage. In: Global Environmental Change 24 (1): 8–19.

New Zealand Herald (n. Y.): Natural disasters. Retrieved from http://www.nzherald.co.nz/natural-disasters/news/headlines.cfm?c_id=68, Accessed 23.02.2017.

Niccolini, D. (2017): Is small the only beautiful? Making sense of `large phenomena´ from a practice-based perspective. In: Hui, A., Schatzki, T. & Shove, E. (Eds.): The nexus of practices: connections, constellations, practitioners. London, New York: Routledge: 98–113.

Nielsen, S. T. (2010): Coastal livelihoods and climate change. In: Verner, D. (Ed.): Reducing poverty, protecting livelihoods, and building assets in a changing climate. Social implications of climate change in Latin America and the Caribbean. Washington D. C.: The World Bank: 123–165.

Nix-Stevenson, D. (2013): Human response to natural disasters. In: Sage Open 3 (3): 1–12. Retrieved from https://doi.org/10.1177/2158244013489684, Accessed 23.02.2017.

Noceda, M. A. (2016): Iberdrola gana un nuevo contrato en México por 350 millones. In: El País, 4/4/2016. Retrieved from
http://economia.elpais.com/economia/2016/04/04/actualidad/1459798938_735498.html, Accessed 23.02.2017.

Norton, A. (1996): Experiencing nature: the reproduction of environmental discourse through safari tourism in East Africa. In: Geoforum 27 (3): 355–373.

Noticias Voz e Imagen (2016): Suspenden los trabajos de la Hidroeléctrica Chicoasén II., 2/22/2016. Retrieved from http://old.nvinoticias.com/329702-suspenden-trabajos-hidroelectrica-chicoasen-ii, Accessed 23.02.2017.

November, V. (2004): Being close to risk. From proximity to connexity. In: International Journal of Sustainable Development 7 (3): 273–286.

Oldrup, H. H. & Carstensen, T. A. (2012): Producing geographical knowledge through visual methods. In: Geografiska Annaler: Series B, Human Geography 94 (3): 223–237.

Olesen, J. E. (2010): Agrarian livelihoods and climate change. In: Verner, D. (Ed.): Reducing poverty, protecting livelihoods, and building assets in a changing climate. Social implications of climate change in Latin America and the Caribbean. Washington D. C.: The World Bank: 93–122.

Oliver Smith, A. (2002): Theorizing disasters. In: Hoffman, S. M. & Oliver-Smith, A. (Eds.): Catastrophe & culture: the anthropology of disaster. Santa Fe: School of American Research: 26–47.

Organisation for Economic Co-operation and Development (OECD) (2013a): OECD reviews of risk management policies: Mexico 2013. Review of the Mexican national civil protection system. Retrieved from http://dx.doi.org/10.1787/9789264192294-en, Accessed 23.02.2017.

Organisation for Economic Co-operation and Development (OECD) (2013b): Estudio de la OCDE sobre el Sistema Nacional de Protección Civil en México. OECD Publishing. n. P. Retrieved from http://dx.doi.org/10.1787/9789264200210-es, Accessed 23.02.2017.

Ortner, S. B. (1984): Theory in anthropology since the sixties. In: Comparative Studies in Society and History 26 (1): 126–166.

Oslender, U. (2002): "The logic of the river": a spatial approach to ethnic-territorial mobilization in the Colombian Pacific Region. In: The Journal of Latin American Anthropology 7 (2): 86–117.

Oswald Spring, Ú. (Ed.): Retos de la investigación del agua en México. Cuernavaca: UNAM.

Otros Mundos Chiapas (OMC) (2008): Synthesis of programs and activities. San Cristobal de Las Casas. Retrieved from http://otrosmundoschiapas.org/index.php/somos, Accessed 04.02.2017.

Otros Mundos Chiapas (OMC) (2014): Mapa intereses territoriales en Chiapas. El extractivismo corporativo de los bienes comunes. San Cristobal de Las Casas. Retrieved from http://otrosmundoschiapas.org/materiales/2014/10/mapa-intereses-territoriales-en-chiapas/, Accessed 04.02.2017.

Otros Mundos Chiapas (OMC) (2015): Informe global 2015. San Cristobal de Las Casas. Retrieved from http://otrosmundoschiapas.org/docs/informeglobal2015.pdf, Accessed 04.02.2017.

Pahl, S., Sheppard, S., Boomsma, C. & Groves, C. (2014): Perceptions of time in relation to climate change. In: WIREs Climate Change 5 (3): 375–388.

Pain, R. (2004): Social geography: participatory research. In: Progress in Human Geography 28 (5): 652–663.

Pedrozo-Acuña, A., Breña-Naranjo, J. A. & Domínguez-Mora, R. (2014): The hydrological setting of the 2013 floods in Mexico. In: Weather 69 (11): 295–302.

Pellicer, C. (1947): El canto del Usumacinta. n. P. Retrieved from http://bibliotecadig-ital.ilce.edu.mx/sites/fondo2000/vol1/paisaje/html/12.html, Accessed 23.02.2017.

Pelling, M. & Manuel-Navarrete, D. (2011): From resilience to transformation: the adaptive cycle in two Mexican urban centers. In: Ecology and Society 16 (2): 11. Retrieved from http://www.ecologyandsociety.org/vol16/iss2/art11/, Accessed 23.02.2017.

Perevochtchikova, M. & Lezama de la Torre, J. L. (2010): Causas de un desastre: Inundaciones del 2007 en Tabasco, México. In: Journal of Latin American Geography 9 (2): 73–98.

Pérez Moreno, M. P. (2012): O'tan - o'tanil : stalel tseltaletik yu'un Bachajón, Chiapas, México. Corazón: una forma de ser-estar-hacer-sentir-pensar de los tseltaletik de Bachajón, Chiapas, México. Master thesis, Facultad Latinoamericana de Ciencias Sociales Quito.

Pérez Sánchez, J. M. (2007): Desarrollo local en el trópico Mexicano. Los Camellos Chontales de Tucta, Tabasco. Master thesis, Universidad Iberoamericana Mexico D.F. Retrieved from http://www.bib.uia.mx/tesis/pdf/014848/014848.pdf, Accessed 16.02.2017.

Perry, R. W. & Quarantelli, E. L. (2005): What is a disaster?: new answers to old questions. Philadelphia: Xlibris Books.

Pfeffer, K., Baud, I., Denis, E., Scott, D. & Sydenstricker-Neto, J. (2013): Participatory spatial knowledge management tools. In: Information, Communication & Society 16 (2): 258–285.

Pichardo González, B. (2006): La revolución verde en México. In: Agrária (4): 40–68.

Pidgeon, N. (2009): Risk, uncertainty and social controversy: from risk perception and comunication to public engagement. In: Bammer, G. & Smithson, M. (Eds.): Uncertainty and risk. Multidisciplinary perspectives. London, Sterling: Earthscan: 349–361.

Pile, S. (1991): Practising interpretative geography. In: Transactions of the Institute of British Geographers 16 (4): 458–469.

Pink, S. (2013): Doing visual ethnography. 3. Ed. London, Thousand Oaks, New Delhi, Singapore: Sage Publications.

Pohl, J. (2008): Die Entstehung der geographischen Hazardforschung. In: Felgentreff, C. & Glade, T. (Eds.): Naturrisiken und Sozialkatastrophen. Berlin: Spektrum Akademischer Verlag: 47–62.

Postel, S. & Richter, B. (2003): Rivers for life: managing water for people and nature. Washington D. C.: Island Press.

Postill, J. (2010): Introduction: theorising media and practice. In: Bräuchler, B. & Postill, J. (Eds.): Theorising media and practice. New York, Oxford: Berghahn: 1–32.

Presidencia de la República (2014a): Programa Nacional de Infraestructura 2014–2018. Mexico D.F. Retrieved from http://cdn.presidencia.gob.mx/pni/programa-nacional-de-infraestructura-2014-2018.pdf?v=1, Accessed 23.02.2017.

Presidencia de la República (2014b): Ley de la Comisión Federal de Electricidad. Diario Oficial 11.08.2014. Mexico D.F.

Puglisi, R. (2014): Algunas consideraciones metodológicas y epistemológicas sobre el rol de la corp oralidad en la pr oducc ión del saber etnográfico y el estatuto atribuido a los sentidos corporales. In: Antípoda. Revista de Antropología y Arqueología 19: 95–119. Retrieved from http://dx.doi.org/10.7440/antipoda19.2014.05, Accessed 20.02.2017.

Quarantelli, E. L. (1998): What is a disaster? Perspectives on the question. New York: Routledge.

Quiroga, P. C. (2015): La investigación basada en la práctica de las artes y los medios audiovisuales. In: Revista Mexicana de Investigación Educativa 20 (64): 219–240.

Ramos Hernández, S., Morales Iglesias, H., Becarios, G., Mota Zaragoza, J. C., Castellanos Zenteno, E., Cossío Pérez, I. G., Díaz Martínez, R. A., Gómez Sarmiento, L. H. & Serrano Ramírez, J. L. (2010): Escenários climáticos para el Estado de Chiapas. Informe final. Fase II. Tuxtla Gutierrez: Universidad de Ciencias y Artes de Chiapas. Retrieved from http://www.semahn.chiapas.gob.mx/portal/descargas/paccch/escenarios_fase_ii.pdf, Accessed 02.02.2017.

Ramos-Gutiérrez, L. & Montenegro-Fragoso, M. (2012): Las centrales hidroeléctricas en México: pasado, presente y futuro. In: Tecnología y Ciencias del Agua. 3 (2): 103–121.

Reckwitz, A. (2002): Toward a theory of social practices: A development in culturalist theorizing. In: European Journal of Social Theory 5 (2): 243–263.

Reckwitz, A. (2003): Grundelemente einer Theorie sozialer Praktiken. Eine sozialtheoretische Perspektive. In: Zeitschrift für Soziologie 32 (4): 282–301.

Reich, K. (2001): Konstruktivistische Ansätze in den Sozial- und Kulturwissenschaften. In: Hug, T. (Ed.): Wie kommt Wissenschaft zu Wissen? Band 4. Einführung in die Wissenschaftstheorie und Wissenschaftsforschung. Baltmannsweiler: Schneider-Verlag Hohengehren: 356–376.

Reichel, C. & Frömming, U. U. (2014): Participatory mapping of local disaster risk reduction knowledge: an example from Switzerland. In: International Journal of Disaster Risk Science 5 (1): 41–54.

Reichenbach, H. (1958): The philosophy of space and time. New York: Dover Publications.

Renn, O. (2008): Concepts of risk: an interdisciplinary review. Part 1: disciplinary risk concepts. In: GAIA 17 (1): 50–66.

Reuber, P. (2012): Politische Geographie. Paderborn: Schöningh.

Reyes Barrón, M. C. (2012): Actores sociales y relaciones de poder. La reconfiguración del territorio frente al proceso de desarrollo local en Catazajá, Chiapas. Master thesis, Universidad Autónoma Chapingo.

Richter, B. D., Postel, S., Revenga, C., Scudder, T., Lehner, B., Churchill, A. & Chow, M. (2010): Lost in development's shadow: The downstream human consequences of dams. In: Water Alternatives 3 (2): 14–42.

Roberts, E. (2013): Geography and the visual image: a hauntological approach. In: Progress in Human Geography 37 (3): 386–402.

Robles García, C. (2015): Proyecta CFE 6 presas sobre el Usumacinta. In: Tabasco Hoy 2015, 6/16/2015. Retrieved from http://www.tabascohoy.com/nota/254379/proyecta-cfe-6-presas-sobre-el-usumacinta, Accessed 23.02.2017.

Rohland, E., Boeker, M., Cullmann, G., Haltermann, I. & Mauelshagen, F. (2014): Woven together. Attachment to place in the aftermath of disaster. Perspectives from four continents. In: Cave, M. & Sloan, S. M. (Eds.): Listening on the edge: oral history in the aftermath of crisis. New York: Oxford University Press: 183–206.

Romero Rodríguez, L. del C. (Ed.) (2011): Tabasco: Entre el agua y el desastre. Expresiones sociales en torno a sus inundaciones. México D.F.: Clave Editorial.

Rose, G. (1997): Situating knowledges: positionality, reflexivities and other tactics. In: Progress in Human Geography 21 (3): 305–320.

Rose, G. (2003): On the need to ask how, exactly, is geography "visual"? In: Antipode 35 (2): 212–221.

Rossing, T. & Rubin, O. (2010): Climate change, disaster hot spots, and asset erosion. In: Verner, D. (Ed.): Reducing poverty, protecting livelihoods, and building assets in a changing climate. Social implications of climate change in Latin America and the Caribbean. Washington D. C.: The World Bank: 63–91.

Roth, R. (2009): The challenges of mapping complex indigenous spatiality: from abstract space to dwelling space. In: cultural geographies (16): 207–227.

Rowles, G. (1978): Prisoners of space: exploring the geographical experience of older people. Boulder: Westview Press.

Ruz, M. H. (2010a): Usumacinta: Agua de encuentros. A manera de introducción. In: Ruz, M. H. (Ed.): Paisajes de río, ríos de paisaje. Navegaciones por el Usumacinta. Mexico D.F.: UNAM: 7–27.

Ruz, M. H. (2010b): "Un lugar verdaderamente deleitable". El pasado virreinal. In: Ruz, M. H. (Ed.): Paisajes de río, ríos de paisaje. Navegaciones por el Usumacinta. Mexico D.F.: UNAM: 79–202.

Ryen, A. (2008): Trust in cross-cultural research. The puzzle of epistemology, research ethics and context. In: Qualitative Social Work 7 (4): 448–465.

Saldaña-Zorrilla, S. O. (2007): Socioeconomic vulnerability to natural disasters in Mexico: rural poor, trade and public response. CEPAL. Mexico D.F.

Sandercock, L. & Attili, G. (2010): Digital ethnography as planning praxis: an experiment with film as social research, community engagement and policy dialogue. In: Planning Theory & Practice 11 (1): 23–45.

Sandoval-Ayala, N. C. & Soares-Moraes, D. (2015): Vulnerabilidad y activos familiares frente a riesgos. Caso de estudio en Ixil, Yucatán. In: LiminaR. Estudios sociales y humanísticos. 13 (1): 56–68.

Schatzki, T. R. (1996): Social practices. A Wittgensteinian approach to human activity and the social. Cambridge: Cambridge University Press.

Schatzki, T. R. (2001a): Introduction: practice theory. In: Schatzki, T. R., Knorr Cetina, K. & Savigny, E. v. (Eds.): The practice turn in contemporary theory. London, New York: Routledge: 10–23.

Schatzki, T. R. (2001b): Practice mind-ed orders. In: Schatzki, T. R., Knorr Cetina, K. & Savigny, E. v. (Eds.): The practice turn in contemporary theory. London, New York: Routledge: 50–63.

Schatzki, T. R. (2002): The site of the social: a philosophical account of the constitution of social life and change. University Park: Pennsylvania State University Press.

Schatzki, T. R. (2003): A new societist social ontology. In: Philosophy of the Social Sciences 33 (2): 174–202.

Schatzki, T. R. (2007): Martin Heidegger. Theorist of space. Stuttgart: Franz Steiner Verlag.

Schatzki, T. R. (2009): Timespace and the organization of social life. In Trentmann, F. & Wilk, R. (Eds): Time, consumption and everyday life: practice, materiality and culture. New York: Berg: 35–48.

Schatzki, T. R. (2011): Where the action is. On large social phenomena such as sociotechnical regimes. Working paper 1. Sustainable Practices Research Group. Retrieved from http://www.sprg.ac.uk/uploads/schatzki-wp1.pdf, Accessed 23.02.2017.

Schenk, G. J. (2007): Historical disaster research: state of research, concepts, methods and case studies. In: Historical Social Research 32 (3): 9–31. Retrieved from http://nbn-resolving.de/urn:nbn:de:0168-ssoar-291428, Accessed 20.02.2017.

Scherer, A. K. & Golden, C. (2012): Revisiting Maler's Usumacinta: recent archaeological investigations in Chiapas, Mexico. San Francisco: Precolumbia Mesoweb Press.

Schetter, C. (2013): Ungoverned territories: the construction of spaces of risk in the 'War on Terrorism'. In: Müller-Mahn, D. (Ed.): The spatial dimension of risk. How geography shapes the emergence of riskscapes. London: Routledge: 97–108.

Schlottmann, A. & Miggelbrink, J. (Eds.) (2015): Visuelle Geographien: Zur Produktion, Aneignung und Vermittlung von RaumBildern. Bielefeld: Transcript Verlag.

Schmid, C. (2005): Stadt, Raum und Gesellschaft. Henri Lefebvre und die Theorie der Produktion des Raumes. München: Franz Steiner Verlag.

Scholl, S., Lahr-Kurten, M. & Redepenning, M. (2014): Considering the role of presence and absence in space constructions. Ethnography as methodology in human geography. In: Historical Social Research 39 (2): 51–67.

Schroth, G., Laderach, P., Dempewolf, J., Philpott, S., Haggar, J., Eakin, H., Castillejos, T., Garcia, J. M., Soto Pinto, L., Hernandez, R., Eitzinger, A. & Ramirez-Villegas, J. (2009): Towards a climate change adaptation strategy for coffee communities and ecosystems in the Sierra Madre de Chiapas, Mexico. In: Mitigation and Adaptation Strategies for Global Change 14 (7): 605–625.

Schurr, C. & Segebart, D. (2012): Engaging with feminist postcolonial concerns through participatory action research and intersectionality. In: Geographica Helvetica 67 (3): 147–154. Retrieved from www.geogr-helv.net/67/147/2012, Accessed 23.02.2017.

Schwarz, R. (1996): Ökonomische Ansätze zur Risikoproblematik. In: Banse, G. (Ed.): Risikoforschung zwischen Disziplinarität und Interdisziplinarität. Von der Illusion der Sicherheit zum Umgang mit Unsicherheit. Berlin: Edition Sigma: 125–131.

Scribano, A. (2013): Expressive creative encounters: a strategy for sociological research of expressiveness. In: Sociology & Culture. Global Journal of Human Social Science 13 (5): 33–38.

Secretaría de Comunicaciones y Transportes (SCT) (2016): Programa General de Protección Civil 2016. México D.F. Retrieved from http://sct.gob.mx/fileadmin/ProteccionCivil/programaGeneralProteccionCivil.pdf, Accessed 20.12.2016.

Secretaría de Desarrollo Social (SEDESOL) (2012): Informe anual sobre la situación de pobreza y rezago social. Chiapas. Retrieved from https://www.gob.mx/cms/uploads/attachment/file/31490/Chiapas_1_.pdf, Accessed 23.02.2017.

Secretaría de Desarrollo Social (SEDESOL) (2013a): Catazajá. Unidad de microregiones. Cédulas de información municipal. Retrieved from http://www.microrregiones.gob.mx/catloc/, Accessed 23.02.2017.

Secretaría de Desarrollo Social (SEDESOL) (2013b): Palenque. Unidad de microregiones. Cédulas de información municipal. Retrieved from http://www.microrregiones.gob.mx/catloc/, Accessed 23.02.2017.

Secretaría de Gobernación (SEGOB) (1917): Constitución politica de los Estados Unidos Mexicanos, que reforma la de 5 de Febrero de 1857. Diario Oficial 05.02.1917.

Secretaría de Gobernación (SEGOB) (1954): Resolución sobre dotación de ejido al poblado Calatraba en Palenque, Chis. Diario Oficial 19.08.1954. Mexico D. F.

Secretaría de Protección Civil (2015a): Escuela de Protección Civil. Retrieved from http://proteccioncivil.chiapas.gob.mx/bienvenida-escuela, Accessed 23.02.2017.

Secretaría de Protección Civil (2015b): Atlas estatal de peligros y riesgos del estado de Chiapas. Retrieved from http://proteccioncivil.chiapas.gob.mx/atlas-estatal-riesgos-peligros, Accessed 23.02.2017.

Secretaría de Relaciones Exteriores (SRE) (2012): Proyecto de integración y desarrollo de Mesoamérica. Libro blanco. Mexico D.F. Retrieved from http://sre.gob.mx/images/stories/doctransparencia/rdc/8lbm.pdf, Accessed 23.02.2017.

Secretaría General del Gobierno (SGG) (2014): Decreto No. 563. Ley de Protección Civil del Estado de Chiapas. Periódico Oficial 138: 6–56. Retrieved from http://www.proteccioncivil.chiapas.gob.mx/documentos/decretos/Decreto-No-563.pdf, Accessed 23.02.2017.

Shove, E., Pantzar, M. & Watson, M. (2012): The dynamics of social practice. Everyday life and how it changes. Los Angeles, London, New Delhi, Singapore, Washington D. C.: Sage Publications.

Silva Herzog, J. (1964): El agrarismo mexicano y la reforma agraria. Vida y pensamiento de México. 2. Ed. Mexico D.F.: Fondo de cultura económica.

Sistema Nacional de Protección Civil (SINAPROC) (n. Y.): Website: Conoce el SINAPROC: Organización. Retrieved from http://www.proteccioncivil.gob.mx/en/ProteccionCivil/Organizacion, Accessed 23.02.2017.

Slater, D. (1997): Spatial politics, social movements. Questions of (b)orders and resistance in global times. In: Pile, S. & Keith, M. (Eds.): Geographies of resistance. London, New York: Routledge: 258–276.

Sletto, B. I. (2009): "We drew what we imagined". Participatory maping, performance, and the arts of landscape making. In: Current Anthropology 50 (4): 443–476.

Slovic, P. (1987): Perception of risk. In: Science, New Series 236 (4799): 280–285.

Smith, K. & Ward, K. (1998): Floods. Physical processes and human impacts. Chichester: Wiley.

Smith, S. J., Pain, R., Marston, S. A. & Jones III, J. P. (2010): Introduction: situating social geographies. In: Smith, S. J., Pain, R., Marston, S. A. & Jones III, J. P. (Eds.): The SAGE Handbook of social geographies. London, Thousand Oaks, New Delhi, Singapore: Sage Publications: 1–39.

Socialwatch (2011): Proyecto Mesoamérica: cambio de nombre, no de empresa. In: Cuadernos Ocasionales 7: 8–10. Retrieved from http://www.socialwatch.org/sites/default/files/CO7_ProyectoMesoamerica_2011.pdf,

Accessed 23.02.2017.

Soja, E. W. (1989): Postmodern geographies. The reassertion of space in critical social theory. London, New York: Verso.

Soja, E. W. (1996): Thirdspace. Journeys to Los Angeles and other real-and-imagined places. Cambridge, Oxford: Blackwell Publishers.

Solís-Castillo, B., Solleiro-Rebolledo, E., Sedov, S., Liendo, R., Ortiz-Pérez, M. & López-Rivera, S. (2013): Paleoenvironment and human occupation in the Maya lowlands of the Usumacinta River, Southern Mexico. In: Geoarchaeology: An International Journal (28): 268–288.

Soriano Roque, D. (1988): Modelos hidrologicos de alarma para la prevencion de inundaciones. Master thesis, Universidad Autónoma de México.

Sousa Santos, B. de (2011): Epistemologías del Sur. In: Utopía y Praxis Latinoamericana. Revista Internacional de Filosofía y Teoría Social 16 (54): 17–39.

Sousa Santos, B. de (2015): Prólogo. In: Leyva Solano, X., Alonso, J., Hernández, R. A., Escobar, A., Köhler, A. & et al. (Eds.): Prácticas otras de conocimiento(s). Entre crisis, entre guerras. San Cristobal de Las Casas: Cooperativa Editorial Retos: 12–22.

Spaargaren, G. (2011): Theories of practices: agency, technology, and culture. Exploring the relevance of practice theories for the governance of sustainable consumption practices in the new world-order. In: Global Environmental Change 21 (3): 813–822.

Sperber, D. (1985): On anthropological knowledge. Three essays. Cambridge, Paris: Cambridge University Press; Editions de la Maison des Sciences de l'Homme.

Stevens, R. P. (1968): Spatial aspects of internal migration in Mexico, 1950–1960. In: Revista Geográfica 69 (1): 75–90.

Strauss, A. & Corbin, J. (1994): Grounded theory methodology – an overview. In: Denzin, N. K. & Lincoln, Y. S. (Eds.): Handbook of Qualitative Research. Thousand Oaks: Sage Publications: 273–285.

Strauss, A. & Corbin, J. (1996): Grounded Theory. Grundlagen qualitativer Sozialforschung. Weinheim: Beltz.

Subcomandante Marcos (1994): Chiapas: El Sureste en dos vientos, una tormenta y una profecia. Ejercito Zapatista de la Liberación Nacional. Selva Lacandona.

Swyngedouw, E. & Williams, J. (2016): From Spain's hydro-deadlock to the desalination fix. In: Water International 41 (1): 54–73.

Tahmiscioğlu, M. S., Anul, N., Ekmeçi, F. & Durmuş, N. (2007): Positive and negative impacts of dams on the environment. In: International Congress on River Basin Management: Vol. 1, 22–24 March 2007, Antalya. Ankara: General Directorate of State Hydraulic Works: 760–769.

Tansey, J. & O´Riordan, T. (1999): Cultural theory and risk: a review. In: Health, Risk & Society 1 (1): 71–90.

Tapia-Silva, F.-O., Contreras-Silva, A.-I. & Rosales-Arriaga, E.-R. (2015): Hydrological characterization of the Usumacinta river basin towards the preservation of environmental services. The International Archives of the Photogrammetry, Remote Sensing and Spatial Information Sciences, Volume XL-7/W3, 2015 36th International Symposium on Remote Sensing of Environment, 11–15 May 2015, Berlin: 1505–1509.

The Guardian: Oroville dam: authorities lift evacuation order for nearly 200,000. Retrieved from https://www.theguardian.com/us-news/2017/feb/14/oroville-dam-evacuation-order-lifted-california, Accessed 02.03.2017.

The New York Times (n. Y.): Disasters. News about Disasters, including commentary and archival articles published in The New York Times. Retrieved from https://www.nytimes.com/topic/subject/disasters, Accessed 02.03.2017.

The Sphere Project (2004): Sphere training package. Geneva.

The World Bank (2012): FONDEN. Mexico's natural disaster fund – a review The World Bank. Washington D. C.

The World Bank (2013): Strengthening disaster risk management in Mexico. 4.09.2013. Retrieved from http://www.worldbank.org/en/results/2013/09/04/disaster-risk-management-mexico, Accessed 23.02.2017.

Thoft-Christensen, P. & Baker, M. J. (Eds.) (1982): Structural reliability theory and its applications. Berlin, Heidelberg: Springer.

Thompson, R. & Poo, M. de (1985): Cronología histórica de Chiapas 1516–1940. San Cristobal de Las Casas: CIESAS.

Thornes, J. E. (2004): The visual turn and geography. In: Antipode 36: 787–794.

Thrift, N. (2009): Space: the fundamental stuff of geography. In: Clifford, N. J., Holloway, S., Rice, S. P. & Valentine, G. (Eds.): Key concepts in geography. London, Thousand Oaks, New Delhi, Singapore: Sage Publications: 85–96.

Tierney, K. J. (2007): From the margins to the mainstream? Disaster research at the crossroads. In: Annual Review of Sociology 33 (1): 503–525.

Tierney, K. J. (2014): The social roots of risk. Producing disasters, promoting resilience. Stanford: Stanford University Press.

Torras Conangla, R. (2012): La tierra firme de enfrente. Lacolonización campechana sobre la Región de Los Ríos (Siglo XIX). Merida: UNAM.

Trahar, S. (2011): Introduction. Travelling in the borderlands or a story of not quite fitting in. In: Trahar, S. (Ed.): Learning and teaching narrative inquiry. Travelling in the borderlands. Amsterdam: Benjamins: 1–13.

Tseng, C.-P. & Chen, C. W. (2012): Natural disaster management mechanisms for probabilistic earthquake loss. In: Natural Hazards 60 (3): 1055–1063.

Tuan, Y. (1979a): Space and place: humanistic perspective. In: Gale, S. & Olsson, G. (Eds.): Philosophy in geography. Dordrecht, Boston, London: D. Reidel Publishing: 387–427.

Tuan, Y. (1979b): Sight and pictures. In: Geographical Review 69 (4): 413–422.

Turner, B. L., Kasperson, R. E., Matson, P. A., McCarty, J. J., Corell, R. W., Christensen, L., Eckley, N., Kasperson, J. X., Luers, A., Martello, M. L., Polsky, C., Pulsipher, A. & Schiller, A. (2003): A framework for vulnerability analysis in sustainability science. In: Proceedings of the National Academy of Sciences of the United States of America 100 (14): 8074–8079. Retrieved from http://www.ncbi.nlm.nih.gov/pmc/articles/PMC166184/pdf/1008074.pdf, Accessed 02.02.2017.

Turner, J. K. (1910): México bárbaro. Mexico D. F.: Costa Amic Editor.

United Nations Development Program (UNDP) (2014): Mexico: country case study. How law and regulation support disaster risk reduction. IFRC–UNDP Series on Legal Frameworks in support of Disaster Risk Reduction. New York. Retrieved from http://www.undp.org/content/undp/en/home/librarypage/crisis-prevention-and-recovery/effective-law---regulation-for-disaster-risk-reduction.html, Accessed 16.03.2017.

United Nations Disaster Assessment and Coordination (UNDAC) (2013): Field Handbook. 6. Ed. n. P.

United Nations Framework Convention on Climate Change (UNFCCC) (1997): Kyoto Protocol to the United Nations Framework Convention on Climate Change. Retrieved from http://unfccc.int/essential_background/kyoto_protocol/items/1678.php, Accessed 04.02.2017.

United Nations International Strategy for Disaster Reduction (UNISDR) (n.Y.): Website: About UNISDR. Who we are. History. Retrieved from https://www.unisdr.org/who-we-are/history, Accessed 26.11.2016.

UNISDR (United Nations International Strategy for Disaster Reduction) (2005): Hyogo framework for action 2005–2015: Building the resilience of nations and communities to disasters. Retrieved from http://www.unisdr.org/files/1037_hyogoframeworkforactionenglish.pdf, Accessed 23.02.2017

United Nations International Strategy for Disaster Reduction (UNISDR) (2008): Indigenous knowledge for disaster risk reduction: good practices and lessons learned from experiences in the Asia-Pacific region. Bangkok. Retrieved from https://www.unisdr.org/we/inform/publications/3646, Accessed 15.02.2017.

United Nations International Strategy for Disaster Reduction (UNISDR) (2009): UNISDR terminology on Disaster Risk Reduction. Geneva. Retrieved from http://www.unisdr.org/files/7817_UNISDRTerminologyEnglish.pdf, Accessed 15.02.2017.

UNISDR (United Nations International Strategy for Disaster Reduction) (2015a): Sendai framework for disaster risk reduction 2015–2030. Retrieved from http://www.unisdr.org/files/43291_sendaiframeworkfordrren.pdf, Accessed 23.02.2017

United Nations International Strategy for Disaster Reduction (UNISDR) (2015b): Mexico marks 30 years of progress. Press release, Geneva, September 21, 2015. Retrieved from https://www.unisdr.org/archive/45854, Accessed 02.03.2017.

United Nations Office for the Coordination of Humanitarian Affairs (UNOCHA) (2013): OCHA in 2014 + 2015. Plan and Budget. New York. Retrieved from https://docs.unocha.org/sites/dms/Documents/OCHA%20in%202014-15%20vF%2072%20dpi%20single%20WEB.pdf, Accessed 23.02.2017.

United States Senate (2012): Clean Energy Standard Act of 2012. Washington D. C. Retrieved from https://www.energy.senate.gov/public/index.cfm/files/serve?File_id=b3580f37-ec8c-4698-a635-3e19f9815b9a, Accessed 4.02.2017.

Universidad Autónoma de México (UNAM) (Ed.) (1992): Atlas Nacional de México 1990–1992. Mexico D.F.

Valadez Araiza, C. (2011): "Ojala y ya no llueva". Organización y percepción social ante las inundaciones en la zona urbana de Ciudad Valles, S.L.P. Master thesis, El Colegio de San Luis.

Vargas Suárez, R. (2015): Reforma energética. De servicio público a modelo de negocios. In: Política y cultura (43): 125–145.

Vega Ó. (1980): Presas de almacenamiento y derivación. Mexico D.F.: UNAM.

Verne, J. (2012): Ethnographie und ihre Folgen für die Kulturgeographie: eine Kritik des Netzwerkkonzepts in Studien zu translokaler Mobilität. In: Geographica Helvetica 67 (4): 185–194.

Verner, D. (Ed.) (2010): Reducing poverty, protecting livelihoods, and building assets in a changing climate. Social implications of climate change in Latin America and the Caribbean. Washington D. C.: The World Bank.

Villegas, P. & Torras, R. (2014): The extraction and exportation of Campeche Wood by foreign colonists. The case of B. Anizan and Co. In: Secuencia 90: 79–93.

Vivó Escoto, J. A. (1964): Weather and climate of Mexico and Central America. In: West, R. C. (Ed.): Handbook of Middle American Indians. Volume one: natural environment and early cultures. Austin: University of Texas Press: 187–215.

Voisin, M. (2012): Evaluation of farmers´ vulnerability from a multidimensional point of view in the coastal plain of Tabasco, Mexico. Master thesis, Norwegian University of Life Sciences.

Vos, J. d. (1987): La contienda por la selva Lacandona. Un episodio dramático en la conformación de la frontera sur, 1859–1895. In: Historias (16): 73–98.

Walsh, K. (2014): Placing transnational migrants through comparative research: British migrant belonging in five GCC cities. In: Population, Space and Place 20 (1): 1–17.

Ward, P. R. B., Räsänen, T. A., Meynell, P. J. & Ketelsen, T. (2013): Flood control challenges for large hydroelectric reservoirs: Nam Theun -Nam Kading Basin, Lao PDR. Project report: Challenge Program on Water & Food Mekong project MK3 "Optimizing the management of a cascade of reservoirs at the catchment level". International Centre for Environmental Management, Hanoi.

Warde, A. (2005): Consumption and theories of practice. In: Journal of Consumer Culture 5 (2): 131–153.

Weichart, P. (2008): Entwicklungslinien der Sozialgeographie. Von Hans Bobek bis Benno Werlen. Stuttgart: Franz Steiner Verlag.

Weichselgartner, J. (2001): Naturgefahren als soziale Konstruktion. Eine geographische Beobach-
    tung der gesellschaftlichen Auseinandersetzung mit Naturrisiken. Dissertation, University
    Bonn.
Weichselgartner, J. (2016): Vulnerability as a concept in science and practice. In: Fekete, A. &
    Hufschmid, G. (Eds.): Atlas of Vulnerability and Resilience. Pilot version for Germany, Aus-
    tria, Liechtenstein and Switzerland. Cologne, Bonn: 18–21.
Weichselgartner, J. & Kelman, I. (2015): Geographies of resilience: challenges and opportunities of
    a descriptive concept. In: Progress in Human Geography 39 (3): 249–267.
Werlen, B. (1993): Gibt es eine Geographie ohne Raum? Zum Verhältnis traditioneller Geographie
    und zeitgenössischen Gesellschaften. In: Erdkunde 47 (4): 241–255.
Wilkerson, S. J. K. (1991): Damming the Usumacinta: the archaeological impact. In: Fields, V. M.
    (Ed.): Sixth Palenque Round Table. Norman: University of Oklahoma Press: 118–134.
Wilkinson, I. (2006): Psychology and risk. In: Mythen, G. & Walklate, S. (Eds.): Beyond the risk
    society. Critical reflections on risk and human security. Berkshire: Open University Press: 25–
    42.
Wilson, A. & Hodgson, P. (2012): Trust, coercion and care: researching marginalised groups. In:
    Love, K. (Ed.): Ethics in social research. Bingley: Emerald Group Publishing: 111–128.
Wisner, B., Blaikie, P., Cannon, T. & Davis, I. (2004): At risk. Natural hazards, people´s vulnera-
    bility and disasters. 2. Ed. London, New York: Routledge.
Wittgenstein, L. (2001 [1953]): Philosophische Untersuchungen: kritisch-genetische Edition.
    Herausgegeben von Joachim Schulte. Frankfurt a. M.: Suhrkamp.
Wood, J. (2005): 'How green is my valley?' Desktop geographic information systems as a commu-
    nity-based participatory mapping tool. In: Area 37 (2): 159–170.
World Commission on Dams (2000): Dams and development. A new framework for decision-mak-
    ing. The report of the World Commission on Dams. London, Sterling: Earthscan.
World Economic Forum (2014): Global Risks 2014. Ninth edition. World Economic Forum. n. P.
Wunderlich, F. M. (2008): Symphonies of urban places: urban rhythms as traces of time in space. A
    study of 'urban rhythms'. In: Näripea, E., Sarapik, V. & Tomberg, J. (Eds.): Koht ja Paik /
    Place and location. Tallin: The Research Group of Cultural and Literary Theory: 91–111.
Zachmann, K. (2014): Risk in historical perspective: concepts, contexts, and conjunctions. In: Klüp-
    pelberg, C., Straub, D. & Welpe, I. M. (Eds.): Risk – A multidisciplinary introduction. Cham,
    Heidelberg, New York, Dordrecht, London: Springer: 3–35.
Zaga, D. (2015): Masked development: exploring the hidden benefits of the Zapatista conflict. Cen-
    tre for Finance and Development Working paper 08. Geneva. Retrieved from http://graduatein-
    stitute.ch/files/live/sites/iheid/files/sites/cfd/shared/working%20papers/CFDWP08-2015.pdf,
    Accessed 02.02.2017.
Zebadúa, E. (2010): Chiapas. Historia breve. Mexico D.F.: El Colegio de México.
Zehetmair, S. (2012): Societal aspects of vulnerability to natural hazards. In: Raumforschung und
    Raumordnung 70 (4): 273–284.
Zurich Insurance Group (2013): Helping society better manage flood risk. Zurich.
Zurich Insurance Group (2014): Helping Mexican communities cope with floods. A case study of
    the Zurich's Flood Resilience Program in the region of Tabasco. Zurich. Retrieved from
    https://www.zurich.fr/de-de/corporate/knowledge/articles/2014/05/case-study-zurichs-flood-
    resilience-program-in-the-region-of-tabasco, Accessed 02.03.2017.
Zurich Insurance Group (2016): Risk nexus. Measuring flood resilience – our approach. n. P. Re-
    trieved from https://www.zurich.com/en/corporate-responsibility/flood-resilience/measuring-
    flood-resilience, Accessed 02.03.2017

SOFTWARE

ArcGIS version 10.1/2012, © ESRI
Atlas.ti version 7.5.10, © 1993–2017 by ATLAS.ti GmbH, Berlin

OTHER SOURCES AND SUPPORT

English corrections of the thesis (spelling and grammar) carried out by Mrs. Debra
Ardelean in March 2017

Interview transcriptions:
Three interviews transcribed by Lorena Valeria Guzmán Wolfhard in 2014
One interview transcribed by Sandra Patricia Alfonso in 2016

ANNEX

[Note of the author: As part of the empirical research process of this thesis, a range of audio and audio-visual recordings have been generated. For reasons of anonymity of research partners and third parties, the original recordings are not added to the annex of the dissertation. Transparency of the research process and the process of analysis of data and information is ensured by presenting the transcripts of interviews and the analysis of selected video sequences in written text.]

ANNEX 1: LIST OF INTERVIEWS CARRIED OUT BY THE AUTHOR DURING EMPIRICAL RESEARCH IN 2014 AND 2015 (NAMES OF PERSONS AND EXACT LOCATIONS ARE ANONYMISED)

Nr.	Interview anonymised	Date	Location	Nr.	Interview anonymised	Date	Location
1	2014_C_1	01.10.2014	Village	31	2014_PR_2	09.08.2014	Village
2	2014_C_2	22.10.2014	Village	32	2014_PR_3	09.08.2014	Village
3	2014_C_3	22.10.2014	Village	33	2014_PR_4	09.08.2014	Village
4	2014_C_4	05.09.2014	Village	34	2014_PR_5	10.08.2014	Village
5	2014_C_5	06.09.2014	Village	35	2014_PR_6	10.08.2014	Village
6	2014_C_6	06.09.2014	Village	36	2014_PR_7	10.08.2014	Village
7	2014_C_7	06.09.2014	Village	37	2014_PR_8	11.08.2014	Village
8	2014_C_8	06.09.2014	Village	38	2014_PR_9	11.08.2014	Village
9	2014_C_9	11.09.2014	Village	39	2014_PR_10	11.08.2014	Village
10	2014_C_10	12.09.2014	Village	40	2014_PR_11	07.09.2014	Village
11	2014_C_11	12.09.2014	Village	41	2014_PR_12	07.09.2014	Village
12	2014_C_12	24.09.2014	Village	42	2014_LB_1	08.08.2014	Village
13	2014_C_13	25.09.2014	Village	43	2014_PTA_1	08.08.2014	Village
14	2014_C_14	30.09.2014	Village	44	2014_P_1	10.09.2014	External
15	2014_C_15	02.10.2014	Village	45	2014_P_2	03.10.2014	Other
16	2014_C_16	06.10.2014	Village	46	2014_P_3	10.10.2014	Other
17	2014_C_17	07.10.2014	Village	47	2014_P_5	08.10.2014	Other
18	2014_C_18	07.10.2014	Village	48	2014_P_6	23.10.2014	Other
19	2014_C_19	20.10.2014	Village	49	2014_S_1	21.07.2014	Other

20	2014_C_20	21.10.2014	Village	50	2014_S_2	19.08.2014	Other
21	2014_C_21	22.10.2014	Village	51	2014_S_3	18.09.2014	Other
22	2014_C_22	22.10.2014	Village	52	2014_T_1	22.08.2014	Other
23	2014_CZ_1	02.09.2014	Village	53	2014_V_1	04.08.2014	Other
24	2014_CZ_2	03.09.2014	Village	54	2014_M_1	02.07.2014	Other
25	2014_CU_1	12.08.2014	Village	55	2014_M_2	03.07.2014	Other
26	2014_CU_2	12.08.2014	Village	56	2014_M_4	04.07.2014	Other
27	2014_CU_3	12.08.2014	Village	57	2014_M_5	31.10.2014	Other
28	2014_CU_4	12.08.2014	Village	58	2014_X_1	12.03.2015	Other
29	2014_CU_5	12.08.2014	Village	59	2015_C_1	02.06.2015	Village
30	2014_PR_1	08.08.2014	Village	60	2015_C_2	02.06.2015	Village
61	2015_C_3	03.06.2015	Village	82	2015_C_24	03.06.2015	Village
62	2015_C_4	03.06.2015	Village	83	2015_C_25	03.06.2015	Village
63	2015_C_5	03.06.2015	Village	84	2015_C_26	03.06.2015	Village
64	2015_C_6	03.06.2015	Village	85	2015_C_27	03.06.2015	Village
65	2015_C_7	03.06.2015	Village	86	2015_C_28	03.06.2015	Village
66	2015_C_8	03.06.2015	Village	87	2015_C_30	21.05.2015	Village
67	2015_C_9	03.06.2015	Village	88	2015_C_31	21.05.2015	Village
68	2015_C_10	04.06.2015	Village	89	2015_C_32	26.05.2015	Village
69	2015_C_11	24.06.2015	Village	90	2015_C_33	26.05.2015	Village
70	2015_C_12	30.06.2015	Village	91	2015_C_34	26.05.2015	Village
71	2015_C_13	30.06.2015	Village	92	2015_C_35	03.06.2015	Village
72	2015_C_14	30.06.2015	Village	93	2015_C_36	24.06.2015	Village
73	2015_C_15	30.06.2015	Village	94	2015_C_37	24.06.2015	Village
74	2015_C_16	30.06.2015	Village	95	2015_P_1	21.05.2015	Other
75	2015_C_17	01.07.2015	Village	96	2015_P_2	10.06.2015	Other
76	2015_C_18	02.07.2015	Village	97	2015_T_1	27.05.2015	Other
77	2015_C_19	02.07.2015	Village	98	2015_T_2	27.05.2015	Other
78	2015_C_20	03.07.2015	Village	99	2015_P_4	21.07.2015	Other
79	2015_C_21	08.08.2015	Village	100	2015_E_1	30.07.2015	Other
80	2015_C_22	08.08.2015	Village	101	2015_V_1	31.07.2015	Other
81	2015_C_23	03.06.2015	Village				

ANNEX 2: LIST OF ARCHIVES, LIBRARIES AND PRIVATE COLLECTIONS CONSULTED AS SOURCES OF SECONDARY DATA AND SCIENTIFIC LITERATURE FROM 2013–2015

Name of Institution and/ or person	Month/Year
Cuernavaca	
Library of researcher Ursula Oswald Spring	07/2014
Emiliano Zapata	
Municipal library	07/2015
Museum of History, Archive	07/2015
Mexico City	
CIESAS Centro, Library	11/2013, 07/2014 & 10/2014
UNAM – Central library	11/2013 & 07/2014
CONAGUA Headquarters, Archive	10/2014
Library at UNAM – Department of Geography	05/2015 & 08/2015
Palenque	
Humedales Usumacinta (NGO)	09/2014 & 06/2015
Library of the local chronicler (Agustin "Chamuchin")	06/2015
Municipal library	06/2015
Municipal Presidency Palenque	07/2015
Procaduria Agraria, Archive	10/2014 & 07/2015
San Cristobal de las Casas	
CIESAS Sureste, Library	07–09/2014
EcoSur, Library	09/2014
UNACH, Library	08/2014
UNICACH, Library	08/2014
UACH, Library	08/2014
San Luis Potosí	
Colegio de San Luis, Library	11/2013
UASLP, Library	11/2013
Tenosique	
Municipal library Tenosique	06/2015
Chroniclers (Ramírez Pavon, Bech Cano & Aldecoa Calzada)	06/2015
Tuxtla Gutiérrez	
Protección Civil Chiapas, Archive & University	08/2014
Archive of the State Congress, Tuxtla Gutierrez	07/2015
State library "Jaime Sabines", Tuxtla Gutierrez	07/2015
Museum of Anthropology and History, Tuxtla Gutierrez	07/2015
INEGI Chiapas: Digital catalogue of data and maps	07/2015

Villahermosa	
Public state library José Ma. Pino Suárez, Library and Archive	06/2015
UJAT, Library of history "José Martí"	06/2015
Regional Anthropology Museum "Carlos Pellicer Cámara"	06/2015

ANNEX 3: GUIDELINE OF QUESTIONS FOR EXPLORATORY INTERVIEWS IN 2014

1. Como viven en su comunidad en tiempo de creciente?
2. Que efectos tiene la creciente en su comunidad/ su casa? Qué cambios hay durante este tiempo?
3. De qué viven ustedes? Cual es su profesión?
4. Se ayudan aqui en la comunidad durante la creciente? Cómo?
5. Viene ayuda de afuera en tiempo de creciente (Protección Civil/Municipio/Gobierno Nacional)?
6. Han pensado una vez en irse de la comunidad a un lugar donde no se inunda?
7. Han escuchado de planes de construir una presa en el Río Usumacinta? Qué piensan de la idea?

Datos estadísticos
8. Nombre
9. Edad
10. Ocupación
11. Número y edad de personas que viven en la casa

English translation of interview guideline

1. How do you live in your village in times of flood?
2. Which effects does the flood have in your community/ in your house? Which changes occur during this time?
3. From what do you live? Which is your profession?
4. Do you help each other in the village during the flood? How?
5. Does help come to the village during flood times (from Civil Protection/Municipality/National Government)?
6. Did you ever think of leaving the village to live in a place that does not get flooded?
7. Did you hear of the plans to construct a dam on the Usumacinta River? What do you think of this idea?

Statistical Data
8. Name
9. Age
10. Occupation
11. Number and age of people who live in the house

ANNEX 4: INTERVIEW SEQUENCES ANALYSED IN CHAPTERS 6, 7 AND 8

Transcripts in Original language (Spanish) and translated into English

Chapter 6

6.3 Flood, flood risk or living with floods

Interview Sequence 1

Interview 2014_C_7: Lines 98–101*
[P1]: Aqui estamos acostumbrados. Y el rio lo tenemos medir. [D1]: El ri...? [P1]: El rio, el rio. Lo tenemos medido. Por ejemplo el tiempo de la inundación, ya sabemos hasta donde va llegar.
Translation into English language
[P1]: Here we are used to it. And the river, we have it measured. [D1]: The ri...? [P1]: The river, the river. We have it measured. For example, in the time of flood, we already know until where it will reach.

Interview Sequence 2

Interview 2015_C_17: Lines 45–50
[P1]: [...] Ah´hay un arroyo que cuando entra el agua, suena como un motor o algo, como "Hiiiii". Cuando está el agua bajita, y corre, este. Como un, este, lo que cae de la altura, como le dicen? Una cascada. Si, así suena el agua.
Translation
[P1]: [...] Ah, there is a stream that when the water enters, it sounds like a motor or something, like "Hiiiii". When the water is low, and it flows, this one. Like an, this, which falls from large height, how is it called? A cascade. Yes, this is how the water sounds like.

Interview Sequence 3

Interview 2015_C_7: Lines 93–97
[P2]: Si. Si. Hace, como cuatro años, hubo una inundación grande que aquí entró el agua a, a la cocina cinco centimetros. A la casa diez, porque la casa es más baja. Aquí nunca nos ibamos al agua. Pero esa vez si.
Translation
[P2]: Yes. Yes. It is about four years ago, there was a large flood that water entered here, into the kitchen five centimetres. Into the house ten, but the house is lower. Here we had never gone into the water. But that time yes.

Interview Sequence 4

Interview 2015_C_14: 8–20; 32–51
[P1]: Bueno pues, para mi, la inundación es algo pues, no muy agradable porque se imagina que todo se crece. Para dondequiera que nos muevamos, tiene que se en un cayuco porque no se puede caminar de otra forma. Y aunque para muchos este, beneficia, a otros no porque por ejemplo el que trabaja, detiene su trabajo muchos meses cuando hay creciente grandes. Que tiene su trabajo, por ejemplo que siembre maíz. Aha, Este y en este tiempo no tienen trabajo los que se ponen, que siembran maíz. Salen a trabajar por fuera. Mhm, salen a trabajar por fuera. [...] Y te, por eso le digo que siiii, a mi me da miedo la creciente, por, porque hay culebras. Aunque a otros les beneficia porque dice: Ya vienen las despensas [riza]. [D1]: [No se escuche] [P1]: Si, para las despensas. Y como es en tiempo de creciente, quien sabe donde sale tanta gente pero sale much gente para las despensas. Aha, y te, pues por otro lado tambien es bonito porque entra mucho pescado acá a las lagunas. Cuando ya va bajando el agua va la gente pesca mucho pescado como le contaba pues. Mucho pescado. Pero tambien es para la gente muy, muy así, como le diría? Este. Tienen problemas porque los animales, los que tienen su ganadito lo tienen que subir a, a lo alto. Y los que no, pues, tienen que alquilar tierras altas tambien, para llevarlos. Y por un lado hace bien pero por otro lado hace mal tambien la creciente.

Translation
[P1]: Well, for me, the flood is something, well, not very pleasant because you imagine that everything rises. Wherever we move, it has to be in the boat because you cannot travel in another way. And although for many well, they benefit, others don´t because for those who work, has to stop his work for many months if the flood is a big one. Who works, for example who sows corn. Aha, and in this time they don´t have work, those who do, who sow corn. They leave to work outside the village. Mhm, they leave to work outside the village. [...] And I, because of that I tell you yeeees, I am afraid of the flood, for, because there are snakes. Although others have benefit because they say: Now come the food supplies [laughter] [D1]: [Not understandable] [P1]: Yes, because of the food supplies. And as it is flood time, who knows from where all these people come out but many people go out for the food supplies. Aha, and I, well on the other side it is beautiful because a lot of fish enters there into the lagoons. When the water level gets lower people fish a lot of fish just as I told you. A lot of fish. But also for the people it is very, very like that, how should I tell you? Like. They have problems because of the animals, those who have cattle they have to bring them to up, to higher grounds. And those who don't, well, have to rent higher lands as well, to bring them there. And on one side it does good but on the other side it does bad as well the flood.

Interview Sequence 5

Interview 2015_C_7: Lines 93–97
[P1]: Porque perdemos ahorita. Si vamos a sembrar ahora, corremos el riesgo que venga la creciente y nos agarre y ya cuando nosotros hicimos el tamal, el maíz, y lo perdemos.

Translation

[P1]: Because now we lose. If we are going to sow now we take the risk that the flood comes and seizes us and then when we prepared the tamal, the corn, and we lose it.

Interview Sequence 6

Interview 2015_C_7: Lines 398–399, 402–403
[P1]: A nosotros el riesgo no nos hace nada, el riesgo, porque como siempre esteee cuando empiezan la lluvias, empieza a crecer [...]. Aca no, aca no, aca la vemos venir y nos protegemos. A nosotros no nos hace nada.
Translation
[P1]: For us, the risk doesn´t do anything to us, the risk, because it is like always, well when the rains start, it begins to grow […]. Here no, here no, here we see it coming and we protect ourselves. To us, it doesn´t do anything.

Interview Sequence 7

Interview 2015_C_22: Lines 359–365
[P23]: En lo que hemos vivido con las inundaciones, pues ya lo vemos con naturalidad porque ya nos hemos acostumbrados a ese, a lo que es la naturaleza, la, las inundaciones porque eso es. Las inundaciones es año con año. Y pues este, ya estamos tan acostumbrados que ya no, ya no nos asustan.
Translation
[P23]: About what we have lived with the floods, well we already see it with naturalness because we have already gotten used to this, to what is nature, the, the floods because that is what it is. The floods it is year after year. And well this, we are already so used to it that we already, already they do not scare us.

Interview Sequence 8

Interview 2015_C_20: Lines 8–18
[P1]: Bueno la creciente, es algunas veces nos sorprende. Pero ya estamos acostumbrados y año con año, la esperamos. Por esto de que empiezan las lluvias. Claro que los huracanes. Y todo el agua que se lleva, pues decir el viento, rumbo a la selva. Baja otra vez por este río caudaloso, que se le llama rio Usumacinta. Y ahí estamos preparados, según va avanzando el agua. Hacia arriba los niveles. Pues nosotros tambien nos estamos preparando por sea caso que sube de nivel.
Translation
[P1]: Well the flood, it is sometimes it surprises us. But we are already used to it and year after year, we expect it. For that when the rains start. Surely that the hurricanes. And all the water that is brought with it, to say the wind, there in the direction of the forest. It goes down again the water of this river abounding in water, that is called river Usumacinta. And here we are prepared, when the water continues. Until up the levels. Well we are also preparing ourselves for the case that the level rises.

Interview Sequence 9

Interview 2014_C_7: Lines 129–135
[P5]: Prepara, no. Aqui estamos acostumbrados, a la fecha que viene, ya estamos al dia, ya sabemos la fecha que...

[P1]: Ya, ya.

[P4]: Estamos acostumbrados a lo que pasa. .. A lo que pasa.

[P3]: Ya estamos al día, ya.

[P1]: Como ahorita ya estamos al dia que, el rio ya vemos que ya va, va subiendo, ya.

[P5]: No nos espanta. Porque nosotros sabemos que ano con ano pasa eso.

Translation

[P5]: Prepare, no. Here we are used to it, to the date that it comes, we are up to date, we already know the date that…

[P1]: Already, already.

[P4]: We are used to what happens…to what happens.

[P3]: We are already up to date, already.

[P1]: Like now we are already up to date, the river we already see that already is, it is rising, already.

[P5]: It doesn´t scare us. Because we know that year after year this happens.

Interview Sequence 10

Interview 2014_C_7: Lines 17–28

[D1]: Y cómo viven acá la creciente?

[P1]: La creciente?

[D1]: Si.

[P1]: Este, se agarra lo mas alto. Cuando la creciente llega hasta la orilla de casa, muchas familias tienen familia donde está mas alto el terreno. Se van par allá. Otros agarran a Zapata. Otros agarran y ponen tabla a la casa. Que ahí andan, a dentro de la casa. Con tablas, tapesco. Los animales las suben arriba del, del tapesco, pero muchos se ahogan. Caen y se ahogan al agua.

Translation

[D1]: And how do you live the flood here?

[P1]: The flood?

[D1]: Yes.

[P1]: The, you get to the highest you can. When the flood reaches the edge of the house, many families have family where the terrain is higher. They go there. Others go to Zapata. Others take and put planks to the house. That there they walk, inside the house. On wooden planks, tapesco. They take the animals up the, the tapesco, but many drown. They fall and drown in the water.

Interview Sequence 11

Interview 2015_C_13: Lines 22–40

[P1]: Los que viven en la orilla, todo se llena de agua.

[D2]: Todo se va al agua? Y qué hacen en ese tiempo, durante?

[P1]: Tenemos que aguantar el agua dentro de la casa.

[D2]: Que hacen los niños dentro de la casa cuando hay creciente?

[P1]: Este [riza]. Se ponen a jugar.

[D2]: Y ahí llegó agua dentro de tu casa?

[P1]: Si. Por ahí por la mata de nona y se llenó.

[D2]: Y como te sientes?

[P1]: Pues a veces se puede llenar.

[D2]: Pero cómo te sientes, estas alegre o triste? Que sientes?
[P1]: Alegre, porque ahi estan estos pescaditos chiquititos.
[D2]: Cómo los agarras?
[P1]: Con un traste [riza]
Translation
[P1]: Those who live at the river bank, everything fills up with water.
[D2]: Everything goes into the water? And what do they do in these times, during?
[P1]: We need to bear the water inside the house.
[D2]: What do the children do inside the house when there is flood?
[P1]: Well [she laughs]. They play.
[D2]: And there the water entered into your house?
[P1]: Yes. From there where there is the nona tree and it filled.
[D2]: And how do you feel?
[P1]: Well sometimes it can fill the house.
[D2]: But how do you feel, are you happy or sad? What do you feel?
[P1]: Happy, because there are those small fishies.
[D2]: How do you catch them?
[P1]: With a plate [laughs].

Interview Sequence 12

Interview 2014_C_7: Lines 52–64
[P1]: En lancha llegas aquí hasta la puerta de la casa.
[D2]: Si?
[P1]: Si.
[D2]: No es muy, muy bajo? Por el motor, no?
[P2]: Ah no, pero se utiliza el remo.
[P1]: Mhm.
[P2]: Andale. Aquí le decimos remo o canalete. Y se utiliza para jalar cuando está bajito. El motor no funciona aquí. Allá atrás si porque está mas hondo, pero por acá no.
[D2]: Mhm.
[P2]: Ya se da vuelta y vuelta. Entras, sales.
Translation
[P1]: In the motor boat you arrive until here, until the door of the house.
[D2]: Yes?
[P1]: Yes.
[D2]: Is it not very, very low? With the motor, no?
[P2]: Ah no, but you use a paddle.
[P1]: Mhm.
[P2]: That's it. Here we call it paddle or canalete. And you use it to paddle when the water level is low. The motor doesn't work here. There outside yes because it is deeper, but here no.
[D2]: Mhm.
[P2]: And you go around and around. You enter, you leave.

6. 4 Livelihood, the "good life" and development

Interview Sequence 13

Interview 2015_C_7: Lines 60–67
[P1]: Aquí te vas a comer una mojarra fresca. Vas alinear un pollo fresco. [D1]: Mhm. [P1]: Y si te lo tienen en el refri, sería por un o dos días. No como lo tienen en SuperChe y todo, donde te cambian el etiquéta y sigue. [D1]: Si. [P1]: Si, es muy diferente.
Translation
[P1]: Here you are going to eat a fresh mojarra (local fish variety). You are going to gut a fresh chicken. [D1]: Mhm. [P1]: And if you have it in the fridge, it would be for one or two days. Not like they have them in SuperChe (supermarket chain) and all, where they change the label and it stays. [D1]: Yes. [P1]: Yes, it is very different.

Interview Sequence 14

Interview 2015_C_1: Lines 114–122
[P4]: Mucha gente dice que [...] (anonymised) es tranquilo. Que les gusta. Porque vivir en una ciudad, la vida es muy rápida. O todo cuesta. Y aquí por lo menos, lo vamos llevando. Porque tenemos una mata de limón, una de naranja, o crecemos maíz, para la tortilla. Sembramos frijol. Calabaza. [D1]: Mhm. [P4]: Y pues en una ciudad es diferente.
Translation
[P4]: Many people say that [...] (anonymised) is quiet. That they like it. Because living in a town, life is very fast. Or everything costs something. And here at least, we keep it going. Because we have a lime tree, one of orange, or we grow corn, for the tortilla. We sow beans. Pumpkin. [D1]: Mhm. [P4]: And well in a town it is different.

Interview Sequence 15

Interview 2015_C_7: Lines 10–16
[P1]: [...] (anonymised) es tranquilo. [...] (anonymised) es una, es una comunidad pues, donde la gente se dedica a trabajar. A sembrar millo, otros a su chambarito de, de tienda. Otros se dedican al Picante, a la Chiva, a la escoba. Y hasta la fecha es un lugar tranquilo. Claro que entran muchos a hacer su, su rollo por acá pero todo tranquilo.
Translation
[P1]: [...] (anonymised) is quiet. [...] (anonymised) is a, it is a community, well, where people are dedicated to working. Sowing millo, others with their little work in, in a shop. Others dedicate to

the picante, to the chiva, to the broom stick. And until today it is a quiet place. Sure that many enter in order to make their, their things here but everything quiet.

Interview Sequence 16

Interview 2015_C_32: Lines 463–467 & 470–473
[D1]: Pero que piensa usted? Durante ese tiempo de creciente, vienen a dar despensa por ejemplo, pero usted dice que eso no es necesario. [P1]: Pues la verdad, lo agarramos porque ellos lo traen. [D1]: Si claro. [P1]: Pero en realidad no es una cosa que nos haga muchísima falta, no? [...] [P1]: porque nos la buscamos como le digo, principalmente como logramos sacar todo lo que podemos de nuestras cosechas de maíz, tenemos para el posole y la tortilla y con mi hija salimos a pescar a anzuelos, con la red, como se pueda y tenemos la comida. Hay oportunidad de ir a matar un venado, lo matamos y tenemos comida, exactamente.
Translation
[D1]: But what do you think? During this time of the flood, they come to bring despensa for example, but you say that this is not necessary. [P1]: Well the truth, we take it because they bring it. [D1]: Yes, sure. [P1]: But in reality it is not a thing that we would lack a lot, no? […] [P1]: because we gather it for us as I tell you, basically how we achieve to take out all we can of our corn harvests, we have for the posole (drink based on corn) and the tortilla (corn bread) and with my daughter we leave to fish with a fishing-hook, with the net, as you can and we have food. There is the chance to go and kill a wild game, we kill it and have food, exactly.

Interview Sequence 17

Interview 2015_C_32: Lines 478–484
[P2]: Pero lo que yo digo, yo para mi que con lo que viene es más que suficiente, porque él lo que dice que no viene jabon, no viene nada, por ejemplo, para limpieza del baño,eso es lo que cice, pero vienen muchas cosas que sí la verdad hace falta. El dice que no mucho hace falta, pero hay gente que verdad lo necesita por aqui, y si le hace bien la despensa porque hay mucha gente como dice hace cosecha de maíz, nosotros tenemos para hacer tortillas todos los días, pero hay mucha gente aqui que no siembra maíz y compran la tortilla.
Translation
[P2]: But what I say, I, for me with what comes it is more than enough, because he what he says there comes no soap, there comes nothing, for example, for cleaning of the bath, this is what he says, but there come many things that indeed is lacking. He says that not much is lacking, but there are people here who truly they need it, and yes the despensa does good to them because there are many people as he says who harvest corn, we have to make tortillas all the days, but there are many people here who don´t sow corn and they buy tortilla.

6.5 Creating and performing collective identities

Interview Sequence 18

Interview 2014_C_10: Lines 128–135
[P1]: Había mucho, de esto, de esto. Las muñecas que se encontraron ahí. De barro, lo trabajaron los mayas. Aqui era en tierra, allá era piedra, porque hubo mucha piedra. Había mucho. Vino una brigada de, de gringos buscando esas cosas.Ahí donde esta el tanque de agua potable, ahi se excavó. Y encontraron varios, de muñecos de esto, varios.
Translation
[P1]: There were many of these, of these. The dolls which were found there. Of clay, they had been made by the Mayas. Here it was in the earth, there was stone, because there were many stones. There was a lot. There came a brigade of, of gringos (US Americans) looking for these things. There by the tank for drinking water, there it was excavated. And they found various, of the dolls of this, various.

Interview Sequence 19

Interview 2014_C_10: Lines 57–72, 75–85
[P1]: El duque se llamaba José María [...] (anonymised) y aquí vivió. Los de Zapata que hicieron lo de la tierra, le pusieron "[...] (anonymised)". [D2]: José María... [P1]: José María [...] (anonymised). Era su apellido, [...] (anonymised). Y como aquí vivió esa, esa familia, de Zapata, que eran dueños de esas tierras, y ahí le pusieron "[...] (anonymised)". Así le quedó, [...] (anonymised). Hasta la fecha. [D2]: Y esa história le fue a usted, le contaron. Es história que fue contada por su papa, o por su abuelo o por, como? [P1]: Un señor que vivió en la casa al lado, un chofer cargaba la história, la leyenda. Y viajando a Tuxtla me la enseñó. Y cuando la, comencí a leer, yo [...] [P1]: Ahí fue cuando yo leí un pedacito, porque es una leyenda larga. [D1]: Aha. [D2]: Y usted dijo que sus abuelos eran españoles. Por parte de su papá o de su mamá? [P1]: De mi mamá. Los papás del padre de mi madre eran, vinieron de españoles. Entonces el papá de mi mamá era bastante guero, todavía. No hablaba muy bien, no hablaba muy bien. Y aquí se casaron. Ya con gente de aquí.
Translation
[P1]: The duke had the name José María [...] (anonymised) and here he lived. Those from Zapata who did this with the land, gave it the name "[...] (anonymised)". [D2]: José María… [P1]: José María [...] (anonymised). It was his last name, [...] (anonymised). And as here lived this, this family from Zapata who were the owners of these lands, and they gave the name "[...] (anonymised)". This is what is stayed, [...] (anonymised). Until the day. [D2]: And this history came to you, they told you? This story was told to you by your father, or by your grandfather or by, how?

[P1]: A man who lived in the house by the side, a driver carried the story, the legend. And travelling to Tuxtla he showed it to me. And when I started to read, I …

[…]

[P1]: There it was where I read a small part because it is a large legend.

[D1]: Aha.

[D2]: And you said that your grandparents were Spaniards? From the side of your father or of your mother?

[P1]: From my mother. The fathers of the father of my mother were, they came from Spanish. So the father of my mother was quite white, still. He didn´t speak very well, he didn´t speak very well. And here they got married. Already with people from here.

Interview Sequence 20

Interview 2015_C_19: Lines 57–64

[P1]: Pero también trae desventaja, porque ahí tenemos que aliviar, sufrir con los animales. Ya sea, la vaca, los caballos, hasta con los aves.

[D1]: Se mueren?

[P1]: Si. A veces se hace un tapesco, se suben pero a veces se duermen, se caen, se le lleva la corriente y se ahogan, se mueren.

Translation

[P1]: But it also brings disadvantage, because there we have to alleviate, to suffer with the animals. Be it a cow, the horses, even with the birds.

[D1]: They die?

[P1]: Yes. Sometimes a tapesco is made, they climb up but sometimes they fall asleep, they fall, the current takes them and they drown, they die.

Interview Sequence 21

Interview 2015_C_11: Lines 1–13

[P1]: Todos estos lindos doce años que ahora tengo,

me han servido para conocer a mi pueblo.

Un pueblo chiquito y lleno de amor

donde su gente es muy trabajadora

y humilde de corazón.

Se oyen los pajarros al amanecer de cada día,

la gente te recibe con alegría.

Pero que tristeza cuando todo comienza a cambiar,

porqué llega otoño y a los pajarros no se oye su cantar.

El río crece y el trabajo del campesino se pierde,

todo lo que era alegría desafortunadamente se entristeze.

Cuando comienzan las lluvias el río sólo crece y crece,

algunos pajarros huyen y no vuelven a verse.

Translation

[P1]: All of these beautiful twelve years that I am now,

have served me to get to know my village.

A village small and full of love

where it´s people is very hard-working and humble of heart.

You can hear the birds at the dawn of each day,

people welcome you with happiness.

But what a sadness when everything starts to change

because autumn comes and the birds are not to be heard any longer.

The river grows and the work of the farmer is lost,

all of which was happiness unfortunately becomes sad.

When the rains start the river only grows and grows,

some birds fly away and are not seen again

Interview Sequence 22

Interview 2015_C_17: Lines 488–497

[P1]: Si, entonces un sacrificio, vamos pasado, pero es la naturaleza. La naturaleza y es que aguantar.

[D1]: Mhm.

[P1]: Si. Hasta que Jesús dice, ven para acá. Y nos vamos por el cielo. [Riza] Mientras estamos sufriendo. Si. Así es.

[D1]: Y qué piensa usted, porqué hay esos cambios del clima, del agua, de la lluvia?

[P1]: Es la naturaleza, la naturaleza.

Translation

[P1]: Yes, well it is a sacrifice, we go, but it is nature. Nature and it has to be endured.

[D1]: Mhm.

[P1]: Yes. Even Jesus says, come here. And we go to heaven. [Laughter] Until then we suffer. Yes. That´s how it is.

[D1]: And what do you think, why are there these changes of climate, of water, of rain?

[P1]: It is nature, nature.

Interview Sequence 23

Interview 2015_C_4: Lines 4–28

[P1]: Bueno, lo que pasa. Aquí lo de la fiesta, acostumbramos a, que para estas fechas, que es el día del Señor de Tila, que es el patrón de aquí de la comunidad, se acostumbra hacer una fiesta. La misa, después de la misa, terminando la misa, la repartición de comida. En algunas ocasiones unos compañeros han donado una rez. Para la comida. Y, otro compañero compraron la tortilla, los guisos, los vasos, platos. La manera de como convivir. El día de mañana ya, que es el día del Señor. Ahí donde nos reunimos en la iglesia y para, de esa manera festejamos al Tila. La fiesta de nosotros, aquí en la comunidad.

[D2]: Se ve que se toma el tiempo para eso. Porque hoy es todo el día la preparación, no?

[P1]: Si, hoy se trabaja prácticamente todo el día. Y mañana, es todo el día. De hecho, la iglesia, tenemos que permanecer desde las dos de la noche hasta que ya, que llegue el padre a empezar la misa, y se retira. La comida y a las dos, tres de la tarde ya vamos a la casa ya descansando en casa ya. Eso es una de las importancias que hay es una costumbre de muchos años ya.

Translation

[P1]: Well, what happens. Here, what is the festivity, we are used to that in these dates, which is the day of the "Señor de Tila", which is the patron of the community, we are used to making a celebration. The mass, after the mass, finishing the mass, the redistribution of the food. In some occasions some fellows have given a cow. For the food. And other fellows bought tortilla, the stews, the cups, plates. The way to live together. The day of tomorrow, which is the day of the Señor. There where we gather in the church and for, in this way we celebrate the Tila. The celebration of us here in the community.

[D2]: One can see that you take your time for that. Because today the preparations are the whole day, isn´t it?

[P1]: Yes, today it is practically worked the whole day. And tomorrow, it is the whole day. In fact, the church, we have to stay from two in the night until then, the father arrives to start the mass and leaves. The food and at two, three o´clock in the afternoon we already go home to relax in the house. This is one of the important things that exists, it is a custom of many years already.

Interview Sequence 24

Interview 2015_C_4: Lines 69–74 & 80–82
[P1]: Todo el que viene, llega de voluntad y a la repartición de la comida, a todos, por igual, no ha y distinción ni nada. Ni alguien en especial que llegue. No, todos. Todo el que llegue será bien recibido y se les da su platillo. […] Para nosotros es un orgullo que vienen a visitar, en esta fecha.
Translation
[P1]: Everyone who comes, comes voluntarily and the repartition of the meal, to all, equally, there is no distinction and nothing. Not even anyone special who would come. No, all. Everyone who comes will be well received and gets his plate. […] For us it is a pride that they come to visit on this date.

Interview Sequence 25

Interview 2014_C_5: Lines 69–78
[P1]: Pero estamos olvidados acá, somos olvidados. [D1]: De quien? Olvidados del gobierno? [P1]: Del gobierno. El que tiene su casita por aqui. Aunque fueramos trabajando por aqui, sembrando millo, sembrando chile, y de eso salimos adelante porque. Y lo que la cosecha nos da, lo invertimos en un 100, 100 bloq. Y por otro año asi vamo comprando otra cosita, hasta que llege un detalle. Ya se iba ya. [06:07] [D1]: Si. [P1]: Asi? Pero aquí del gobierno? No hay ninguna ayuda. No es que ese companero, ese companero tenga su casa por el gobierno? No. No
Translation
[P1]: But we are forgotten here, we are forgotten about. [D1]: By whom? Forgotten by government? [P1]: By the government. He who has his house over here. Although we worked here, sowing millo, sowing chile and from this we got through because. And this of the harvest give to us, we

invest it in 100, 100 blocks. And in the other year like that we go on buying another thing until we come to a certain point. And it goes.

[D1]: Yes.

[P1]: Like that? But from government? There is no help at all. It is not that this fellow, this fellow would have got his house because of the government. No, no.

Interview Sequence 26

Interview 2015_C_1: Lines 276–291
[P5]: No cambiaría. No, no. [Otras mujeres hablando, no se entiende] [D2]: Pero como va la canción? Alguien se la sabe para que la cante? [P5]: No, parece que es así: Este es mi gente que por nada dejo aunque yo iría sufrir igual. [P6]: Ahh si. [P5]: Aunque a veces pues, aunque a veces nos aruinemos y entre nosotros, ¿verdad? Pero tenemos eso de que si a alguien le pasa una desgracia. Pues, yo, mi persona, yo lloro por ellos. Podemos tener una tal diferencia, eh, sea lo que sea, pero una desgracia que le pase a un compañero, yo lo siento. [P3]: Se coopera. [P5]: Siii.
Translation
[P5]: … I would not change. No, no. [Other women talk; it is not understandable] [D2]: But how is the song? Somebody knows it so that you can sing it? [P5]: No, it seems that it is like that: This is my people, which I would not leave for anything even if I suffered the same. [P6]: Ah, yes. [P5]: Even though sometimes, well, although sometimes we ruin and among us, right? But we have this that if to someone a misfortune happens. Well, me, my person, I cry for them. We can have such or such difference, eh, whatever it may be, but a misfortune that happens to a fellow, I feel it. [P3]: It is cooperated. [P5]: Yeeees.

6.6 Retelling and re-performing history

Interview Sequence 27

Interview 2015_C_10: Lines 146–184
[P1]: Si. No sabía decir, si fue población maya o no. […] (anonymised) era, tal vez vinieron como, vamos como ahorita, aquí hay una ranchería maya, y otra maya, y otra. Pero, tal vez, vivieron aqui, pues, porque como dejaron muñecos. [D2]: Aha. [P1]: como muchas trabajaron. [D2]: Si, si.

[P1]: Y aquí pasaban hasta la costa del Atlantico. Toda esta parte pasaron.

[D1]: Mhm.

[P1]: Como toda esta gente. Porque eran mas fuertes que nosotros. Aquí en las ruinas de Palenque. En el Palacio de las leyes, como lo llaman, son de tan alto, cada paso y ahi subían con carga, con piedras para trabajar ahí. Eran más que nosotros. Nosotros una piedrita, ya no aguantamos. Era muy grande esa gente. Cuando descubrió, que se trabajón en el 49, descubrieron muchas cosas. Tumba hay, esceleto de esta gente. Era de este tamaño. Se llamaba Cali, cali. Era grande esa gente.

[D2]: Mhm, mhm.

[P1]: Era bonito. Pero eso es de otra éra ya, de otra éra. Es muy antiguo. No sé porque desapareció esa gente, los mayas. Por qué? Antes del diluvio o después del diluvio?

[D2]: Pues aquí seguimos.

[P1]: Como?

[D2]: Aquí seguimos los mayas.

[P1]: Ahhh?

[D2]: Si, están los Ch´oles, los Tseltales, los Tsotsiles, los de Tabasco ahí estan los, no recuerdo pero igual, estan los mayas peninsulares. Entonces, seguimos. No, pues hubo una población antes, que se dispersó.

[P1]: Cómo se acabaron?

[D2]: Pues, ellos también migraron a otros lugares.

Translation

[P1]: Yes. I could not tell if it was Mayan population or not. [...] (anonymised) was, maybe they came like, well like now, here there is a Mayan ranch, and another Maya one, and another one. But maybe they came, well because as they left dolls.

[D2]: Aha.

[P1]: As they elaborated many.

[D2]: Yes, yes.

[P1]: And like that they passed until the coast of the Atlantic. All that part they passed.

[D1]: Mhm.

[P1]: Like all these people. Because they were stronger than us. Here in the ruins of Palenque. In the Palace of the laws, as they call it, they are so tall, every step and there they stepped up with loads, with stones to work there. They were more than we are. We one small stone, we cannot stand it. This people was very tall. When they discovered, what they worked in 49, they discovered many things. There are tombs, skeletons of the people. They were of this size. It was called cali, cali. This people was tall.

[D2]: Mhm, mhm.

[P1]: They were beautiful. But this is from another era yet, from another era. It is very old. I don´t know why this people disappeared, the mayas. Why? Before the diluvium?

[D2]: Well here we remain.

[P1]: What?

[D2]: Here we remain the Mayas.

[P1]: Ahhh?

[D2]: Yes, there are the Ch´ol, the Tseltal, the Tsotsil, those from Tabasco there they are, I don´t remember but it´s the same, there are the peninsular Mayas. So, we remain. No, well there was a population before, which dispersed.

[P1]: How did they end?

[D2]: Well, they also migrated to other places.

Interview Sequence 28

Interview 2015_C_10: Lines 97–121

[P1]: Pues, entre lo que mi papá me decía que el todavía alcanzó una parte de eso. Cortaba mucho tinto y lo sacaba. Se usa para una pintura. El tinto da una pintura muy bonita. Quien sabe cuantos mas sacaron. Si, sacaron mucha madera. El tinto, lo... y ..., los trozos, cuadrados. Quien sabe que sacaron de ahí. Pintura y otras cosas. Tinta, no?

[D2]: Mhm.

[P1]: Mas adelante no se, que otra madera. Pero el tinto, si. El tinto si, lo explotaron mucho. Pues había mucho tinto. El tino, lo cortaba uno, lo agarra en la calle, hasta quedo morado la mano.

[D2]: Ah, si.

[P1]: Este tinto, lo estan trozando ahí. Quien sabe que otras cosas más, porque ellos saben trabajar, ellos sabía trabajar. Cuantas cosas mas. Trabajaban, sabían trabajara pues. Y ya tiene su rato.

[D2]: Ya tiene su rato.

[P1]: Quien sabe que años sería. No mucho, quizas no mucho. Como esa gente de ahí se acordaba que vivió aquí el señor [...] (anonymised). Como los dueños de la tierra, como se acordaban de que aquí vivió esa familia. No está muy viejo. José María [...] (anonymised).

Translation

[P1]: Well, among what my father told me that he still lived part of that. He cut a lot of tinto and took it out. It is used for a paint. Tinto gives a very beautiful paint. Who knows how many more they took out. Yes, they took out a lot of wood. Tinto, they, and, the pieces, square-cut. Who knows what they took out? Paint and other things. Tinta, isn´t it?

[D2]: Mhm.

[P1]: Further ahead I don´t know which other woods. But tinto, yes. Tinto yes, they exploited it a lot. Well, there was much tinto. Tinto, it was cut by one, he takes it to the street, until the hand gets coloured.

[D2]: Ah, yes.

[P1]: This tinto, it´s what they cut into pieces over there. Who knows which other things, because they know how to work, they knew how to work. How many other things. They worked, they just knew how to work. It is some while ago.

[D2]: It is some while ago.

[P1]: Who knows which year would it be? Not much, maybe not much. As these people from there they remembered that her lived Mister [...] (anonymised). Like the owners of the land, as they remembered, here lived this family. It is not very old. José María [...] (anonymised).

Interview Sequence 29

Interview 2014_C_14: Lines 419–438

[P1]: Este el apellido vino de Belice trajeron al señor este Graf lo trajeron José como se llamo mi padre lo trajeron de para el corte de madera.

[D1]: Sí

[P1]:Que había mucho mucha selva aquí todo era selva entonces para cortar a este era como capataz el que mandaba a los trabajadores y y lo trajeron vino con su esposa Juana Kupp irlandesa de origen Irlandés.

[D1]: Mhm irl o holandesa?

[P1]: Ir irlan

[D1]:Irlandesa, ahh.

[P1]: Y este pero se murió la señora al al al al tener su niño una niña se fue y el señor se fue y se crió la muchacha la niñita se crió aquí los mismo patrones la criaron porque a la la señora la dejo ahí enterrada ahí entre las Cerves? [no se entiende bien] No sé si en Campeche o aquí en Tabasco o acá en Chiapas pero si ahí quedo.

[D1]: Aha.

[P1]: Y se crío la muchacha y ya luego ya de ahí este buscó un hombre y ya nació mi padre. Pero no llevan el apellido del papá sino de la abuela mi abuela y ya siguió el apellido Graf porque si no sería si lo hubiera reconocido el padre se hubiera terminado ahí el apellido Graf no existiera Graf aquí. Esta muy cerca de pero ya era de color el señor ya era de antillano ahí de del Caribe ya era negro ya no vino güero. [Risas]Ya no vino güero por eso es que todos estamos así ya medio medio negritos.

Translation

[P1]: This, the family name came from Belize, they brought Mister Graf, they brought José as was my fathers´ name, they brought him for the cutting of timber.

[D1]: Yes.

[P1]: That there was a lot, lot of forest here, everything was forest so in order to cut it he was like very able, who directed the workers and, and they brought him, he came with his wife Juana Kupp, Irishwoman, of Irish origin.

[D1]: Mhm, Irish or Dutch?

[P1]: Ir, Irish…

[D1]: Irishwoman, ahh.

[P1]: And this, but the lady died when, when, when she had her child, a girl, she died and the sir was, and he raised the girl, the little girl he raised here, the patrons raised her because the lady he left her buried there between the Cerves? [not understandable] I don´t know if in Campeche or here in Tabasco or there in Chiapas but yes, there she stayed.

[D1]: Aha.

[P1]: And the girl was raised and, and later she found a man and so my father was born. But they didn´t take the last name of the father but of the grandmother, my grandmother and so stayed my last name Graf because if it wasn´t, if the father had accepted, the last name Graf would have ended there, Graf wouldn´t exist here. It is very close to but he was coloured the man he already was from the Antilles, there fro, from the Caribean, he already was black, he yet didn´t come white [laughs]. He yet didn´t come white that is why all we are like this, half, half black.

Interview Sequence 29

Interview 2014_C_14: Lines 45–73
[P1]:Y… (Sonido de gallo) la dueña era la hermana de don Ovidio, Virginia Jasso pero quedo infestada y al morir él era este el que seguía ¿no?... [D1]: Sí

[P1]: pues comenzó a explotarlo pues mientras él buscaba maneras de arreglarlo… pues tuvo muy buena política ese señor. Fue senador federal de la República y ahí [...]. Entonces, este…aquí tenía que trabajar ¿no? Para fomentar algo. Allá atrás hay una loma de la misma [...] (anonymised). La derrumbaron y sembraron zacate los y sembraron pasturas. Luego este comenzaron a venir unas gentes a abrir aquí para sus trabajadores. Había unos que se llamaban Afol Rodolfo Pechi, Layo Pechi, Juan Guzmán este… varias gentes. Luego vinieron otros otra gente los Mendoza y ya esos son familia de nosotros. Papá de mi mujer… Mendoza, Manuel Mendoza y en fin…otros… y vino mi papa también, Belisario Hernández con mi abuelo… en fin…ya era bastante grande la comunidad. Y si se dieron idea que, que había que pelear. Entonces este estaba Emiliano Zapata con Porfirio Díaz con este… sí, Emiliano Zapata. El que estaba gestionando las tierras, peleando las tierras para los campesinos. Y aquí había un presidente Municipal de Zapataque se llamaba Juan Marinez, apoyaba mucho a los campesinos. Y que dicen a a pelear a pelear y todo. Y lo van garantizando… (tos) van ganando aquí… pero costó un pleito grande, el señor como tenía dinero (gritos fuertes de niños jugando) traía los soldados… y la gente tenía miedo. Claro que tenía miedo. El señor decía que se van porque si no les voy a meter bala pues metele plas.. pues metele pla y aquel presidente los apoyaba a ellos. Muy bien y papeles y oficios antes no es que no era campesino ¿no?. Prolectariado le decían al cam …

[D1]: Proletariado.

[P1]:… al trabajador así como nosotros ahorita. Proletariado. Después vino el agrario.

[D1]: ¿Vino la?

[P1]: El agrario, el agrario.

Translation

[P1]: And the owner was the sister of Ovidio, Virginia Jasso but she got sick and upon dying he was the one who came, no?

[D1]: Yes.

[Pa]: Well he started to exploit it while he was looking for ways to improve it. Well, he had good politics this mister. He was federal senator of the Republic and there [...] Then, this, he had to work here, no? In order to foster something. There, behind there is a small hill of [...] (anonymised) itself. They tore it down and sowed Zacate they and they sowed meadow. Later, well, it started to come some people to open here for their workers. There were some who were called [names] various people. Later there came others, other people the [name] and yet those are family of us. Father of my woman. [Name] and finally, others and my father came as well, [name] with my grandfather. Finally, the community already was quite big. And yes they gave themselves the idea that, that it had to be fought. So, this was Emiliano Zapata with Porfirio Díaz with this, yes, Emiliano Zapata. He, who procured the lands, fighting the lands for the farmers. And here there was a Municipal president of Zapata who was called [name], he supported the farmers a lot. And they said to, to fight, to fight and everything. And they go to assure it, they win here, but it costed a grand case, the mister as he had money, he brought soldiers, and the people was afraid. Of course they were afraid. The mister said that they should leave <because if not, I will put bullet.> <Well put it then, put it then.> And the president supported them. Very good and papers and office, before no, it is that they were farmers, no? Proletariat they called the farmers.

[D1]: Proletariat.

[P1]: to the worker like to us now. Proletariat. After that came the agrario.

[D1]: It came the …?

[P1]: The agrario, the agrario.

Interview Sequence 30

Interview 2014_C_14: Lines 85–90
[P1]: Tropera de luchadores, trabajadores, peleadores para la tierra y todo. Hasta que vino el entonces era un ingeniero para medir. Midieron todas las tierras lo que era [...] (anonymised). Eran 600 hectáreas. [D1]: 600 hectáreas. [P1]:Y le dieron la mitad, 300 a [...] (anonymised) al grupo y las otras 300 se las dejaron a ese señor. A don Ovidio Jasso.
Translation
[P1]: Troop of fighters, workers, roughnecks for the land and everything. Until then came, there was an engineer to measure. They measured all the lands of what was [...] (anonymised). It was 600 hectares. [D1]: 600 hectares. [P1]: And they gave half, 300, to [...] (anonymised) to the group and the other 300 they gave to this mister. To Don Ovidio Jasso.

6.7 Fighting for land and the importance of "la tierra"

Interview Sequence 31

Interview 2015_C_17: Lines 331–340
[P1]: Por eso nos quedamos aquí. Por eso alguna gente dice que, que sufrimos porque queremos. Porque hubieramos ido allá. Pero que dió un cambio de uno a otro. Que no era para más de casa esa. Está muy bonito, porque ahí se quedaron esta familia [nombre], ahí tienen un terreno. Ese no lo vendieron, se quedaron allá. Y está muy bonito la tierra. Pero tienen que estar viajando a ver sus animales. Tienen ganado allá. Siempre están viajando.
Translation
[P1]: That is why we stayed here. That is why some people say that we suffer because we want to. Because we had gone there. But what gave a change from one to the other. It was not for more than houses that one. It is very beautiful, because there stayed this family [name], there they have their land. Those did not sell it, they stayed there. And it is very beautiful the land. But they have to travel in order to see their animals. They have cattle there. They are always travelling.

Interview Sequence 32

Interview 2015_C_17: Lines 368–402
[P1]: Nosotros estamos inundado y en ese tiempo casi no auxiliaban en nada. Por lo más que nosotros solicitabamos, despensas, no, no, no. No nos auxiliaban. Entonces nos fuimos a Tuxtla, yo con [nombre], otro señor que se llama [nombre], los tres nos fuimos. Estuvimos allá como cinco dias, pidienso audiencia. Y no nos daban la, no nos daban la audiencia. Nos quedamos de dia a noche, ese señor tiene mucho trabajo. Y iba a descansar y este, o quizas, porque lo solicitaba, o bien porque estaba ahí porque [no se entiende]. Hasta que le digo a mi compañero: Cuando entre ese grupo, te vas con ellos. Ah bueno, y cuando vino un grupo que tenía audiencia, entraba, y después de que hablaban con el gobernador, con ese señor Juan Sabines, que en paz descanse, le habló. Y comprende ustedes que era un poco, este, su vocabulario así de grocería. Que pasa, porque no me hablaban, que están inundados. Que humildes que están acá. Y cuanto tiempo que

están acá? Tenemos tres días que estamos acá. Mañana voy a ver si puedo ir con ustedes. Y digo, y dice señor: Si, estamos esperando. Y si vino hasta acá. Estaba aquí con un pantalón blanco. Y yo le
puse un cayuco ahí para que llegara a la escuela, la que está destruida. Que era una parte donde no queríamos que se cayó. No, pa aquí me voy y se enbatió hasta arriba. Y estuvo ahi con nosotros. Fue que nos, le pedimos que nos die esa tierra. Pero manda el ingeniero y dice que no se veía, que lo queríamos para la agricultura sino para vivir. Ya nos dió ese terreno que le digo.

Translation
[P1]: We are flooded and in this time they almost didn´t help in nothing. For as much as we asked for food supply, no, no, no. They didn´t help us. So we went to Tuxtla, me with [name], other mister whose name is [name], the three of us went. We were there like five days, asking for an audience. And they didn´t give us the, they didn´t give us the audience. We stayed there, day and night, this mister has a lot of work. And he went to take a rest, or maybe, because he asked to, or well […] Until I say to my fellow: <When this group enters, you go with them>. Ah well, and when a group came that had an audience, he entered and after they had talked to the governor, with this mister Juan Sabines, who in peace may rest, he talked to him. And understand that he was a little, well, his vocabulary was like with rudeness. <What happens, why don´t you tell me that you are flooded? How humble you are there. And how long have you been here?> <We have been here for three days.> <Tomorrow I will see if I can come with you.> And I say, he said: <Yes, we are waiting.> And he came here. He was here with white trousers. And I gave him a cayuco so that he could get until the school building, the one that is destroyed. That was a part where we didn´t want that he fell. <No, here I go.> And he went in until up. And he was here with us. It was that he, we asked him to give us this land. But he sent the engineer and said that it wasn´t seen, that we wanted it not for agriculture but to live. Yet he gave us this terrain that I say.

Interview Sequence 33

Interview 2015_C_17: Lines 565–569 & 581–591
[P1]: El gobernador es que dicen que es el dueño de esa finca. Finca Nueva Esperanza, muy famosa. Porque criaron aqui era puro ganado fino. Puro ganado bonito. Como era gente rica, que ahí criaron puro ganado fino. [...] Pero el problema que habido que ya el gobernador sabe, tal vez no conoce pero ya sabe. Este, había un pajarral. Un pajarral donde dormía muchíssimas pajarros de cierta clase. Y tumbaron esos árboles. Entonces los pájaros se hecharon huir. Y no sé quien los renunció pero corrieron esos pájarros. Entraron esos pájaros silvestres. Hay alguien que los cuida tambien.

Translation
[P1]: The governor is who they say is the owner of this finca. Finca Nueva Esperanza, very famous. Because they raised here, it was purely the finest cattle. Only beautiful cattle. As they were rich people, who raised only finest cattle here. […] But the problem that was there that the governor already knows, maybe he is not aware but he already knows. This, there was a pajarral. A pajarall where so many birds of a certain class slept. And they fell these trees. So the birds went to flee. And I don´t know who indicted them but they chased these birds. These wild birds entered. There is somebody who takes care of them, as well.

Interview Sequence 34

Interview 2015_C_6: Lines 107–122
[D1]: Y no hay formas como se pueden preparar? [P1]: Nosotros, tenemos un terreno communal como lo nombramos nosotros aquí, donde los ejidatarios venimos a pastar nuestro ganadito. Pero no podemos subir más de diez cabezas de animales por persona y el que le sobra, pues tiene que ya rentar algun rancho por ahí. Antes lo rentaban ahí, que ahora es del gobierno. Si, nos daban pasture ahí. Pero hoy, ya pues, es de un hombre que…
Translation
[D1]: And are there no other ways how you can prepare? [P2]: We, we have a communal terrain as we call it us here, where the ejidatarios can come to pasture our cattle. But we cannot put more than ten heads of animals per person and what remains, well, you have to rent a ranch over there then. Before those from there they rented, which is now from the government. Yes, they gave us pasture there. But today, well, it is owned by a man who […].

6.8 Interacting with government

Interview Sequence 35

Interview 2015_C_6: Lines 130–144
[P2]: Yo pienso que este, el presidente no nos apoya. En el presidente está todo. A mi me dijó ahí una vez en el secretariado de gobierno, que iba a Tuxtla. Que todo este iba con el president. Si el presidente lo apoya y no hay nada. [P1] a [P3]: Vamos por la tienda, tío? [P2]: El que tiene, mhm. [P1]: Voz, voz y voto para dar fé de que, de, de, de lo que se está pidiendo, sea realidad. [P2]: Si, exactamente. [P1]: Ellos, ellos son los indicados para … Es como un juez que dice yo los libero. Y por que? Porque yo considero que no hay culpa. Al igual son los presidents municipales. Todos los que son jefes. Ellos dice: Este va, este no.
Translation
[P2]: I think of this, the president does not support us. In the president is everything. He told me here once in the government secretariat, that he would go to Tuxtla. That all this would go with the president. If the president does not support it, there is nothing. [P1] to [P3]: Should we go to the shop, uncle? [P2]: He who has, mhm. [P1]: Voice, voice and vote to give faith that, that, that, that what is asked for, would be reality. [P2]: Yes, exactly. [P1]: They, they are the indicated for … He is like a judge who says: I liberate you. And why? Because I consider that there is no guilt. The same are these municipal presidents. All of those who are bosses. They say: This one goes, this one not.

Interview Sequence 36

Interview 2015_C_6: Lines 50–54 & 64–72

[P1]: A la verdad que hace el gobierno que algun día mirara a la, pusiera su, su, la mira hacia acá. Con un bordo carretero como está en Jobál, no nos fueramos a Pique. [...] Vienen las, este, vienen los ayuntamientos y con el esperamiento que nos van apoyar y no nos dan nada, lo último. No nos dan nada. Que nos vienen a traer? Una despensita de equivalente a unos 100 pesos o 150 pesos. Ya ellos dicen: "Ya [...] (anonymised) fue beneficiado con tantos miles de pesos". Y eso es una mentira. Es una mentira eso.

Translation

[P1]: To tell the truth, what the government does, that one day it would look to, it would put its, the gaze towards here. With a dam and a street like it is in Jobál, we would not go to "Pique". [...] There come the, these, there come the councils and with the hope that they would support us and they don´t give us anything, the least. They don´t give us anything. What do they come to bring? A small food supply equivalent to 100 pesos or 150 pesos. Already they say: "[...] (anonymised) was benefitted with how many thousands of pesos". And this is a lie. This is a lie.

Interview Sequence 37

Interview 2014_: Lines 226–229, 234–237 & 246–248

[P1]: Porque busca uno ahí un amigo, amigo político de los buenos. ¿No es cierto? Y allá tenía yo un amigo que trabajaba en ¿en qué oficina trabajaba este señor? ¡Ah! era delegado de programación y presupuesto. Me fui a Tuxtla buscando un crédito para comprar unas tierras...
[...]
[P1]: Era amigo de nosotros. Llegabamos ahí cada rato, ya ya lo conocíamos a el mucho [risa].
[D1]: Sí, sí.
[P1]:Pero pues no paso a más. No pasó a más.
[...]
[P1]:Tengo, tengo amigos políticos si ahí, que quiero visitarlos. Creo que uno de los más allegados al gobierno para candidato de gobierno municipal de Palenque se llama [nombre]. Un amigo bueno, un amigo bueno.

Translation

[P1]: Because one looks there for a friend, a political friend of the good ones. Isn´t it? And there I had a friend who worked in, in which office did this mister work? Ah! He was delegate of program and budget. I went to Tuxtla looking for a credit to buy some lands…
[…]
[P1]: He was a friend of ours. We went there every now and then, yet, yet we knew him a lot [laughs].
[D1]: Yes, yes.
[P1]: But well, it didn´t proceed to more. It didn´t proceed to more.
[…]
[P1]: I have, I have political friends there, which I would like to visit. I think one of the closest to government for candidate of the municipal government of Palenque is called [name]. A good friend, a good friend.

Interview Sequence 38

Interview 2015_C_10: Lines 372–416

[D2]: Porque pasó algo o porque occurrió algo?

[P1]: Por lo que se está viendo.

[D2]: Mhm.

[P1]: Se está viendo. Todo lo que se está viendo. Se ve y se oye aquí de tantas cosas. Hace como tres años, dos años, la balacera allá bajito de [...] (anonymised). Allá como a dos o trés kilómetros. Una

banda de narcos y una banda de gobierno. Pasó la balacera.

[D1]: Mhm.

[P1]: Pero prestación dura. Por qué? Por lo que estamos hablando. ...todavía de esta brigada. Bien armada. Le hecharon balas. Murieron mucha gente.

[D1]: Eso hace cuanto tiempo?

[P1]: Hace como cuatro años.

[D1]: Cuatro años.

[P1]: Aha. Pasaron los soldados, el gobierno. Y se topa por allá con los maleados. Ellos, estaban bien, bien armados. Y eso fue un balazo. [no se entiende bien]. Y ahorita, este gobierno, hace como una semana, una semana [riza], dos partidos. Ahi vienen a la última casa de aquí. El verde con el PRI. A mi me molestaba mucho. Si yo era [...] ellos hubieron matado a alguien. Le cortaron aquí, aquí, a uno de los verdes. Y ahí arriba, en la entrada a la carretera estaba la gente así. Un Pri-ista. Estaba esperando el verde que se iban a matar todos.

[D1]: Por qué?

[P1]: Ahí voy. Ahí voy. Pero a parte de ese personal, es gente de, es gente extranjera. Gente maleante com dije yo. Estaban ahi, dando cinco mil pesos a cada uno, les arman, y vamos hacer este trabajo. Ahí voy. Ahi está la camioneta, del verde.

[D2]: Ah, todavía siguen.

[P1]: Hee?

[D2]: todavía siguen ahí una camioneta de...

[P1]: Si, del PRI o del Verde. El PRI, iban a quitarles todo. Eso ya es típo guerra.

[D2]: Si.

[P1]: México, es en guerra. México es en guerra.Un enfrentamiento así, y ya es duro, ya es duro. Es guerra, estamos en guerra.

Translation

[D2]: Because something happened or something occurred?

[P1]: Because of what one can see.

[D2]: Mhm.

[P1]: One can see. Everything that you can see. You can see and hear so many things here. It´s like three years ago, two years, the shooting there down of [...] (anonymised). There like at two or three kilometres. A gang of narcos and a gang of government. The shooting happened.

[D1]: Mhm.

[P1]: But strong performance. Why? For what we are talking about...still of this brigade. Well-armed. They fired bullets. Many people died.

[D1]: This is how long ago?

[P1]: It´s like four years ago.

[D1]: Four years.

[P1]: Aha. The soldiers passed by, the government. And they clash there with the bad guys. Them, they were well, well-armed. And this was a shooting. And now, this government, it is like one week ago, one week [laughter], two parties. There they come, the last house from here. The Green with the PRI. I was very disturbed. If I had […] they would have killed someone. They cut him here [indicated on his body], here, to one from the Green. And up there, in the entrance to the highway there were the people like this. A PRI-ista. He was waiting for the Green that they all were killing each other.

[D1]: Why?

[P1]: This is where I go. This is where I go. But apart from this personal, this people from, this people from, foreigners. Bad people as I said. They were there, giving five thousand pesos to each of them, they arm them, and let´s do this work. This is where I go. There is the pickup, from the Green.

[D2]: Ah, they are still there?

[P1]: Hee?

[D2]: They are sitll here with the pickup from...

[P1]: Yes, from PRI or from the Green. The PRI, they were about to take everything from them. This is already type of war.

[D2]: Yes.

[P1]: Mexico is in war. Mexico is in war. A confrontation like this, and it is already hard, it is already hard. This is war, we are in war.

Chapter 7

Video sequence 1

Video 1 (Minutes 09:40–09:51), transcribed in Interview_2015_C_22: Lines: 134–135
[VM1]: Ya va empezar a tener una ya tina de agua en el fuego ya.   [P1]: Ya cuanto menos.
Translation into English language
[VM1]: Already you will start to have a tub of water yet on the fire.   [P1]: Yet the least.

Video sequence 2

Video 1 (Minutes 10:03–10:07) transcribed in Interview_2015_C_22: Lines 136–137
[VM1]: Charlando, esperando los cerditos.   [P1 & P2]: [rizas]
Translation into English language
[VM1]: Chatting, waiting for the piggies.   [P1 & P2]: [laughter]

Video sequence 3

Video 1 (Minutes 10:34–10:56) transcribed in Interview_2015_C_22: Lines 143–147
[P1]: Aprovechale la cámera.   [VM1]: Estoy aquí grabando.   [P2]: Si me presta esta [no se entiende]   [VM1]: Quitate, quitate de ahí, la hormiga, quitate, quitate.   [P1]: Agarrale de ahí.
Translation into English language

[P1]: So you take advantage of the camera.

[VM1]: I am here, recording.

[P2]: If you would lend me this [not understandable]

[VM1]: Stay away, stay away from there, the ant, stay away, stay away.

[P1]: Take it from there.

Chapter 8

8.1 Social practices performed by civil protection actors

Interview sequence 1

Interview 2014_T_1: Lines 18–29, recorded in English

[P1]: If I don't attack the root of the flood, in 10 or more years the effect will maybe even be stronger. This is the part I try to tackle in my work. This is what I try to make the people understand. The problem is: Because here there is a lot of water, more water than before on the rivers. We have already explained it to the people and they understand that it is not our guilt or our responsibility. It has happened with earthquakes. People say, you have to help me, it is your obligation. It is my work to help, that is why I come and help. But it is not my fault that your house is flooded. Your house has been in bad conditions, there was poor maintenance, it was constructed poorly. A lot was missing. I will support you but you have to understand that the problem came from your side.

Interview sequence 2

Interview 2014_T_1: Lines 64–83, recorded in English

[P1]: Then we arrive there and see the house was built poorly, in a place where it should not be, there is no allowance to build in that place. So the people have to understand this part and it is difficult for them to understand.

[D1]: And how was it in Catazajá?

[P1]: Catazajá is a region that floods every year, so up to 6 months of the year the region can be flooded. They live practically at the side of the river and due to the saturation of water that is present in the ground, already with small rainfalls, the water rises and stays flooded. The problem is that earlier administrations did not give a lot of support, they didn't come there. So people lived fine, they were used to this situation. They put up their things, they went in their boat and everything was fine. This is no solution either but people don't want to live in any other place. If you tell them to be removed, they say no. So what can I do? The zone is a zone of lagoon, a giant lagoon, and this cannot be changed. So what do we do? The earlier administration, the state government gave too much assistance to these people. Even if the level of water was not high, they already gave blankets, food, all that they wanted. They installed programmes so that people would be fine and lots, lots of money for them. So what the people do is, they get used to it. And now anything that happens, they say <Listen, I have water. Listen, it already rained. I want to have food support>.

Interview sequence 3

Interview_2015_P_2: 20–36

[D1]: Y ahorita que es o va empezar la temporada de lluvias...

[P1]: Si.

[D1]: ...como, que, como cambia su trabajo? Cambia su trabajo diario?

[P1]: Si, si, ya existe un plan de contingencia., verdad?

[P2]: Si.

[P1]: Ya existe un plan de contingencia.respecto a las lluvias, las tormentas tropicales. No te queda por ahí? Para que lo vea. Lo vi ayer? Queda por acá.

[P2]: Ah, ya. No tengo por aquí...

[P1]: Es un manual, verdad? Para estar preparado.

[D1]: Si.

[P1]: Si.

[D1]: Muchas gracias. Y esto es elaborado en Tuxtla? O aquí?

[P1]: No, aquí. O sea mas o menos respectivo a la región, vberdad?

[D1]: Aha.

[P1]: Con los problemas de aquí del municipio.

[D1]: Si.

[P1]: Cada año se hace. Se va actualizando.

Translation

[D1]: And now that there is or will begin the rainy season...

[P1]: Yes.

[D1]: ...how, what, how does your work change? Does your daily work change?

[P1]: Yes, yes, there already exists a contingency plan, isn´t it?

[P2]: Yes.

[P1]: There is already a contingency plan concerning the rains, the tropical storms. Don´t you have one over there? So that she could see it. I saw it yesterday. It is over there.

[P2]: Ah, yes. I don´t have it over here.

[P1]: It is a manual, right? To be prepared.

[D1]: Yes.

[P1]: Yes.

[D1]: Thank you. And this is elaborated in Tuxtla? Or here?

[P1]: No, here. Or let´s say more or less concerning the region, right?

[D1]: Aha.

[P1]: With the problems of here the municipality.

[D1]: Yes.

[P1]: Every year it is made. It is updated.

Interview sequence 4

Interview_2015_P_2: 110–126

[P1]: Yo tengo aquí una guia, de como se hace un analisis de riesgo, pero es una simple guia. Este. Atlas de peligros y [no se entiende] del municipio de Palenque. Este. Verdad?

[D1]: Si.

[P1]: Yo voy colecionando informaciones y voy hacer mi dictamen. Asi por ejemplo el grado de sismicidad, si en esta zona es bajo no? De una 4, pongo una 4 en mi lista. Eso es un ejemplo.

[D1]: Si.

[P1]: En cuestiones hidro–meteorológica ya sea en las zonas bajas, inundables, o zonas altas de deslaves. Y eso va aqui. Y sobre esto me voy orientando.

[D1]: Mhm.

[P1]: Pero en sí, hace falta un atlas de riesgo personificado de aquí del municipio de Palenque.

[D1]: Aha, aha.

[P1]: O sea, tenemos este. Es del departamento de identificación de riesgos, de Tuxtla. Y yo me baso en este.

[D1]: Aha. Si.

[P1]: Pero se necesita más información.

Translation

[P1]: I have here a guide of how a risk analysis is made, but it is a simple guide. This. Atlas of dangers and […] of the municipality of Palenque. This. Right?

[D1]: Yes.

[P1]: I go collecting information and go making my survey. Here for example the level of seismicity, yes in this zone it is low, isn´t it? Of around 4, I put 4 in my list. This is an example.

[D1]: Yes.

[P1]: In hydro-meteorological questions in the low zones, flood-prone, or high zones with landslides. And this goes here. And on this I keep orienting.

[D1]: Mhm.

[P1]: But in itself, there is a lack of a risk atlas, personalised of the municipality of Palenque.

[D1]: Aha, aha.

[P1]: Or let´s say, we have this. This is from the department of risk identification in Tuxtla. And I base myself on this.

[D1]: Mhm.

[P1]: But there is the need for more information.

Interview sequence 5

Interview_2015_P_2: 48, 51–60

[P2]: No, no se hizo refugio porque solo fue un encharcamiento.

[P1]: Entonces un encharcamiento es tal vez menos de un metro. [...] O sea, menos de un metro. Ya pasando un metro, inundación.

[D1]: Aha, aha.

[P1]: Si, si.

[P2]: Y son afectaciones indirectas. Ya afectaciones directas es cuando...

[D1]: ...entra el agua en las casas.

[P2]: ...en las casas.

[D1]: Indirectas es mas como, como por ejemplo, destruye las milpas y eso?

[P2]: Asi es.

[D1]: Eso es como un efecto indirecto que ya la gente pierde parte de su cosecha.

[P2]: O el camino.

Translation

[P2]: No, there was no shelter made because it only was an encharcamiento.

[P1]: So an encharcamiento is maybe less than a metre. [...] Or let it be less than a metre. When it passes over one metre, flood.

[D1]: Aha, aha.

[P1]: Yes, yes.

[P2]: And these are indirect effects. Yet, direct effects is when...

[D1]: ...water enters into the houses.

[P2]: ...into the houses.

[D1]: Indirect is more like, like for example, it destroys the milpa or something like that?

[P2]: That´s how it is.

[D1]: It is like an indirect effect that people loose part of their harvest.

[P2]: Or the street.

Interview sequence 6

Interview 2014_T_1: Lines 100–105 & 110–112, recorded in English

[P1]: Many people was given the opportunity to build a house, some people actually received a house like this...

[D1]: Like [name of a person]´s house.

[P1]: Yes, exactly. But the government said <I will give you a house, for sure. But now you cannot live in your first house anymore.> People many times want to live in their old house during the year and when the flood comes they want to move to their small house. [...] They want a new house but they want to continue living like they did before. And they will continue to have the same problem, year after year after year.

Interview sequence 7

Interview 2014_T_1: Lines 90–97, recorded in English

[P1]: So we have to find a way to support them but make them understand that this is a support, the solution is that you move away from your house. You are bad in the place where you live. You have to find a dry place. But there, they have no dry places, no elevated place, everything is at the water level. So there is a lot of work to do there, a lot of work. For me one ideal form to live there would be to build houses that are resistant to the water that comes every year. Elevated houses and such. But the people don't want to. They give all kinds of reasons, for their animals, etc.

Interview sequence 8

Interview_2015_P_2: 170–200

[P1]: Hay una zona donde debería hacer el gobierno. Como aquí es inundable, seria un proyecto productivo nada más. Uso habitacional vamos a pasarlo a la zona segura. Pero que si pierdan sus tierras.

[...]

[D1]: Si, creo que es la más grande preocupación de las personas. Que una vez que se vayan a vivir en otros lados que tambien van a perder sus tierras.

[P1]: Si, ese es el detalle. O sea, ahí, es un feómeno social.

[D1]: Mhm.

[P1]: Habría que estudiarlo a fondo para ver una solución. Porque ahorita se estudia, antes no. Porque en los años de Juan Sabines, ya hubo ciudades rurales.

[D1]: Mhm. En los 80s?

[P1]: No, apenas. Hace como 10 años de aca.

[D1]: Ah.

[P1]: O sea, ya viene esa parte de inversión por parte del gobierno. De que por ejemplo dejen sus propiedades, se ha logrado ya en Chiapas y donde se construyen ciudad, tienen suficiente. O sea, hay todo para hacerlo.

[D1]: Mhm.

[P1]: O sea, yo ya he llegado a Tuxtla. Ya me he dado cuenta.

[D1]: Y son autosustentables? Así que tienen tambien...

[P1]: Ciudades sustentables, o rurales.

[D1]: ...milpa, o? Tienen tierra para hacer agricultura?

[P1]: Si. Y lo que el gobierno pretende hacer es invertir en zonas seguras.

[D1]: Si, si.

[P1]: Porque cada desgracia que hay, hay que meterle mas dinero. Y es más caro.

Translation

[P1]: There is a zone where the government should make. As here it is flood-prone, it would only be a production project. Settlement use we will transfer it to the safe zone. But that yes they lose their lands.

[…]

[D1]: Yes, I think that this is one of the major concerns of the persons. That once they leave to live in another place that also they will use their lands.

[P1]: Yes, this is the detail. Or let´s say it is a social phenomenon.

[D1]: Mhm.

[P1]: It would be necessary to study it in detail in order to see a solution. But now it is studied, before not. Because in the years of Juan Sabines, there already were rural cities.

[D1]: Mhm. In the 80s?

[P1]: No, recently. About 10 years ago.

[D1]: Ah.

[P1]: Or let´s say, there already comes a part of investment from government. That for example they leave their properties, this has been reached in Chiapas already and where they construct a city, they have enough. Or, there is everything to make it.

[D1]: Mhm.

[P1]: Or I have already gone to Tuxtla. I have already noticed.

[D1]: And are they self-sufficient? So that they also have…

[P1]: Sustainable or rural cities.

[D1]: …fields, or? Do they have land to do agriculture?

[P1]: Yes. And the government what they try to do is invest into safe zones.

[D1]: Yes, yes.

[P1]: Because every misfortune there is, it has to be put more money. And it is more expensive.

8.2 Social practices performed by government actors

Interview sequence 9

Interview_2014_P_2: 4–16
[P1]: [...] yo me acerque ahí el 31 de julio fue mi primera fecha que yo fui a [...] (anonymised) no conocía tampoco por indicaciones de de mi jefa la la mamá del gobernador. La mamá del gobernador visitó esa comunidad y de ahí regresó iba a ayudar ahí y les llevó a Tinaco porque les metieron una solicitud de que ellos no tienen agua potable. Entonces ella le llevó a donar sus tinacos, si para almacenar sus aguas. Cuando la señora se presentó, la señora Leti Coello se llama. Este... se presentó con todos ellos les dijo: miren mi delegado que soy yo, les dijo, mi delegada va estar pendiente de ustedes. Y a mí cuando me dan una tarea hasta que la sacó ¿no?...   [D1]: Claro.   [P1]: Entonces yo dije bueno si a ella le preocupa esta comunidad yo tengo que quedar este con un trabajo muy bien plantado aquí para que la gente se sienta bien y ojalá y todas las comunidades yo las pudiera visitar y pudiera hacer el trabajo.
Translation
[P1]: I approached there the 31st of July this was my first time that I went to [...] (anonymised) I didn't know either, for instructions of, of my boss the, the mum of the governor. The mum of the governor visited this community and returning from there she was to help them there and she brought tinaco because they had put an application that they had no drinking water. So, she brought them to donate their tinacos, yes in order to store their waters. When the lady presented herself, the lady Leti Coello, that's her name. Well, she presented to all of them and said <Look, my delegate>, that is me, she told them, <my delegate will be here to take care of you>. And to me, if you give me a task, until I fulfil it, no?   [D1]: Sure.   [P1]: So I said to myself, good, if she cares about the community I have to do this with a well–planted work here so that the people feel well and hopefully and all the communities I could visit them and could do the work.

Interview sequence 10

Interview_2014_P_2: 82–87
[P1]:Y les gusta depender del gobierno y eso es triste. A mí me gustaría hacer un trabajo de cambiar un poquito la mentalidad. Por eso decía yo si trabajaran en el proyecto y que ellos vieran esto no es: hoy si tengo ganas de hacer una escoba y mañana no! Es un trabajo diario. Y van a tener una alternativa, me gusta esta tierra para cosechar pero no para vivir. Puede ser: tengo mi casa en Zapata y vengo y cosecho aquí cuando hay inundación me voy.
Translation
[P1]: And they like depending on the government and this is sad. I like to do a work to change a little bit the mentality. That is why I said if they work in the project and that they would see, this is not <Today if I feel like making a broom stick and tomorrow no.> It is a daily work. And they will have an alternative, <I like this land to harvest but not to live>. It can be: <I have my house in Zapata and come and harvest here, when the flood comes, I leave>.

8.3 Other actors linked to social practices of flood management in the case study region

Interview sequence 11

Interview_2014_S_1: Lines 212–226 & 237–244
[P1]: Pero. La parte importante es que, no se necestia mas electricidad en México, tiene capacidad suficiente por lo menos para los proximos 30 anos. Ehm, se deberian mejorar la eficiencia de companías de las casas del sector publico etcetera. Eso no se esta haciendo. No estas buscando alternativas a mejorar la tecnología que existe. Si?   [D1]: Si.   [P1]: Y se puede reducir la demanda de energía en Mexico. Y entonces este 40% de exceso se aun ahorro y no sea necesaria finalmente no? Ehm, con todos los tratados que Mexico ha firmado del cambio climatico tambien, las represas se vuelven una muy buen excusa de inversion Porque las represas son considerada como energías límpias.   [D1]: Si.   [P1]: Y eso significa entonces que reciben bonos de carbono...   [D1]: Mhm, para la construccion de represas.   [P1]: Para la construccion de represas. Y reciben financiamiento del banco mundial, de companias incluso que quieren comprar bonos de carbono y decir que esta en un periodo de transicion. Si?   [...]   [P1]: La capacidad enorme de retencion de agua, la produccion de electricidad. Estas son las más daninas para el medio ambiente.Existen otros tamanos y ramos pero lo mas, lo importante es la producción de metan, en por materia organica en descomposicion.   [D1]: Que está abajo del agua.   [P1]: Abajo del agua. Si. Es decir, tu no produces carbono, produces metan y mientras mas grande es el proyecto, mas hectareas de materia organica tienes abajo del agua produciendo metano por decenas de anos.
Translation
[P1]: But. The important part is that, in Mexico it is not needed more electricity, it has enough capacity for at least the next thirty years. Ehm, it would have to be improved the efficiency of companies of the houses of the public sector etcetera. This is not being done. You don't look for alternatives to improve the technology which exists. Yes?   [D1]: Yes.   [P1]: And you can reduce the demand for energy in Mexico. And so this 40% of excess is saved and it would not be financially necessary, isn't it? Ehm, with all the agreements that Mexico has signed of the climate change also, the dams become a very good excuse of investment. Because dams are considered as clean energy.   [D1]: Yes.   [P1]: And this means well that they receive carbon credits.   [D1]: Mhm, for the construction of dams.   [P1]: For the construction of dams. And they receive financing from the World Bank, of companies as well that these want to buy carbon credits and say this is a period of transition. Right?   [...]

[P1]: The enormous capacity of water retention, the production of electricity. These are the most harmful for the environment. There exist other sizes and types but more, the important is the production of methane, in the organic material in decomposition.

[D1]: That is below the water.

[P1]: Under the water. Yes. This means, you don´t produce carbon, you produce methane and the larger the projects, the more hectares of organic material is below the water that produces methane for decades.

Interview sequence 12

Interview_2014_S_1: Lines 571–574

[P1]: Si, exacto. Pero bueno, todos los gobiernos, incluyendo a Estados Unidos, estan financiando la construccion de infraestructura. Nuevas carreteras, nuevos puertos, nuevos aeropuertos. Entonces hay de todo, la verdad. Pero si, si creo que continuan con esa misma logica de construir energía para que se puedan instalar mas companias en la zona.

Translation

[P1]: Yes, exactly. But well, all the governments, including United States, finance the construction of infrastructure. New highways, new harbours, new airports. So there is everything to tell the truth. But yes, yes I think that they continue with this same logic of constructing energy so that they can install more companies in the zone.

ANNEX 5: FULL ANALYSIS OF SELECTED PHOTOGRAPHS AND VIDEO SEQUENCE SNAPSHOTS WITH THE DOCUMENTARY METHOD USED IN CHAPTER 7

Photograph 1

2.1.1 Planimetric composition

2.1.3 Scenic choreography

2.1.4 Relation of sharpness/blurriness

1. Formulating interpretation (following Panofsky): WHAT can be seen on the photo?
1.1 Pre-iconographic interpretation
A muddy pathway surrounded by green grass and vegetation. The mud path is very wet, has some small holes filled with water. The path leads towards a house made out of stone in the upper part of the picture/in the background of the picture. Besides the house there is a palm tree.
1.2 Iconographic interpretation
The picture was taken in a tropical area. This can be recognised in the dense vegetation and the palm trees. The one-storey house made out of bricks (concrete) and with an aluminium roof gives hints that the people who live here do not have a lot of money. However, they are also not poor people, because these would not be able to afford aluminium roof and concrete, but would live in a house made out of wood or bamboo.
2. Reflecting interpretation
2.1 Iconic interpretation

2.1.1 Planimetric composition

The main line in the picture goes from the right front of the picture to the centre and then turns right. It points towards the house in the upper middle part of the picture. The centres of attention are the house and the water holes in the mud.

2.1.2 Perspective projection

Long-shot. Picture taken from a standing position. The view is directed slightly to the ground. The focus of the picture lies on the water hole in the pathway in the centre of the photography.

2.1.3 Scenic choreography

The picture is parted by the pathway into a left part of the picture and a right part. Both parts are characterised dominantly by green vegetation. In the left part there is more grass in the foreground of the picture, on the right side there is more grass in the centre of the picture with some shrubs towards the right side of the picture. The house is in the background of the picture but located at the middle of the upper part of the picture. A palm tree stands at the left side of the house. Other trees are found to the left and the right of the house, whose tops are beyond the picture. Over the house the sky can be seen and it is largely covered with clouds.

2.1.4 Relation of sharpness/blurriness

The foreground of the picture is sharp, especially the green grass on the lower left part of the photography and the earth holes filled with water in the right front and in the centre of the picture. The left and the right parts of the picture are blurry. The house and the palm tree and the sky are blurry too. The brightest parts of the picture are the sky and the house as well as the water holes, in which the white of the clouds reflects. The upper right and upper left part of the picture are in dark green, due to tree vegetation and the shadow it casts on the grass. The foreground and centre grass is in brighter green.

2.2 Iconological-iconic interpretation (Identification of habitus/meaning)

The picture is dark as the sky is covered with clouds. The mud and the water are very important topics in the picture. The house is an important topic in the picture, too. One imagines that the producer of the picture shows the activity of walking towards the house and having to pass the mud. The house is quite far away from the producer of the picture, he/she has to walk the muddy path, but he/she wants to reach that house It is a distant position and at the same time a challenging position, as the way clearly leads towards the house but it is not an easy way to walk.

Photograph 2

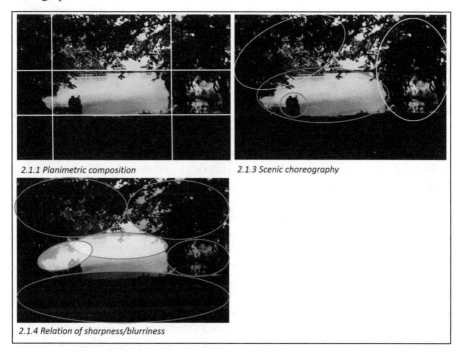

2.1.1 Planimetric composition   2.1.3 Scenic choreography

2.1.4 Relation of sharpness/blurriness

1. Formulating interpretation (following Panofsky): WHAT can be seen on the photo?

1.1 Pre-iconographic interpretation

People in a boat arriving at or leaving the river bank of a big river or water body. There are three people in the small boat, two of them sitting, one standing, with the back to the camera. This man wears a hat. He has his hands akimbo and looks into the water. All three people wear short sleeves. They have coloured skin. The boat has a motor. In front of the picture you can see the earth ground. It lies in the shadow of trees that can be seen around. The water has a brown colour; some shrubs are in the water. The movement of the water can be seen to the right of the boat in circles on the water surface. In the background part of the water body, a horizontal reflection of some parts of the water show a movement of water On the other side of the water body trees and shrubs can be seen, the vegetation is very dense and green. The sky is filled with white clouds.

1.2 Iconographic interpretation

The picture was taken in a tropical area. This can be recognised in the dense vegetation and the red-brown colour of the water. The man standing seems to be the driver of the boat, as he stands in the boat and looks into the water and the place next to the motor is left empty. He seems to have control over the boat and the people seem to trust him.

2. Reflecting interpretation

2.1 Iconic interpretation

2.1.1 Planimetric composition

The picture is divided by two parallel horizontal lines, one that shows the line of the ground in the foreground and the other one is the horizon line on the other side of the water body. The two vertical lines divide the picture into two areas of tree vegetation on the right and on the left side of the picture and the water body in the middle of the picture. Like this, a rectangular with slightly round side lines is formed in the centre of the picture which is the centre of attention: The water body, the boat and the people.

2.1.2 Perspective projection

Long-shot. Picture taken from a standing position. The view is directed slightly downwards, with a focus on the boat and the water.

2.1.3 Scenic choreography

The focus in this scene is put on the people in the boat and on the water body. This lies in the lower and left central part of the photography. The other objects like trees, shrubs and earth ground surround the water and the people, but play only a minor role. The trees of the vegetation give a homogenous picture of a round form that places the water body and the people in a circle-like formation of the scene.

2.1.4 Relation of sharpness/blurriness

The largest part of the picture lies in the shadow and is dark. The water body is the exception. The main part of the water body reflects the white colour of the sky and is bright. Only a small part of the water body lies in the shadow. The water has a brown colour. The boat and the people in the boat are partly in the sun, partly in the shadow. The ground in the foreground of the picture lies in the shadow and it is hardly possible to recognise objects on the ground, except for tree leaves, soil and small wood pieces.

2.2 Iconological-iconic interpretation (Identification of habitus/meaning)

The main focus of attention is created in this picture by the contrast of light and darkness. The water and the people in a boat are given the main attention in this picture. There is a peaceful atmosphere in the photography. The only thing that can be recognised to be in movement in the picture is the water which is flowing. The water surface is however without waves. The man standing with his arms akimbo in the boat is the person directing the boat and the way of the two passengers. While he stands upright, expresses in a dominant gesture and looks into the water, the two other people sit in the boat and appear comparably small. The two persons sitting in the boat are women. They look towards the river bank where they have arrived. They sit in a position with their arms and hands relaxed. It appears to be a routine activity. There is no excitement in the people. The boat driver appears to be quite sure of what he does, he is in a routine. The women are waiting for his action and calling.

While the spectator has questions about where the people might come from or where they go to, it is clear that you can see a routine activity and that people are used to this environment and activity in the water. It is a moment of waiting, waiting for the observation of a man, who knows what he is doing. All subsequent activity depends on his pausing and his observation of the water or something inside of the water.

There is a close connection of the man and the water. He knows the movements of water, the animals that live in the water. He can identify dangers and good things in the water. He is a patient person, not stressed but very calm and discrete in what he does.

Photograph 3

2.1.1 Planimetric composition

2.1.3 Scenic choreography

2.1.4 Relation of sharpness/blurriness

1. Formulating interpretation (following Panofsky): WHAT can be seen on the photo?

1.1 Pre-iconographic interpretation

A group of cows with different colours stand on a meadow. There are big cows and two baby cows. One white cow in the centre of the picture looks into the direction of the camera; the other cows look into other directions. They are on a green and partly yellow meadow and behind a fence made with wooden sticks and wire. In the foreground of the picture, a road can be seen. In the background of the picture there are trees and shrubs. There are red pigmentations on the upper right and upper left of the photography, which look like stains from the process of exposure.

1.2 Iconographic interpretation

The cows on the meadow seem to be part of the village life and economy. The people in the community might own these cows and live from the trade with cattle. As the grass in this tropical area is not purely green but also a little yellow, it seems to be dry season.

2. Reflecting interpretation

2.1 Iconic interpretation

2.1.1 Planimetric composition

The picture is mainly composed of rather horizontal lines. The road is one line, the bottom line of the fence is another one and the end of the meadow and beginning of the shrubs and trees is the next horizontal line. One shrub and one fencing post make for two short vertical lines in the centre of the picture, breaking the purely horizontal orientation. In between these two vertical lines lies the centre of attention, the white cow and the two baby cows.

2.1.2 Perspective projection

Long-shot. Picture taken from a standing position on the roadway. The view is directed slightly downwards, with a focus on the cows and the green meadow.

2.1.3 Scenic choreography

The cows, the major objects on the picture are all located on one virtual horizontal line in the middle of the picture. In the front are only small shrubs in the left part and the fence posts in the right part of the picture. In the left part of the picture there are two cows in a group, in the middle there is one cow and two baby cows in a group and on the right there are three cows in a group, while one of them is lying on the ground and only the head can be seen on the photography. While the cows in the centre of the picture look towards the camera, the other cows look into the direction of the cows in the centre.

2.1.4 Relation of sharpness/blurriness

The cows in the centre of the picture are sharp, while the rest of the picture is not very sharp. The road is also sharp. The bright parts of the picture are the centre of the picture, especially the two white cows (baby and adult cow), the yellow parts of the meadow and the road. The dark parts of the picture lie in the background, in the trees which have a dark green colour and have a lot of shadow.

2.2 Iconological-iconic interpretation (Identification of habitus/meaning)

The picture shows cows as they live in the area of the village. People rear cattle as a livelihood activity. They also care for the reproduction of the cows. The cows are not fat; you can see the ribs and some other bones of the cows under their skin. The picture looks quite peaceful, but the cows looking in the direction of the camera seem to take notice of the people and seem to put their ears into a position directed towards the direction of the camera holder as well.

The cows are kept in an area with a fence, so it seems important to protect the cows from getting stolen or from walking away from their owner´s land. They seem to be precious to their owners. As all the cows´ heads are left and the bodies are right, it seems like they are walking in one line in the left direction of the picture. The white cow in the middle has turned the body in such a way that it has the camera in front and can watch the camera holder with both eyes and listen with both ears and the body can react directly.

The picture also represents an adult-child relationship, as the central objects in the picture are the white cow and the white baby cow and the brown baby cow behind the white one, the young standing right beside the adult cow.

Photograph 4

2.1.1 Planimetric composition

2.1.3 Scenic choreography

2.1.4 Relation of sharpness/blurriness

1. Formulating interpretation (following Panofsky): WHAT can be seen on the photo?
1.1 Pre-iconographic interpretation
A group of young people walking along a muddy pathway. One girl jumps over a muddy part in the middle of the pathway, another girl wants to follow her. The other youths from the group are already a little bit ahead on the way, two of them turn around his head to look in the direction of the girl jumping. All people of this group wear white shirts with blue collar. In the foreground of the picture you can see small holes in the pathway filled with water. In the middle of the photography there is the group of young people in movement, in the background there are also other youths. In the background you can see a grey stone house and some trees, some of them palm trees of banana trees. To the left and to the right of the pathway there is green vegetation. On the left side, close to the house there is a fence.
1.2 Iconographic interpretation
A group of young students walking along a way on something like an excursion. They have a common goal, as they all walk in the same direction. They all seem to know each other as they walk very close to each other. The walk takes place in a village, in a rural area in the tropics. Their clothes and bags show that they are modern and aware of style. One boy wears a baseball cap turned around (the shield part to the back of his head) and they wear sports shoes. The girls wear tight jeans and the boys wear black trousers.
2. Reflecting interpretation
2.1 Iconic interpretation
2.1.1 Planimetric composition

The central lines in this photograph are composed by the pathway, which takes large parts of the picture. The lines of the right and the left border of the pathway meet on the top of the picture, but are broken by the young people who walk on the pathway. The lines also direct to a girl in the centre of the picture. The horizon line of the picture is in the background of the picture and is not very dominant. It is the line where most trees spring off from the ground.

2.1.2 Perspective projection

The photograph is a long-shot. The person who took the photo is in a standing position. Between him/her and the young people there is some distance. He/She might have waited some time for them to walk the pathway in order to have them all on the picture. He/She stands at the right side of the pathway and makes the photography directed slightly to the left, thereby not only capturing the pathway and the youths, but also the house.

2.1.3 Scenic choreography

There is a clear dominance of water and mud in the foreground of the picture. In the centre of the photograph, there is a group of ten young people. There is one girl in the centre of the picture, who jumps in the middle of the pathway. The other youths walk on the border between the muddy pathway and the green grass. The girl raises the attention of the spectator due to her central position [golden ratio], her movement and due to the fact that the front of her body is directed towards the camera. All the other persons of the picture are directed to the camera with their back or the side of their body. The girl to the right of the girl in the centre is the second person arising attention in the picture, she is directed towards the other girl and stretches out one of her arms, which indicates that she might follow the girl on her jump. Some of the other youths look into the direction of the pathway, some look down or to their side onto the pathway. Two other people, one girl and one boy, turn their heads and look in the direction of the girl or in the direction of the camera. The house, the fence and the trees in the background of the picture.

2.1.4 Relation of sharpness/blurriness

The pathway and the holes filled with water in the foreground of the picture are the parts sharpest in the picture. The two girls in the centre of the photograph are also sharp, while the other youths and the background of the picture is blurry. The water holes reflect the white colour of the sky and are among the brightest part of the photography. The shirts of the youths are also among the brightest parts of the picture. The white shoes of the girl in the centre also attract the eye of the spectator. The green vegetation and especially the larger trees in dark green colour in the background of the picture are the darkest parts of the picture.

2.2 Iconological-iconic interpretation (Identification of habitus/meaning)

The central point of interest in this picture is the muddy pathway and the group of youths walking on it or alongside it into one common direction. The girl in the centre of the photography jumps into the middle of the pathway, on a very muddy part of the pathway. Her body position shows that she is feeling comfortable. She wears white shoes, which are clean, although the group has already walked some part of the muddy pathway. The youths seem to know where and how to walk, without getting dirty. The white shirts that all of them wear shows unity of the group. The trousers, bags, caps and hairstyle show the individuality of each person. Style seems to be important to them, more to some then to others. The muddy pathway appears like an obstacle for a person walking, but the youths due to their body performance and position seem familiar with the situation. Most of them walk on the right or left side of the pathway, only the one girl jumps into the centre. It appears that she wants to follow another girl which walks on the left side of the pathway. That other girl seems very calm. Some youths seem to take notice of the camera and look into the direction of the camera, some seem not to. As the body gestures of the youths are individual, it seems that they all feel familiar and at ease with the situation of walking in a group. They all seem to know each other and trust each other and the excursion or walk they do seem to be fun to them.

Photograph 5

2.1.1 Planimetric composition   2.1.3 Scenic choreography

2.1.4 Relation of sharpness/blurriness

1. Formulating interpretation (following Panofsky): WHAT can be seen on the photo?

1.1 Pre-iconographic interpretation

A field with agricultural plants like corn and plantain. In the foreground there is a part if a muddy pathway with a hole filled with water. Then there is some small strip of green grass, behind it there is a fence of wooden sticks and wire. Behind that wire there are crop plants, in green, yellow and brown, maybe ripe for harvest. In the background there are high trees in dark green and a part of the sky blue, a part filled with round and white clouds.

1.2 Iconographic interpretation

A field in the tropic with subsistence, small-scale agriculture.

2. Reflecting interpretation

2.1 Iconic interpretation

2.1.1 Planimetric composition

The main lines in this picture are horizontal: The ground line of the pathway and of the ground of the crop field. The upper tree line in the background of the picture. The main vertical lines are made of the crop (corn or millo) in the centre of the photography. Two other vertical lines are produced by the wooden planks of the fence and a large tree in the left part of the picture in the background.

2.1.2 Perspective projection

The photography is taken from a standing position, the look slightly to the ground. The producer of the picture stands on a pathway or other plane area next to the field.

2.1.3 Scenic choreography

The crops are located in the centre of the picture, almost all the part from the left to the right in the central part. In the right lower corner of the picture there is the hole filled with water and some muddy strip of pathway. In the upper left part of the picture there is one line of large trees reaching higher than the end of the photography. In the right upper corner of the picture, there is the sky filled with bright white clouds which have a round form. The three fence sticks in the centre, in front of the crops create a visual borderline, although it is a relatively open fence. The young crop plants, surrounded on one side by the fence, on the other side by the high trees appear in a protected position. The fence stick in the centre of the picture is the central part of the picture.

2.1.4 Relation of sharpness/blurriness

The sharpest part of the picture is the central stick of the wooden fence. The green grass and the muddy pathway and water hole are sharp as well. The crops are less sharp. Less sharp are the other plants on the right side of the photography as well as the large trees. The brightest part in the picture is the white cloud in the upper right corner. The crops in light yellow are the second brightest part. The large trees in the background of the picture are the darkest part.

2.2 Iconological-iconic interpretation (Identification of habitus/meaning)

Agriculture is one of the central aspects of this picture. It is presented in a way that shows that the crops are protected by the village members. There are clear fences and borders of agricultural land. The producer of the picture stands relatively close to the crops at the same time he/she has a good overview of the whole part of land. Doing agriculture on a small scale, without high fencing or other technology is an everyday activity in the village. There are diverse crops. The young crops are taken special care about. The picture shows that there is a lot of rain. The big white clouds and the water hole on the ground indicate this.

Photograph 6

2.1.1 Planimetric composition

2.1.3 Scenic choreography

2.1.4 Relation of sharpness/blurriness

1. Formulating interpretation (following Panofsky): WHAT can be seen on the photo?
1.1 Pre-iconographic interpretation
A one storey building located at the side of an earth road. The house is run-down, many windows are broken, the largest part of the roof is missing. The house has had a blue and white painting but is almost completely washed out. There are trees growing from within the building reaching over the building structure. The earth road is in red colour. On the ground right to the building there lies a corrugated sheet. In the upper part of the picture there are two electric wires in front of a sky covered with white clouds.
1.2 Iconographic interpretation
It is a scene in the tropical area, with its red earth and big white clouds and palm trees. The building seems to be a large community building, a public one, not a private house.
2. Reflecting interpretation
2.1 Iconic interpretation
2.1.1 Planimetric composition
The vanishing point of the mainly horizontal lines in this picture lies to the left outside of the picture itself. The only vertical lines made in this picture are the trees and the closed doors of the house. The door in front of the picture attracts main attention of the spectator.
2.1.2 Perspective projection
It is a central projection and a long-shot picture. The producer of the photography has taken the picture from a standing position, on the road, directed slightly to the left. The upper line and the lower line of the house give an impression of the great length of this building.

2.1.3 Scenic choreography

The main object/body in this photography is the house. Right of the house there is the corrugated sheet, it seems to be scattered on the floor, not given special attention. The part of the road on the left side of the picture shows that the building is located at a central location in the village. The only other object in the picture are the trees, diverse tree species growing from inside the house and one large palm tree to the right of the house. The central object of the house structure is the right door, which is closed.

2.1.4 Relation of sharpness/blurriness

The sharpest part of the picture is the door of the building and the part of the building in the right part of the picture. The left part of the building, the road, the trees and the sky are blurry. The bright parts of the picture are the sky filled with white clouds and the white part of the building. The darkest parts of the picture are the left side, with trees and the right side of the picture, the head of the building and the vegetation to the right side of the building.

2.2 Iconological-iconic interpretation (Identification of habitus/meaning)

The building appears to be at a central location in the village. It might have been in use by the community in the past. But it is not used any longer. The doors are closed, the roof missing and objects scattered around the building. It appears that the building has not been used for a long time since there are some high trees growing from the inside of the building. The right door of the building is put a focus on in this picture, maybe it was used frequently by the people of the community in earlier times. The building takes a large part of the picture so it seems to be part of everyday experience of the producer of the picture. The kind of windows indicates that it might have been a public building like a community hall or school. The wires that pass over the building show the connection to electricity in the community. This wire follows the path of the road in this picture. It appears as if there were other buildings along the road as well, which are connected to the electricity grid. Electricity is an important service in the community. It makes a contrast to the run-down building, just as the road does.

Photograph 7

2.1.1 Planimetric composition

2.1.3 Scenic choreography

2.1.4 Relation of sharpness/blurriness

1. Formulating interpretation (following Panofsky): WHAT can be seen on the photo?

1.1 Pre-iconographic interpretation

In the foreground of the picture there is a large water body. On the right side there are shrubs and trees, some of them inside the water. Some wooden sticks and treetops reach out of the water. In the background you can see a line of earth and vegetation, like the other bank of the river or lake. In the upper part of the picture there is the open sky, filled with clouds. There is a clear lower frontline of the clouds, they seem to have one main direction. The vegetation and the clouds reflect on the water.

1.2 Iconographic interpretation

The picture shows a flood situation, where the banks of a river are flooded so that the vegetation is submerged in the water. The photography is taken in a tropical area where there are big clouds filled with a lot of rain.

2. Reflecting interpretation

2.1 Iconic interpretation

2.1.1 Planimetric composition

There are two main lines in the picture, which are horizontal lines: The line of the water surface against the river bank on the other side of the water body and the line of the cloud front. There is one vertical line which is a thin tree that stands in the water and its reflection in the water.

2.1.2 Perspective projection

It is a long-shot. The producer of the picture seems to be standing in the water or kneeling down directly at the river bank. The water line is slightly below the natural horizon line of the producer of the picture so he/she has directed the view of the camera slightly downwards.

2.1.3 Scenic choreography

The water body makes for the largest part of the picture. It is the central object. The sky, especially the frontline of the clouds is the other central object. The cloud front produces an impression of movement. The trees in the water are another focus of the picture [golden ratio?]. The small stretch of land between the water and the clouds in the background part of the picture create an impression of narrowness, of the reduction of space. But it is this part of the picture where it is possible to have a very far look. The trees inside the water show also the change of the extension of the water. One wooden stick reaching out of the water bends down the upper part, which might show that this tree or shrub is already dead. There is one circle on the water, that might have been produced by a rain drop or by an animal inside the water.

2.1.4 Relation of sharpness/blurriness

The front of the picture, especially the tree inside the water and its reflection on the water surface are the sharpest part of the picture. The objects in the background and the sky are blurry. The brightest part of the picture is the part of the clouds which is white.

2.2 Iconological-iconic interpretation (Identification of habitus/meaning)

The picture presents a scene from the flood in the rainy season. It shows the trees that are partly submerged in the water, the dominance of the water and of the clouds. Although there is a lot of reflection on the water body and white clouds, the picture has a dark impression. The lower line of the cloud front with its dark grey colour makes a threatening impression. The spectator has the impression to be in the middle of the water, surrounded by water and clouds.

The second level of the picture is the narrow space between the water and the clouds on the central left part of the picture. Here, a new extension of space opens up, which is not a vertical extension but a horizontal extension to a place far away. The eye of the spectator is directed towards this space filled with different layers of clouds.

Video sequence 1 (Video 1 Minutes 09:40–09:51)

1. Formulating interpretation

1.1 Pre-iconographic interpretation

A man who stands beside a metal bud which is placed on a roast over a fireplace. Another man leans on a tree. Trees and grass surround the scene and in the back part a river bed can be seen.

Additional info screenshot 2: The man bends down and places a piece of wood in the fire.

Additional info screenshot 3: The man walks towards the camera.

Additional indo screenshot 4: The man bends down to the left just in front of the camera. You can see a tool in the back pocket of his trouser.

1.2 Iconographic interpretation

The picture was taken in a tropical area. This can be recognised in the dense vegetation and clothing of people. There is a fast river bed to be recognised in the back of the photograph. The man in the centre of the photograph wears a base-cap, used shirt and trousers and rubber boots. This underlines his activity of working, which he does in lighting up a fire.

2. Reflecting interpretation

2.1 Iconic interpretation

2.1.1 Planimetric composition

The main line in the picture is a horizontal line created by the line where the soil meets the river. A vertical line is drawn from a tree in the right half of the picture. The centre of attention is in the middle of the picture on a round water bud and a man standing beside it.

2.1.2 Perspective projection

From long-shot to close-up. Video taken from a standing position. The view is directed slightly to the ground. The focus of the picture lies on the water bud in the centre of the photography and a man who moves from the centre of the picture (screenshot 1) to the front of the picture (screenshot 4).

2.1.3 Scenic choreography

The picture is parted into a left and a right part by the water bud and fire place. In the right part of the picture activity takes place, while in the left part there is no activity. While in the first screenshot there are two people in the picture, in the other screenshots the person leaning on a tree is not present any more.

2.1.4 Relation of sharpness/blurriness

There is a high contrast between shade and light in the video. While most parts of the pictures are in the shade, which is created by trees, rays of the sun can be distinguished on the ground and the sky above the river in the back part of the pictures is illuminated and characterised by strong contrast of light.

2.2 Iconological-iconic interpretation (Identification of habitus/meaning)

The screenshots of this video sequence show a high contrast of light and darkness which is produced by the shade of the trees from the sunlight. The high light contrast shows that the video was recorded at daytime, around noon. The screenshots show a central activity in the village. The man lights a fire on which a metal bud is placed. He is active, while another man leans on a tree. The man is between 45–60 years of age. The last screenshot provokes a feeling of closeness and intimacy, as the camera is very close to the man who bends down.

Video Sequence 2 (Minutes 10:03–10:07)

1. Formulating interpretation

1.1 Pre-iconographic interpretation

Two women leaning on a table in an open space, below a tree. On the table, a plastic cup is placed upside-down. The women wear T-Shirts and each have a small towel over their left shoulders. The atmosphere is friendly as the women smile and look happy.

Additional info screenshot 1: Both women smile, while woman 1 looks in front of her smiling with her mouth open, woman 2 looks into the camera, with her mouth closed.

Additional info screenshot 2: Both women look in front of them, slightly smiling with their mouths closed

Additional info screenshot 3: Woman 1 looks into the camera with a more serious look, while woman 2 looks at her hand, smiling with her mouth open and smiling.

Additional indo screenshot 4: Woman one looks in front of her, with a more serious expression in her face while woman 2 looks straight at the camera with her arm leaning on the table and her hand in front of her mouth.

1.2 Iconographic interpretation

The picture was taken in a tropical area. This can be recognised in the dense vegetation and cloth-ing of people. Women lean on a table in a bodily position that indicates waiting and relaxation. Both women look into the same direction and from time to time look into the camera.

2. Reflecting interpretation

2.1 Iconic interpretation

2.1.1 Planimetric composition

Two main lines can be recognised in the snapshots, one is a horizontal line in the lower part of the photograph, which is the line of the table on which women rest with their arms. The other main line is a vertical line formed by the trunk of a tree behind the women. The centre of attention is in the middle of the picture on the two women.

2.1.2 Perspective projection

Medium shot. Video taken from a standing position. The view is directed to the front on eye level with the two women. The focus of the picture lies on the faces of the women in the centre of the picture.

2.1.3 Scenic choreography

The picture is parted into a main vertical extension which is the tree in the back of the women and into a rectangular field in the lower central part of the picture which is filled by the torso and heads of the women.

2.1.4 Relation of sharpness/blurriness

Large parts of the pictures are in shade. While screenshot 1 is very shady, especially the women, in the other screenshots the women are more light. The brightest parts of the pictures are the sky behind the tree leaves, the shirt and earring of one woman, the white plastic cup on the table and the teeth of the women to be seen in screenshots 1 and 3. The cup in the lower centre of the picture is sharpest, after that the women are focused on while the other objects on the picture are less sharp.

2.2 Iconological-iconic interpretation (Identification of habitus/meaning)

The screenshots of this video sequence show a strong focus on the two women. The video maker closely observes the action of the two women. The closeness and intimacy is created by the me-dium shot in which the facial expressions of the women can be seen clearly. Observation of facial and bodily expressions is a dominant feature of the video sequence.

Video sequence 3 (Minutes 10:34–10:56)

1. Formulating interpretation

1.1 Pre-iconographic interpretation

Various men stand at the back of a red transporter. They prepare to open the door of the trans-porter. One man on the left of the picture holds a yellow rope while other men are close to the truck working with their hands to open the door.

Additional info screenshot 1: The man in the left part of the picture looks towards the camera and has his mouth open, speaking in the direction of the camera while the two other men stand with their backs to the camera. In the background of the picture a woman with black hair and a pink T-Shirt can be recognised.

Additional info screenshot 2: The camera slightly turns to the left. The man on the left still looks towards the camera. In the background behind him, a woman with black T-Shirt and blond hair can be recognised.

Additional info screenshot 3: The camera slightly turns back to the right again. Another man appears in this picture close to the camera with his back to the camera. Another man walks toward the group of men on the right. Hands of men work on the left and the right part of the doors of the transporter.

Additional indo screenshot 4: The yellow rope and the rope that has been removed from the transporter lie on the ground. A man gathers them from there. The door of the transporter has been opened. In the right part of the picture a small fire can be seen, which provokes some smoke.

1.2 Iconographic interpretation

The picture was taken in a tropical area. This can be recognised in the dense vegetation and clothing of people. While people in the right part of the picture wear casual clothes adapted for hard work, the man on the left wears better clothes and elegant shoes. While most people on the pictures have tanned skin and black hair, the woman in the left background part of screenshot two has comparably light skin and blond hair. Many people stand and work together in this sequence.

2. Reflecting interpretation

2.1 Iconic interpretation

2.1.1 Planimetric composition

The main lines in the pictures are vertical lines which are produced by the man in the left part of the picture and the men in the right part of the picture. One horizontal line is formed by the lower part of the door of the transporter and the ground in the background of the picture.

2.1.2 Perspective projection

Medium long shot. Video taken from a standing position. The view is directed to the transporter, while it is slightly directed downwards as well.

2.1.3 Scenic choreography

The picture is parted into the left and the right part separated by the transporter. Main changes and movement take place in the right part of the picture. The group of men on the right part of the picture changes from one snapshot to the other.

2.1.4 Relation of sharpness/blurriness

The main focus of sharpness lies on the transporter. While the camera turns to the left and the right, the focus of sharpness slightly changes. While the men who stand in front of the transporter are sharp, the other persons who are farer away or closer to the camera are blurry.

2.2 Iconological-iconic interpretation (Identification of habitus/meaning)

The screenshots of this video sequence show a strong focus on the transporter and the men that stand around it. The order of people in space divides the man who gives instructions (man on the left) from people who follow and exert the instructions (men on the right). The bodily position and gestures reveal a hierarchy between the men. The gesture and facial expression of the man on the left towards the camera are quite aggressive (snapshot 1) and dominant (snapshots 1 & 2), while the other people do not take notice of the camera.

ANNEX 6: SONGS AND POEMS USED FOR ANALYSIS

Legend of "El Hombre Pez" (the Fish Man)

Leyenda sobre los origenes de Catazajá

Las antiguas bocas de la tradición cuentan que, en las lejanas albas de la humabidad/humanidad, la lahuna de Catazaja era navegable durante todo el año y en ella bullía la multiforma vida de pececillos polícromos y demás especies comestibles. La laguna azul de las olas lamía las márgenes rodeadas por el tupido y verde biombo de la selva, donde el palo tinto y otras maderas preciosas señoreaban. Poblada esta laguna un terrible monstruo mitad hombre mitad pez, al cual los mayas de Palenque, año tras año tributaban la doncella más hermosa. Sellendo así el vasallaje se podia pescar, cortar madera y extraer tintes.

El ritual se cumplía de este modo: La virgen era ofrentada sola en un cayuco ebandonado en las orillas; venía el hombre pez y se la llevaba consigo a las aguas más profundas (este lugar se conoce hoy simplemente como la poza).

Algunas veces los tributarios, quizá cansados por la sumisión, no realizaban la ofrenda. Entonces el hombre pez, soberbia y furia, no permitía la pesca. Esto hacía sufrir mucho a los maayas, cuya alimentación básica la constituía el pescado.

Pasaron algunos años. Cierta vez e pueblo maya, cansadísimo por el tributo estalló en rebeldía. Fueron convocados los flechadores más carteros de la comarca.. Estos se presentaton al lugar convenido. Allí se organizaron, en tácticas de combate, para atacar el hombre pez. El monstruo opuso resistencia tenaz, pero los hábiles guerreros lograon ensartarle varias flechas. Mortalmente herido huyó buscando refugio. En su huída manchó con su sangre los arenales (este sitio puede citarse ahora, todavía colorado, frente a Punta Arenas).

No consiguió esconder su agonía ni ocultar su cuerpo sin vida, que de inmediato se convirtió en un montón de piedras (este girón puede contemplarse en estos tiempos, llamado el Piedral, a un costado de Paraíso).

A partir de la muerte del hombre pez, las aguas cambiaron su curso. La laguna se secó. Esto hizo creer a los mayas en un castigo divino. Entonces clemencia y aua a sus dioses. Estos concedieron el preciado líquiso solamente por temporadas (así es como se origina la estación de la creciente y la estación de la sequía). De esta realidad climatica surge la palabra Catazajá, germinada del dulzón de la lengua maya: AGUAS QUE VIENEN, AGUAS QUE VAN.

Voces antiguas dicen que ahora el espíritu del hombre pez vaga en forma de fuego sobre las aguas de la laguna o en el pasto durante las secas.

Copied by hand by the author of this study from the original document owned by members of case study village P (anonymised) on 10. August 2014

Song "Sin Fortuna" by Gerardo Reyes

Yo nací sin fortuna y sin nada, desafiando al destino de frente,
hasta el más infeliz me humillaba, ignorándome toda la gente,
y de pronto mi suerte ha cambiado, y de pronto me ví entre gran gente.

Ví a esa gente sentírse dichosa, frente a un mundo vulgar y embustero,
gente hipócrita, ruin, vanidosa, que de nada le sirve el dinero,
que se muere lo mismo que el probre, y su tumba es el mismo agüjero.

Ahora voy por distintos caminos, voy siguiendo tan solo al destino,
y entre pobres me siento dichoso, sí amando doy mí amor entero,
con los pobres me quito el sombrero, y desprecio hasta el más poderoso.

Soy cabal y sincero les digo, he labrado mí propio destino,
yo le tiendo la mano al amigo, pero al rico jamás me le humillo.

Hablado: Yo nunca tuve el calor de un beso, mis pobres viejos trabajaban tanto,
que nunca tuvieron tiempo para eso. y así crecí sin ignorar el llanto.
No fuí a la escuela, yo apendí de grande, para esas cosas no alcanzaba un probre,
las letras no entran cuando se tiene hambre, ní hay quien te tienda la mano si eres pobre.
por eso vuelvo a éste pueblo viejo, donde la vida me trató tan mal,
ésta es mi gente que por nada dejo, aunque volviera yo a sufrir igual.

Soy cabal y sincero les digo, he labrado mí propio destino,
yo le tiendo la mano al amigo, pero al rico jamás me le humillo.

Source: http://www.musica.com/letras.asp?letra=948820 (Accessed 26.12.2016)

ANNEX 7: MEMOS REFERRED TO IN THE ANALYSIS

Chapter 6: Memo_I_150603, anonymised

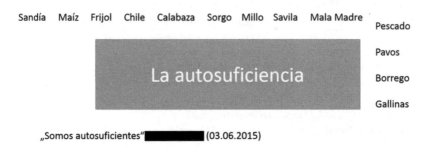

Sandía   Maíz   Frijol   Chile   Calabaza   Sorgo   Millo   Savila   Mala Madre

Pescado

Pavos

Borrego

Gallinas

„Somos autosuficientes" ▮▮▮▮▮ (03.06.2015)

Chapter 8: Memo_I_141014

Instituto Estatal de Proteccion Civil y Escuela de Proteccion Civil, Chiapas
— Manejo Integral de Riesgos y Prevención
— Ahora viene un Frente Frio 7 hacia la región, Protección Civil alerta el norte del Estado de Chiapas

Discusion con Director de Proteccion Civil Chiapas
— Problema principal de desastres naturales en Chiapas es pobreza
— Pobreza de Conocimiento y de Educacion
— "Cost-Benefit" de Reubicación: pequenas comunidades si, grandes no
— Nueva Ley de Prot. Civil
— Pp5
— Cifras de pobreza de Chiapas
— Grijalva y Usumacinta
— Desarrollo y Resiliencia (Cree en concepto de desarrollo)

Discusion sobre las despensas: Yo articulo mi vista crítica, la expectativa que cree en las comunidades; El dice que por los derechos humanos es necesario dar ayuda y despensa a la gente en las comunidades; El prefiere darles despensas que no dar y ser acusado por falta de apoyo y disrespecto de derechos humanos. Le interesa mi investigación y me pide una platica virtual en los proximos meses via skype. Pensamos en un intercambio con Escuela de Prot Civil y Institute of Risk Enineering and Civil Protection así como un intercambio y enlace con BBK y THW. Preucapacion del Director sobre lo de Ebola. El piensa, que hay muchos enemigos de EEUU y estos van a intentar de introduciry dispersar Ebola en EEUU. Eso va afectar a Mexico muy fuertemente.